Heating and Cooling Essentials

Instructor's Manual

by

Jerry Killinger

LaDonna Killinger

Publisher

The Goodheart-Willcox Company, Inc.

Tinley Park, Illinois

International Standard Book Number 1-56637-967-9

2 3 4 5 6 7 8 9 10 03 06 05 04

Contents

Chapter Resources

Introduction

Overview

Heating and Cooling Essentials provides the information and activities students need to develop the knowledge, skills and attitudes required of service technicians in the heating, ventilation, air conditioning, and refrigeration industry. The textbook and accompanying lab manual form a comprehensive overview of heating and cooling systems for use by first-year students in HVAC programs at vocational schools, technical schools, or community colleges. The instructional materials are also suitable for apprenticeship programs or adult continuing education classes, and can serve as an excellent reference for technicians already working in the field. Just enough engineering principles are included to explain essential troubleshooting and service procedures.

Heating and Cooling Essentials was developed in actual classrooms and training programs. It is organized into areas of study that progress from basic to complex concepts and systems. However, it is flexible and may be studied in the sequence that best accomplishes your program goals.

Students should have ample opportunity to fully acquaint themselves with the tools and equipment encountered in the world of work. Then, they should be given the opportunity to apply newly learned skills on existing equipment in the school laboratory. The textbook and lab manual allow for such a shift between instruction and hands-on experience.

The introduction of tools, fasteners, tubing, pipe, soldering and brazing, and mathematics in the first several chapters is intended to help students develop identification, computation, and basic hands-on skills early in the program. Because classroom instruction initially will progress more rapidly than laboratory (shop), lab manual activities enable the instructor to "gain some ground" teaching the theory and basic principles of refrigeration and air conditioning before students become involved with more complex training equipment and tools.

Vocational/technical education must be a real-life learning experience through which the student can develop the skills necessary to obtain gainful employment. Proficiency in manipulative skills is only one part of the training. A successful employee must also possess a good work attitude. This can be accomplished only by providing the student with an atmosphere for developing self-confidence and interpersonal skills.

Using the Textbook

The 33 chapters of the **Heating and Cooling Essentials** textbook contain familiar features such as Learning Objectives, Important Terms, and Test Your Knowledge review questions. These features provide a simple and effective framework for presenting the subject matter. Once students are comfortable with the format of the textbook, they are able to concentrate strictly on the content of the chapters. This makes learning easier, more effective, and more fun.

An outstanding feature of this textbook is the way complicated processes are described in easy-to-understand terms. Explanations are further clarified with carefully selected photographs, diagrams, and other illustrations. Since the textbook describes processes in such great detail, readings should be assigned out of class. The students will then be prepared for in-class discussions and able to comprehend additional content presented through discussions, lectures, demonstrations, and other classroom resources.

Out-of-class study is especially important in vocational courses where students spend considerable time in the laboratory on "hands-on" activities. It not only saves valuable classroom time but also provides students with the experience of researching and securing answers to questions. Out-of-class reading and research also minimizes the tendency of some students to depend on the instructor for direct assistance in finding solutions to all their problems.

Each chapter of the textbook is divided into several areas:

- Objectives.
- Important Terms.
- Body of the Text.
- Summary.
- Test Your Knowledge (review questions).

The text also includes a thorough Glossary of Technical Terms and an Index.

Learning Objectives

Each chapter begins with Objectives that tell students what they are expected to know at the conclusion of the chapter (outcome-based objectives). A measurable action verb introduces each objective; the verb suggests a method of self-evaluation for the student.

When introducing a new chapter, ask for a volunteer to read the objectives aloud in class. Briefly discuss the topics students will encounter in the chapter. Previewing helps them focus on the major points to be covered. Briefly scan the chapter with the students, pointing out important illustrations and key points to remember. Locate the Test Your Knowledge review questions at the end of the chapter.

Important Terms

A list of key words introduced in the chapter is located in the Important Terms section, immediately following the objectives. A thorough understanding of the terms is important to ensure comprehension of the material within the chapter. In the body of the text, these terms are printed in ***boldface italic*** type for emphasis and usually are defined immediately within context. All Important Terms are also defined in the Glossary of Technical Terms at the back of the book.

Body of the Text

The body of the text is divided into informal sections using various head levels, similar to the head levels in an outline. The first-level heads (the largest) are the major divisions of the body, and are identified by a numeric prefix consisting of the chapter number and a sequential number — for example, 10.1, 10.2, 10.3, and so on. The second level heads (the next largest), extend that numeric identifier with an additional sequential number —10.1.1, 10.1.2, 10.1.3, and so on. Second-level heads are further subdivided by third-level heads, which are unnumbered. In a few instances, fourth-level heads are used.

In the body of the text, all *safety* information is printed in red type for emphasis. The material may relate to personal safety or may be a warning that equipment damage can result from certain practices or situations.

Illustrations, in the form of line art and photographs, are used frequently throughout the text. The illustrations are referenced in the body of the text and accompanied by descriptive captions that enhance the topics being presented. When students encounter a figure reference, they should immediately locate the illustration, read the caption, and look carefully at the drawing or photograph, taking note of the details. This helps to reinforce the information presented in the text. Color and callouts are used extensively in the illustrations to help clarify technical concepts and simplify comprehension. A color key facilitates reading illustrations and understanding system operations.

Summary

The Summary provides a synopsis of the information in the body of the text. Generally, the Summary includes information relating to the first-level heads. Some instructors may choose to use it as an introductory overview of the chapter.

Test Your Knowledge

These end-of-chapter questions may be used as either an in-class or out-of-class assignment to review the subject matter within a chapter. A variety of true/false, multiple choice, and completion questions are included.

Glossary of Technical Terms

The Glossary of Technical Terms provides detailed descriptions of technical terms appearing in the textbook and commonly used in the trade. Although many of these terms are defined in the text, the Glossary of Technical Terms provides a convenient reference for all content areas.

Index

The Index is a comprehensive reference to the topics included in the text. If you cannot locate a term when using the Index, refer to a closely related term for a secondary entry.

Using the Lab Manual

The **Heating and Cooling Essentials Lab Manual** is designed to enhance student achievement and comprehension of the material presented in the textbook. The lab manual uses a variety of activities, often "hands-on," to reinforce the concepts, practices, and techniques described in the text.

The general scheme of the lab manual is quite simple. It is divided into 33 sections, corresponding with the textbook chapters. Each of the 33 sections has between two and seven activities for completion by the student. Some can be completed in a short period of time; others will require a full class period or longer.

Any activity requiring tools or materials opens with a list of the needed items. A step-by-step procedure is described, and space is provided to check off each step upon completion. Lines are provided to the right or below the question when a written response is required. Answers can easily be checked against the appropriate answer key in the Instructor's Manual. When appropriate, space is provided a the end of an Activity for your initials and the date.

Note: It is a good practice to review the shop safety rules before allowing students to work on the activities. (Specific safety recommendations are provided at the beginning of each activity, when appropriate.) A review will reinforce the importance of safety precautions and allow you to ascertain whether students understand basic safety rules and policies.

Using the Instructor's Manual

The **Heating and Cooling Essentials Instructor's Manual** is a teaching tool that contains suggestions for teaching HVAC safely, effectively, and efficiently. The Instructor's Manual includes answers to the textbook review questions, lab manual questions, and quizzes. It also contains visual masters to supplement the material in the textbook and lab manual. These masters can be reproduced as student handouts or overhead transparencies.

Chapter Resources

Chapter resources in the Instructor's Manual include learning objectives, listings of instructional materials, teaching strategies, and answers to all textbook review questions, lab manual questions, and quizzes.

Learning objectives

Each chapter of the text includes objectives to help you focus on the important concepts and skills presented in that chapter.

Instructional materials

The materials available for each chapter of **Heating and Cooling Essentials** are listed. Review the appropriate instructional materials before teaching a chapter. Supplementary materials should be prepared in advance for use at the appropriate time.

Teaching strategies

Suggestions and strategies for effectively using the textbook, lab manual, and visual masters are provided. The strategies should be tailored to fit the needs of the curriculum. Other activities and strategies can be developed from ideas presented in the chapter resources. Maintain a file of ideas for future use. The teaching strategies in this Instructor's Manual share these characteristics:

- *Reinforcement.* It is generally accepted that the more ways a student is exposed to a given concept, the greater the chances for his/her understanding and retention of the material. A variety of learning experiences have been designed to meet the reinforcement needs of your students.
- *Extending.* The teaching strategies are directed to students at varying levels of ability and encourage highly motivated students to extend their learning experience. These activities allow students to relate text information to other experiences, particularly life skills.
- *Enrichment.* Enrichment activities are designed to help students learn more about topics introduced in the text. These types of activities, such as research and surveying, give students the opportunity to enrich their learning through more in-depth study.
- *Reteaching.* Studies have shown that students respond differently to various teaching methods and techniques. Therefore, these materials provide suggestions for several strategies for teaching the concepts in the text. This allows you to choose a different strategy to reteach students who showed a low response to a previous strategy.
- *Higher-order Thinking Skills.* Higher-order thinking skills encourage students to become involved in activities beyond memorization. They include analysis, troubleshooting, and problem-solving. Students must extract concepts from the text, combine them with other concepts, and develop creative solutions. Many of the activities in the lab manual involve some degree of problem-solving and synthesis. Students are required to utilize information found in the text and apply it to the activities. The process allows students to

develop skills in critical and creative thinking, problem-solving, and decision-making.

Answers keys

The correct answers to all textbook review questions, lab manual questions, and quizzes are included in the chapter resources.

Visual masters

For each chapter, four to seven reproducible visual masters are included in the chapter resources, immediately following the answer keys. These masters can be used to create overhead transparencies or handouts. Visual masters can also be projected using an opaque projector.

Chapter quiz masters

The reproducible quiz masters are designed to supplement the questions in the textbook and the lab manual. They are located at the end of the chapter resources, following the visual masters. Quiz masters can be used as chapter quizzes or as study guides for chapter examinations. In some cases, it may be beneficial to project an overhead transparency of the quiz while the students work at their desks. In this case, be sure to enlarge the quiz on a photocopy machine before making a transparency so it can be read by those sitting farthest from the projection screen.

Developing a Course of Study

A course of instruction may require a two-, three-, or four-year study plan. The course of study often includes traditional academic subjects, such as communications, humanities, social sciences, and physical sciences that are taught by other instructors. There are four types of programs providing courses of study in the HVACR field. These include:

- *Occupational.* All instruction is performed in the refrigeration laboratory and a classroom. One instructor teaches both technical and related subjects. The school often includes evening classes for adults attending part-time. The occupational course of study is normally completed in about two years or 1800 school hours.
- *Trade.* Half of the student's time is devoted to laboratory experiences; the other half is spent in academic classes taught by other instructors. Trade courses usually require two years of study with about 1200 hours of laboratory work.
- *Technical.* One-fourth of the student's time is spent in the refrigeration laboratory; the other three-fourths is spent in academic subjects. This is a two-year program with about 600 hours of refrigeration laboratory experience. It is designed to prepare students for entry-level jobs as technicians.
- *Pre-engineering.* One-sixth of the student's time is spent in the laboratory, and the other five-sixths is devoted to studying communications, humanities, social sciences, physical sciences, and other academic subjects. The pre-engineering course of study is designed to lead into a college engineering degree program.

Basic Course Outline

A basic outline for a suggested course of study is given on the following pages. Because the textbook is the heart of any program, the outline was derived by selecting some tasks and objectives from each textbook chapter. The course of study is intended for the first year of a two-year vocational/technical or college training program. The outline can be adjusted to suit individual needs, available laboratory equipment, or methods of instruction. State curriculum guidelines may require certain modifications or additions. The ultimate goal of any course of study is to develop a systematic program that incorporates good instructional techniques on a daily basis. A course outline is the first step toward accomplishing professional instruction on a day-to-day basis.

Basic Course Outline

Orientation
- Tour the laboratory.
- Assign work areas and lockers.
- Issue safety glasses and mark new tools.
- Organize the laboratory.
- Develop rules and procedures.

Hand Tools (Chapter 1)
- Identify types of hand tools.
- Select the proper hand tool for a particular job.
- Demonstrate proper use of wrenches.
- Demonstrate proper use of pliers.
- Describe the do's and don'ts of using each hand tool.

Fasteners (Chapter 2)
- Recognize different thread forms.
- Demonstrate proper use of taps and dies.
- Identify screw types and sizes.
- Correctly use bolts, nuts, and washers.
- Identify and use masonary anchors.
- Demonstrate proper use of various fasteners.

Working with Copper Tubing (Chapter 3)
- Identify types and sizes of copper tubing.
- Show how to properly cut and ream copper tubing.
- Demonstrate proper use of tube bending tools.
- Show how to make proper flared connections.
- Show how to make proper swaged connections.
- Identify various fittings and sizes.

Working with Pipe (Chapter 4)
- Identify various sizes of steel pipe.
- Identify various sizes of plastic pipe
- Identify type and size of various fittings.
- Show how to cut threads on steel pipe.
- Demonstrate the correct way to solvent weld plastic pipe.
- Describe how to anchor and hang pipe.

Soldering (Chapter 5)
- Demonstrate proper use of an air-fuel torch.
- Describe proper safety with air-fuel torches.
- Demonstrate the connection and operation of pressure regulators.
- Discuss how to select proper alloys and fluxes for copper tubing.
- Show how to solder soft and hard copper tubing.

Brazing and Flame-cutting (Chapter 6)
- Connect and operate oxyacetylene equipment in a safe manner.
- Select appropriate brazing alloys and fluxes.
- Demonstrate a carburizing, oxidizing, and neutral flame.
- Show how to braze copper tubing with various brazing tips.
- Show how to braze copper tubing with a multiflame tip.
- Set up, adjust, and use an oxyacetylene torch for cutting metal.

Mathematics for Technicians (Chapter 7)
- Demonstrate how to read a ruler.
- Make linear measurements.
- Estimate the area of the laboratory.
- Calculate the volume of a piston cylinder.
- Convert temperature readings from Fahrenheit to Celsius.
- Record time sheet information accurately.
- Calculate profit margin on selected parts.

Basic Thermodynamic Principles (Chapter 8)
- Describe the ways heat is transferred.
- Describe how a substance changes its physical state.
- Explain how sensible, specific and latent heats differ.
- Explain how the British Thermal Unit is calculated.
- Name the five latent heats.
- Define saturated conditions, superheat, and subcooling.
- Calculate a ton of refrigeration.

Temperature and Pressure (Chapter 9)
- Identify four types of thermometer scales.
- Explain the difference between absolute pressure and gauge pressure.
- Describe vacuum measurement.
- Explain Boyle's, Charles', and Dalton's Gas Laws.
- Identify the high-pressure gauge and the compound gauge on a manifold gauge set.

Basic Refrigeration Cycle (Chapter 10)
- Draw a diagram of the basic refrigeration cycle.
- Describe the condition of the refrigerant in each component.
- Explain the purpose of each component in the cycle.
- Illustrate the two division points in the system that separate high and low pressure.
- Identify variations in basic system components.

Basic Course Outline (continued)

Other System Components (Chapter 11)
- Distinguish between condensing and evaporating units.
- Demonstrate proper use of service valves.
- Show how to change a commercial filter-drier.
- Demonstrate system pumpdown.
- Show how to install and properly use a gauge manifold set.

Refrigerants (Chapter 12)
- Use a gauge manifold set to check refrigeration system pressures.
- Use temperature-pressure charts.
- Distinguish disposable cylinders from refillable cylinders.
- Show how to check systems for the type of refrigerant.
- Demonstrate how to determine superheat on a commercial system.
- Identify and explain various refrigerant applications.

Refrigerant Recovery and Recycling (Chapter 13)
- Recognize CFCs, HCFCs, and HFCs.
- Use recovery cylinders properly and safely.
- Recognize recovery and recycle equipment.
- Demonstrate correct methods for recovering refrigerant and recharging a commercial system.

Zeotropic Blends (Chapter 14)
- Determine temperature glide.
- Read and use a temperature-pressure card for blends.
- Find bubble point and dew point.
- Appropriately select and use lubricants for blends.
- Perform system retrofit to a blend.

Troubleshooting Refrigerant Problems (Chapter 15)
- Use soap bubbles for leak detection.
- Use a halide torch for leak detection.
- Use an electronic detector for leak detection.
- Demonstrate how to predict and correct low-side pressures.
- Discuss how to recognize and eliminate causes of high head pressure.
- Show how to use electronic scales and graduated cylinders for charging.

Working with Refrigerant Controls (Chapter 16)
- Install and adjust an automatic expansion valve system.
- Install a capillary tube.
- Install and adjust a thermostatic expansion valve.
- Draw a multiplexed evaporator system and explain its advantages.
- Show how to change an orifice on an air conditioning system.

Special-purpose Valves (Chapter 17)
- Discuss the proper use of hand valves.
- Demonstrate how to adjust EPR and CPR valves.
- Demonstrate how to adjust head pressure control valves.
- Demonstrate how to adjust hot gas bypass valves.
- Describe how to check operation of cylinder unloaders.
- Draw a system using a desuperheating TEV.

Troubleshooting Flow Controls (Chapter 18)
- Troubleshoot and adjust an AEV system.
- Troubleshoot a capillary tube system.
- Troubleshoot a TEV system.
- Troubleshoot an orifice system.
- Troubleshoot a solenoid valve system.

Compressors (Chapter 19)
- Dismantle and reassemble an open-type reciprocating compressor.
- Dismantle and reassemble a semihermetic reciprocating compressor.
- Cut open and examine a hermetic reciprocating compressor.
- Dismantle, clean, and reassemble a water-cooled condenser.
- Show how to align pulleys and replace V-belts.

Compressor Lubrication and Accessories (Chapter 20)
- Show how to add oil to a compressor.
- Show how to rem ove oil from a compressor.
- Describe and demonstrate installation of vibration eliminators.
- Describe how to interpret data plate information.
- Dismantle and reassemble oil pump and separator.

Water Chillers (Chapter 21)
- Explain the operating principles of vapor-compression-type chillers.
- Describe the operation of a centrifugal chiller.
- Explain temperature difference in a closed loop.
- Describe the operation of a low-pressure chiller.
- Explain the operating principles of absorption-type chillers.
- Explain the operating principles of water-ammonia absorption-type refrigerators.

Basic Course Outline (continued)

What Is Electricity? (Chapter 22)
- Demonstrate proper use of test instruments.
- Discuss sizes, types, and uses of conductors.
- Describe how to identify and properly use fuses and circuit breakers.
- Show how to correctly connect wires to plugs, receptacles, and switches.
- Demonstrate how to safely connect portable power tools.

Power Transmission and Circuits (Chapter 23)
- Discuss how to select and replace fuses and circuit breakers.
- Identify common voltages and discuss their uses.
- Show how to make good terminal connections.
- Demonstrate the proper method for connecting a GFCI.
- Show how to select and properly connect a transformer.

Induction Motors (Chapter 24)
- Show how an ohmmeter can be used to identify hermetic motor terminals.
- Troubleshoot induction motor problems.
- Demonstrate how to properly connect multispeed and dual-voltage motors.
- Dismantle and reassemble five types of induction motors.
- Test and install motor capacitors.

Electromagnetic Control Devices (Chapter 25)
- Install and troubleshoot solenoid valves.
- Connect and troubleshoot relay circuits.
- Connect and troubleshoot contactors and line starters.
- Describe how to select and connect a potential relay.
- Describe how to read and use pictorial and schematic diagrams.

Motor Controls (Chapter 26)
- Install and adjust primary motor controls.
- Install and adjust safety controls.
- Troubleshoot thermostatic and pressure-type motor controls.
- Demonstrate how to adjust a pumpdown cycle.
- Troubleshoot oil pressure safety control.

Defrost Cycles (Chapter 27)
- Show how to check operation and wiring connections of various timeclocks.
- Operate and adjust an "off cycle" defrost system.
- Operate and adjust an "off time" defrost system.
- Operate and adjust an electric defrost system.
- Operate and adjust a hot gas defrost system.
- Demonstrate how to connect and troubleshoot defrost termination thermostats.

Ductwork (Chapter 28)
- Identify the components of the supply and return air duct systems.
- Show how to assemble rectangular duct.
- Distinguish between the proper uses of round pipe and rectangular duct.
- Lay out and properly size the ductwork for an extended plenum system.
- Install floor diffusers and return grilles.
- Use mastic and fiberglass reinforcing membrane to seal joints in a duct system.

Gas Heat with Air Conditioning (Chapter 29)
- Identify the components of a gas heating system.
- Show how to adjust and check operation of burner and fan/limitcontrols.
- Connect and check high-voltage wiring circuits.
- Connect and check low-voltage wiring circuits.
- Show how to wire a two-speed fan motor.

Oil Heat with Air Conditioning (Chapter 30)
- Demonstrate how to properly adjust a fuel/air mixture.
- Change a nozzle and adjust electrodes.
- Connect and check high-voltage wiring circuits.
- Connect and check low-voltage wiring circuits.
- Check the operation and placement of safety controls.

Electric Heat with Air Conditioning (Chapter 31)
- Identify components in an electric heating system.
- Connect and check high-voltage wiring circuits.
- Connect and check low-voltage wiring circuits.
- Troubleshoot an electric heating system.
- Troubleshoot a two-stage electric heat system with air conditioning

Heat Pumps (Chapter 32)
- Identify and explain the purpose of major heat pump components.
- Describe the operation of the reversing valve.
- Check operation of different types of defrost systems and controls.
- Show how to read and understand heat pump schematics.
- Troubleshoot a heat pump system.

Customer Relations (Chapter 33)
- Demonstrate how to shake hands properly.
- List ways individuals can improve their appearance.
- Demonstrate how to begin and end service calls properly.
- Cite examples of good (and poor) customer relations.
- Demonstrate how to deal with angry customers.

Instructional Guides

The course outline breaks the course of study into chapters, or specific blocks of study. Each block of study, in turn, is assigned a specific time frame in which to be accomplished. This information can be used to develop more specific day-by-day instructional guides describing exactly how the material will be presented and the order the instruction will follow. Depending upon the instructor's needs and chosen methods, these instructional guides may take the form of course content sheets, strategy sheets, lesson plans, and job sheets. Brief descriptions of each type of guide follow. Samples are provided and may be used as models for teacher-developed materials.

Course content sheets

A course content sheet is rather specific and is needed for each block of study. Each laboratory task is listed in the sequence it is performed. A specific time allotment (in hours) is stipulated for related instruction and laboratory instruction. The content sheet shows the direct relationship between laboratory experiences and related (classroom) instruction. See page 13 for a sample course content sheet.

Strategy sheets

A strategy sheet describes how each block of study will be managed, instructed, and graded. It is a logical follow-up to the content sheet. See page 14 for a sample strategy sheet.

Lesson plans

The lesson plan describes the exact sequence and method of instruction. Lesson plans are numbered and must be taught in the proper sequence since several lesson plans may be required to complete one block of study. Lesson plans are a readily available, organized guide to instruction, serving as a reminder of essential material to cover. A completed set of lesson plans often becomes a teacher's essential tool for maintaining program continuity and is also ideal for use by substitute teachers. Lesson plans help ensure a smooth and accurate program each day. See page 15 for a sample lesson plan.

Information sheets

An information sheet is often attached to a lesson plan to provide essential information for teaching that lesson. Information sheets are also helpful for constructing tests and quizzes. See page 16 for a sample information sheet.

Job sheets

Job sheets are guides for hands-on activities developed by the instructor to supplement those provided in the **Heating and Cooling Essentials Lab Manual**. Like the activities in the lab manual, the job sheets guide the student through a hands-on project in a step-by-step manner. A job sheet should inform the student of the tools needed, describe safety procedures, clearly identify each step, and list the order in which steps should be performed. Completed lab manual activities and job sheets are graded, and the student's progress is noted on a progress chart (see below). See page 17 for a sample job sheet.

Progress Chart

A progress chart is frequently used in the laboratory to assign students to a sequence of activities and maintain a record of each student's progress. The activities listed on the chart may include both numbered activities from the **Heating and Cooling Essentials Lab Manual** and customized job sheets developed by the instructor (see above). The chart is maintained by the instructor and posted in the laboratory. See page 18 for a sample progress chart. The chart contains an empty square for each activity, which can be assigned and then marked "completed" or "repeat." Grades should never be placed on the chart; rather, they should be recorded in a separate and private gradebook.

The progress chart aids the instructor in several ways. It:

- Automatically assigns students to projects.
- Eliminates the need to answer the question, "What do I do next?"
- Identifies high and low achievers.
- Indicates which students need help or assignment to another group.
- Creates healthy competition among students.
- Shows fairness and impartiality.

The progress chart also helps the student in several ways. It:

- Indicates what is expected each day.
- Avoids standing in line to ask the instructor for an assignment.
- Tells what other students are doing.
- Develops organization and self-discipline.
- Motivates individuals to work harder to compete with others.

The progress chart may be constructed for use on a weekly or monthly basis, or it may cover a complete grading period. It should include enough appropriate activities to keep the students involved in more projects than can be completed in the allotted time. Unfinished projects are then carried over to the next chart. High-achieving students often are asked to assist classmates who work at a slower pace, providing a rather uniform advancement for the entire class.

Course Content Sheet

Time Allotment:
Related: 22.5 hours
Laboratory: 45 hours
Total Hours: 67.5
Total Days: 15

Block 16
Working with Refrigerant Controls

Objective: Upon completion of instruction, the student will have a good working knowledge of all refrigerant controls and how they relate to the system.

Laboratory (Shop)		**Related (Classroom)**	
Task:		**Lesson Plan #:**	
16.1	Dismantle and reassemble TEV and AEV trainers.	1.	Purpose and location.
16.2	Operate and adjust AEV settings on trainer.	2.	Operation of AEV.
16.3	Operate and adjust superheat on TEV trainer.	3. 4.	Operation of TEV. Superheat and the TEV.
16.4	Operate and adjust refrigerant charge on capillary trainer.	5. 6.	How capillary tube functions. Capillary tube systems.
16.5	Install TEV, and set superheat.	7. 8.	How to install TEVs. How to adjust superheat.
16.6	Recover, evacuate, and recharge capillary tube system.	9. 10. 11.	Principles of evacuation Vacuum and dehydration. Charging capillary tube system.
16.7	Install filter-drier.	12.	Purpose of filter-driers.
16.8	Change orifice on AC system.	13.	Sizes and types of orifice.
16.9	Troubleshoot TEV system.	14.	Troubleshooting TEV system.

Sample Course Content Sheet

Strategy Sheet

Block 16
Working with Refrigerant Controls

Management Strategy: The student will be furnished complete related instruction regarding the different types of refrigerant controls and their application to the system.

Instructional Strategy: There will be 15 days of instructional and laboratory experiences allowed for this instructional block. Each group of three or four students will rotate the trainers, depending upon how the groups complete the previous assignment. Faster students will be permitted to proceed while slower groups complete unfinished assignments. At this time, it may be necessary to reassign some students to another group in order to compensate for learning difficulties. The instructor will explain and demonstrate the use and purpose of the trainers, and the students must follow and complete the job sheet for each task.

Evaluation Strategy:

Related: The student will be required to orally identify various refrigerant controls and explain how they are adjusted. About every four days of instruction, the student must pass a pencil and paper test with at least 70% accuracy.

Laboratory: Each task in this area must be completed to the satisfaction of the instructor. Job sheets will be evaluated on an individual basis. Student progress will be recorded on the Progress Chart.

Sample Strategy Sheet

Lesson Plan #12

Block 16
Working with Refrigerant Controls

Title: Purpose of filter-driers.

Task 16.7: Install a filter-drier.

Goal: To acquire the knowledge and skill required to change filter-driers.

Lesson (Performance) Objective: Upon completion of instruction, the student will be able to properly change a filter-drier and fully explain its purpose in the system.

Instructional Materials and Equipment: Various sizes of filter-driers (flare-type, sweat-type, residential, commercial); gauges; and hand tools.

References: *Heating and Cooling Essentials* by Jerry and LaDonna Killinger; training brochures from Sporlan, Alco catalog.

Introduction: Filter-driers perform a valuable function and are found in most systems. Located in the liquid line and suction line, they are direction-sensitive and replaced often. Flared and sweat connections are the most common and must be leakproof.

Outline of Objectives (Steps)	Key Points
A. Purpose of filter-drier	A. Filters and cleans refrigerant and oil
B. How the filter-drier works	B. Molecular sieve vs core
C. Location in system	C. Liquid line vs suction line
D. How to size a filter-drier	D. Tubing connection vs capacity
E. Flare vs sweat connections	E. Advantages of each type
F. When to replace	F. Identify need to replace
G. How to replace	G. Critical how-to points
H. Install gauges	H. Why gauges are required
I. Pumpdown and eliminate vacuum	I. How to replace
J. Leak testing	J. Safety

Application: Student will be issued a filter-drier to be properly installed in a trainer system. Gauges must be installed and the system pumped down. Release enough refrigerant to eliminate the vacuum to zero (1 psig to 2 psig). Remove and replace the filter drier. Release refrigerant and leak test.

Evaluation: Students will be graded on the replacement of a filter-drier based on their use of tools, communication of knowledge, practice of safety, and participation. Progress can be recorded on the Progress Chart.

Sample Lesson Plan

Information Sheet for Lesson Plan #12

Block 16
Working with Refrigerant Controls

Lesson Title: Purpose of filter-driers.

Task 16.7: Install a filter-drier.

Essential Information:

1. A properly sized filter-drier does not restrict refrigerant flow. A restriction causes pressure drop across the filter-drier, and the drier becomes cold. A restricted filter-drier indicates a dirty system because moisture does *not* restrict a filter-drier. When a filter-drier is full of moisture, further moisture passes through unrestricted.
2. Factory seals should not be removed until the last minute. When a filter-drier is opened to the atmosphere, it will absorb moisture from the air.
3. Do not attempt to reactivate a used filter-drier. A filter-drier retains trapped oil, acid, and moisture. It should be discarded when no longer effective.
4. Do not open the system when it is in a vacuum. Air rushes in and defeats the purpose of the filter-drier. When changing a flare-type filter-drier, use the system refrigerant to eliminate the vacuum after a pumpdown. The system is pressurized to about 1 psig or 2 psig before removing the old filter-drier. A small amount of refrigerant is lost by this minor purging procedure, but it is accepted practice and perfectly legal. Air does not enter the system because there is no pressure difference between the inside and the outside.
5. When a capillary tube system is opened for repairs, the filter-drier should always be replaced. A single drop of excess moisture or a foreign material will plug tiny orifices.

Sample Information Sheet

Job Sheet

Block 16
Working with Refrigerant Controls

Name ______________________________

Date______________________________

Grade ______________________________

Task 16.7: Replacing a filter-drier on a capillary tube system

Safety: Safety glasses are required. Beware of burns and pinch spots. Many procedures in this exercise can be dangerous to you and others. Exercising good safety procedures is part of your grade.

Equipment and Materials: Gauge manifold, adjustable wrench, valve key wrench, refrigerant recovery unit, recovery cylinder, oxyacetylene outfit, brazing rod, wet rag, new filter-drier, vacuum pump, nitrogen cylinder with regulator, and soap solution.

Procedure:

1. Install gauge manifold and connect the recovery unit to the center hose on the gauge manifold. Use good recovery procedures to remove refrigerant to the recovery cylinder. Save this refrigerant for later recharging into the system. Do not purge refrigerant to the atmosphere.
2. Connect the nitrogen regulator to the manifold center hose, and sweep nitrogen through the SSV at about 2 psig. Permit nitrogen to escape to the atmosphere through the DSV. This removes all air and moisture, and prevents oxidation during the brazing procedure. Oxidation is a major cause of sludge and oil contamination.
3. Carefully use the oxyacetylene torch to remove the old filter-drier. Install a new filter-drier. These procedures can be dangerous and require skill. Use care to avoid plugging the capillary tube with alloy when brazing.
4. After the filter-drier is properly installed and cooled with a wet rag, close the DSV. Adjust the nitrogen regulator to 300 psig, and use a soap solution to check for leaks. When the system is leakproof, close the cylinder valve on the nitrogen, and release all system pressure (nitrogen) to the atmosphere.
5. Connect the vacuum pump to the gauge manifold, and properly evacuate the system. Now is a good time to return the oxyacetylene torch, nitrogen cylinder, and recovery unit to proper storage. Try to keep your work area organized and clear of unnecessary objects. A cluttered work area indicates confusion, causes accidents, and results in sloppy work.
6. When the system is properly evacuated, recharge the refrigerant vapor into the SSV until the system pressure and cylinder pressure are equalized. Now turn the refrigeration system on, and complete the charging process into the SSV as a vapor. Charge the system until the suction and discharge pressures are correct. Have the job approved by your instructor.

Sample Job Sheet

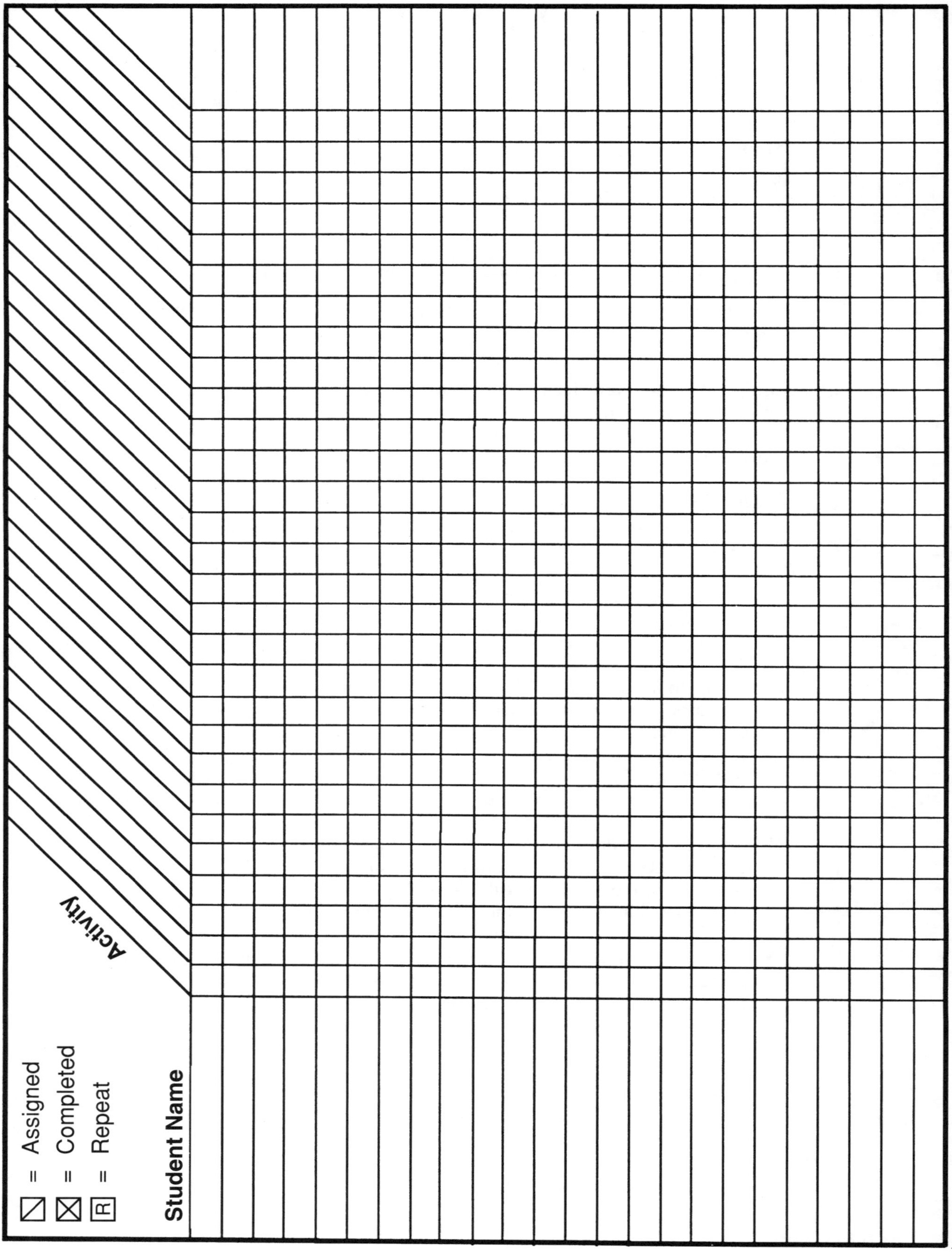

Sample Progress Chart

Teaching Methods

While the textbook is the heart of the program, it should be supplemented with lectures, overhead transparencies, audiovisual materials, demonstrations, speakers, the lab manual, and other resources to stimulate and maintain student interest. Employ a variety of teaching methods to present the subject matter. Teaching methods will vary according to the instructor's experience and approach, but one premise remains constant: learning takes place when students "*hear, see, and do.*"

Lectures

Students typically retain only about 10% to 15% of what they hear in a lecture presentation that relies solely on the spoken word. Therefore, lectures should be short, informative, and interesting. Analogies are helpful since associating new material with something familiar improves comprehension and retention. Verbal cues, such as, "Remember, this is going to be on the test," help capture the students' attention and increase retention. Occasional anecdotes bring information to life and help students make the connection between theory and "real world" situations.

Too many instructors place all the responsibility on the student: he or she is expected to listen carefully, behave in class, and study diligently. A good instructor is well-prepared and uses multiple techniques to make the material interesting and understandable. Students respond favorably to good instructors and negatively to poor instructors. Be friendly, sincere, considerate, and above all, *be prepared.*

Audiovisual Materials

Good audiovisuals, when combined with a well-prepared and interesting lecture, greatly increase information retention. The use of overhead transparencies is preferred to drawings on the blackboard. Hastily made drawings often lack quality and contain errors. Both time and student attention is lost while waiting for the instructor to draw on the board.

Overhead transparencies

A picture is worth a thousand words, and overhead transparencies are an easy and inexpensive way to add visuals to your presentation. Multiple but simple transparencies are better than one or two complicated ones. Keep transparencies simple by resisting the temptation to include too much information on a single transparency. Make sure the text or figures can be clearly seen by students seated in the back of the room.

Transparencies are easily made from any drawing or photograph. Illustrations can be reduced or enlarged with a photocopier before making the transparency. A computer with graphic capability is an excellent tool for making graphic representations and then projecting them on transparencies.

The use of multiple transparencies along with the lecture method is probably the most common and effective teaching method.

Films, videotapes, and slide programs

If available, these audiovisual materials can be excellent supplements to your program. They give the students a different viewpoint and another voice. Such audiovisuals should be selected with care (always preview them before use) and should be pertinent to the program at the time of showing. Avoid programs that are boring, too long, or inappropriate. Never use audiovisuals simply to give yourself a break. Students recognize such duplicity and will react negatively to it.

Demonstrations

Demonstrations break up the monotony of the lecture/transparency presentation and greatly enhance the learning process. Demonstrations should always be given for complex procedures and when the possibility of injury exists. The instructor can make the demonstration alone but should involve members of the class whenever possible. Passing components around for students to see and touch also is in the realm of demonstration. An effective demonstration can serve as the "stepping stone" between classroom theory and hands-on laboratory experiences.

Hands-on Learning Experiences

Such activities typically occur in the laboratory and should be structured to provide maximum reinforcement of classroom instruction. Hands-on experiences provide the student with the necessary skills to enter the world of work. Hands-on projects should always be pertinent and challenging; they should never consist of boring activities designed simply to keep the students occupied.

The **Heating and Cooling Essentials Lab Manual** includes a number of hands-on activities relevant to each chapter of the textbook. Additional projects can be designed in each area of study, if desired, to challenge students and build skills. A project may require only a few minutes to complete, or it may involve two or three hours of effort by the student. Simple projects may not require a demonstration, while complicated projects may require not only a step-by-step demonstration but supervision and guidance as well.

Providing for Individual Differences

Students should be permitted to progress in accordance with their individual abilities, skills, and needs. With the increased emphasis on mainstreaming special needs students into vocational programs, classes are likely to include individuals who have various mental or physical impairments. Gifted students, with their own special needs, are also likely to be enrolled. Some special needs students may have physical limitations that require individualized teaching strategies or adaptation of the classroom or laboratory. Others may exhibit poor reading, writing, and computational skills, lack of motivation, short attention span, disruptive behavior, or other problems requiring specialized learning programs or teaching methods.

Special needs students usually need more time and assistance than other students. The curriculum may require occasional adjustment to accommodate both students with special needs and gifted students who may need additional challenges. A goal should be to provide challenges to every student in your class, regardless of ability. By meeting challenges based upon individual abilities, students will develop self-confidence and maintain a high level of interest in your program. Your grading system must also use a variety of criteria for evaluation because of the differing levels of abilities. It should also reflect the multiple learning styles of students: some are poor performers in the classroom but excel at hands-on activities in the laboratory; others may be "all thumbs" in the laboratory but excel in the classroom.

Pairing students of differing abilities, such as a gifted student and a student with a learning disability, is often mutually beneficial. The gifted student gains experience and maintains interest by helping another student; the student with special needs receives additional instruction that can enhance understanding and help him or her satisfactorily complete required activities. Pairing may also be beneficial in situations where some students have limited proficiency in English.

Basic Interdisciplinary Skills Chart

The Basic Interdisciplinary Skills Chart on page 21 identifies the chapters in **Heating and Cooling Essentials** that encourage the development of basic skills.

These interdisciplinary skills are an integral part of most vocational programs. The academic areas addressed include reading, writing, mathematics, science, social studies, communication (other than reading and writing), and problem-solving.

Classroom Management

The classroom should be a comfortable place for bringing students together for a specific period and type of instruction. Most classroom instruction involves theory, identification of parts, safety procedures, and other essential information needed to perform specific skills. Lectures, transparencies, demonstrations, and other methods are used to introduce material, explain terminology, and make students aware of service procedures. The classroom is also a place to prepare students for hands-on experiences. Such preparatory training is essential to prevent injuries and damage to tools and equipment.

Students are normally permitted to select their own seating at the beginning of the program. However, it should be explained that the instructor reserves the right to change the seating arrangement whenever the need occurs. Classroom instruction should begin at the appointed time. Any delay in starting time teaches bad habits. Therefore, the instructor should be in the classroom and prepared to begin well before the actual starting time. You cannot teach good work habits by setting a poor example.

Students should be encouraged to participate in the learning process by asking questions anytime. Questions are an indication of how the instruction is going. Failure of the students to ask questions may signal poor instruction or an atmosphere that discourages student involvement.

Do not teach to only the fastest learners or to the students who ask questions. Ask for a student to volunteer a different opinion, or ask a reticent student to answer a question. Go beyond teaching—help students learn. Students will become more interested and involved, and will pay closer attention if they never know when their turn will come to answer a question.

When students are not paying attention or lose interest, stop teaching and ask questions. This encourages your students to think about the instruction taking place. Call on individuals by name, and ask questions pertinent to the material just presented. Ask questions that can be easily answered. Never tell students they are wrong. Nobody likes to be wrong, especially in front of a group. Find ways to offer compliments, and do not criticize. Avoid making a "big deal" of student inattention; instead use some tricks to regain their attention, and proceed with the instruction. Maybe you expect too much, and a little humor or a change of direction is needed to relieve the stress or monotony.

Pausing periodically for oral questions on the material being covered is beneficial to both you and the

Basic Interdisciplinary Skills Chart

Chapter Number	Reading	Writing	Mathematics and Science	Social Studies	Problem-solving	Communication (other than reading/writing)
1	√	√	√			√
2	√	√	√	√	√	√
3	√	√	√	√	√	√
4	√	√	√		√	√
5	√	√	√		√	√
6	√	√	√		√	√
7	√	√	√		√	√
8	√	√	√			√
9	√	√	√	√		√
10	√	√	√			√
11	√	√	√			√
12	√	√	√	√		√
13	√	√	√	√		√
14	√	√	√			√
15	√	√	√		√	√
16	√	√	√			√
17	√	√	√			√
18	√	√	√		√	√
19	√	√	√			√
20	√	√	√			√
21	√	√	√			√
22	√	√	√	√		√
23	√	√	√			√
24	√	√	√		√	√
25	√	√	√		√	√
26	√	√	√		√	√
27	√	√	√		√	√
28	√	√	√		√	√
29	√	√	√		√	√
30	√	√	√		√	√
31	√	√	√		√	√
32	√	√	√		√	√
33	√	√			√	√

students. It becomes a time to stop and evaluate the learning process. Stopping points help reinforce important facts and uncover any weaknesses before it is too late. It also gives each student an opportunity to express lack of understanding and request another explanation.

Student Orientation

The first few class periods should include a student orientation and program organization period. Good organization at the beginning of the program can prevent many future problems. This period of introduction and indoctrination will set the tone for the entire program. The instructor should provide a guided tour of the entire facility and point out areas pertinent to the students' daily routine. A tour of the laboratory should include a brief but interesting explanation of the various equipment and trainers the students will be using. Students should be informed of areas where safety precautions must be taken, as well as any areas that are off limits. Make the students feel comfortable by being friendly, but let them know exactly where they stand. This orientation period should get the program off to the right start by showing that you are an instructor who is friendly yet firm.

Classroom Behavior Control

Experts advise the instructor always maintain control of the learning environment, which is sometimes easier said than done. The instructor should set an example of the type of behavior expected by being *fair, firm,* and *friendly.* It is important to maintain a balance among these three important qualities.

Being *fair* is probably the most critical of the three qualities. Students know when they are being ignored, singled out, or given special treatment. Students identify with their peer group: when one of them is being mistreated, the entire class will react with strong resentment. These feelings of resentment are very difficult to overcome. When you make a mistake or discover you are wrong, admit it! This simply shows that you are human, like everyone else, and that we all learn from mistakes.

Being *firm* is strictly a matter of judgment. Firmness, to one person, is seen as bullying by another. If you are too firm, you are perceived as mean or bossy. This often results in rebellion against your authority. If you are not firm enough, students may take advantage of what they perceive as a weakness and engage in disruptive behavior. Being *too friendly* also can be mistaken for weakness, with the same results. Treat your students the way you want to be treated. Respect them, and they will respect you.

Disruptive students in the classroom should be dealt with quickly and firmly. These "problem students" are often merely seeking attention, so try giving them opportunities to get attention in a positive manner. Give sincere compliments for good behavior and correct answers, since positive reinforcement works better than negative reinforcement. If these tactics fail, the student should be taken aside for a private discussion. The conversation should be two-way. Be a good listener, and ask questions. Show honest concern, and enlist the student in finding a solution to his or her disruptive behavior. Obtain a commitment from the student to improve; assure him or her of your help and cooperation. Likewise, warn the student that continued misbehavior will be dealt with harshly. The student must make the choice of which course to follow.

All students appreciate having rules and guidelines that establish appropriate and inappropriate behavior. How you enforce these rules determines your style of firmness. It is recommended that you start the year being a firm (but fair) disciplinarian. Let your students know that you will enforce all rules, regardless of the person who breaks them. As you establish control and learn the students' personalities, you can "loosen up." Never start by being easy and then trying to tighten up. This change in discipline policy often results in chaos.

A strong desire to help students is easily discerned and respected. However, little respect will be shown to an instructor who merely dispenses information. Continuously let your students know that you like them, but maintain an effective learning environment through proper organization and control. Remember, not all students are alike, so different techniques may be needed to resolve a problem. A good instructor must wear different hats for different students while making clear that he or she is maintaining a fair and even-handed approach. Most classroom problems develop because the instructor is unaware of what is taking place, is unprepared, or simply does not care.

Student Involvement

Getting students actively involved in the implementation of their program is usually desirable. The degree of involvement will vary from classroom to classroom, depending upon the type of program and the instructor's style of teaching. Election of officers and committee members is a fairly typical approach.

A temporary election of class officers should take place within one or two days of the beginning of the program. Either the students or the instructor may select the length of time until permanent elections take place (normally two or three weeks). During this time, students have an opportunity to get to know one another and see what each position involves. Officers to be elected will vary with each instructor. Suggested positions are President, Vice President, Secretary, Treasurer, Safety Committee (3 members), and Discipline Committee (3 members).

Duties performed by these elected officers vary, but it is strongly suggested that you give as much authority and power as possible to the officers while providing a high level of counseling and support. Delegation of authority and decision-making responsibilities in some areas provides a learning experience for the students and relieves the instructor to perform teaching duties. However, the instructor should always be available to *any* student for counseling, advice, or assistance of any kind.

The individual officers (President, Vice President, Secretary, Treasurer) perform duties typical of officers in most organizations, but they should be guided by written "job descriptions" provided by the instructor. The instructor must counsel and guide the officers, especially the President, in the performance of their duties.

The two committees play specific roles in the effective management of a program. The *Safety Committee* develops safety rules and regulations for use by all students working in the laboratory. The members report safety violations and hazards to the President or the instructor, depending upon the program.

Where implemented, the *Discipline Committee* typically has authority to decide individual punishment and enforce rules. The instructor must counsel and guide this committee until the members gain experience and confidence in their duties. To prevent favoritism and excesses, the instructor must retain the authority to accept or reject Discipline Committee decisions. Any rejections of decisions should be promptly and carefully explained to the committee.

In order to enforce the rules, Discipline Committee members should set personal examples of proper behavior. For this reason, you may wish to attempt to get "problem students" elected to the Discipline Committee. This would force such students to examine their own behavior and its consequences.

Laboratory Involvement by Students

The appointed position of *Shop Supervisor* is an important one in maintaining proper laboratory operation, since he or she is responsible for the day-to-day operation of the laboratory. A new Shop Supervisor is designated each week, working from the class roster on which students are normally listed in alphabetical order by last name.

The instructor issues orders to the supervisor, and the supervisor decides how and by whom the order is carried out. The instructor must support the supervisor's decisions, right or wrong. This procedure frees the instructor from some time-consuming daily tasks, permitting more time for instruction.

The opportunity of being the "boss" for one week is a valuable learning experience. Students are accustomed to receiving orders but are rarely given the opportunity to issue them. They need to experience both sides of the work environment. Students quickly discover that being fair is not easy and that somebody must do the unpleasant but necessary tasks.

All students should be taught how to properly answer the laboratory telephone; it is one of the duties they will perform as supervisor. Another duty of the supervisor is to have the entire laboratory cleaned up by quitting time. He or she must decide when all activity must stop and cleanup must begin. The supervisor assigns students to such duties as sweeping, mopping, and cleaning the restroom. He or she also gives a grade for each student on the day's performance. These grades must be explained to the instructor's satisfaction. The instructor, in turn, will grade the supervisor's daily performance.

Some students have natural leadership ability; others have none. It is the instructor's duty to help each supervisor become a successful leader. A student's first experience as a supervisor is the most difficult, with each succeeding term of duty becoming easier. The experience results in fewer disciplinary problems because students know how to both issue and accept orders. Delegation of authority enables students to learn from each other in a controlled environment.

A *Toolroom Supervisor* may also be selected weekly by the alphabetical system. The supervisor is responsible for the proper operation and maintenance of the laboratory tools, and is the only student permitted in the toolroom. He or she must issue all tools and maintain records to ensure their return, as well as keep a daily inventory of supplies used and needed. The Toolroom Supervisor is graded by the Shop Supervisor.

Testing and Grading

A written test should be administered in the classroom at least weekly, or more often if the material being covered is somewhat complicated. Written questions should be simply and clearly phrased, and structured so only one answer is possible. Frequent written quizzes consisting of a few questions are better than occasional examinations with large numbers of questions. Continual quizzes help keep the students active, interested, and involved, especially if they get good grades. Frequent testing also serves to reduce absenteeism. It must be remembered that test results are actually an indicator of the quality of instruction. Good student grades are the result of good instruction.

Student grades for work in the *laboratory* should be issued daily and be based upon the following criteria:

- Proper use of tools and equipment.
- Ability to take direction and accept criticism.
- Proper use of materials and supplies.
- Safe work habits.
- Quality of work.
- Interest in work.
- Quantity of work.
- Shop attitude.
- Finished in required time.
- Cleanup procedures.
- Ability to work with others.
- Attendance.

The student's daily laboratory grades are normally averaged to obtain a weekly grade.

Trade Associations

The organizations listed represent various facets of the HVACR field and are excellent sources of information. Some offer publications or educational materials you may find useful. Electronic mail or worldwide web (www) addresses are provided for computer access.

Air Conditioning Contractors of America (ACCA)
1712 New Hampshire Avenue, NW
Washington, DC 20009
(202) 483-9370
info@acca.org

Air Conditioning and Refrigeration Institute (ARI)
4301 North Fairfax Drive, Suite 425
Arlington, VA 22203
(703) 524-8800
ari@dgsys.com

American Boiler Manufacturers Association (ABMA)
950 North Glebe Road, Suite 160
Arlington, VA 22203
(703) 522-7350
www.abma.com

American Frozen Food Institute (AFFI)
2000 Corporate Ridge, Suite 1000
McLean, VA 22102
(703) 821-0770
asfi@pop.dn.net

American Gas Association (AGA)
1515 Wilson Boulevard
Arlington, VA 22209
(703) 841-8400
www.aga.com

American National Standards Institute (ANSI)
11 West 42nd Street
New York, NY 10036
(212) 642-4900
www.ansi.org

American Pipe Fittings Association (APFA)
111 Park Place
Falls Church, VA 22046-4513
(703) 538-1786
apfahq@aol.com

American Society of Heating, Refrigerating and Air Conditioning Engineers, Inc. (ASHRAE)
1791 Tullie Circle, NE
Atlanta, GA 30329
(404) 636-8400
ashrae@ashrae.org

American Society of Mechanical Engineers (ASME)
345 East 47th Street
New York, NY 10017-2392
(212) 705-7722
www.asme.org

American Solar Energy Society Inc. (ASES)
2400 Central Avenue
Boulder, CO 80301
(303) 443-3130
ases@ases.org

American Welding Society (AWS)
550 NW LeJeune Road
Miami, FL 33126
(305) 443-9353
www.aws.org

Association of Home Appliance Manufacturers (AHAM)
20 North Wacker Drive
Chicago, IL 60606
(312) 984-5800
www.aham.org

Commercial Refrigerator Manufacturers Association (CRMA)
1200 19th Street, NW, Suite 300
Washington, DC 20036-2401
(202) 857-1145
www.sba.com

Compressed Gas Association, Inc. (CGA)
1725 Jefferson Davis Highway, Suite 1004
Arlington, VA 22202
(703) 412-0900
www.cganet.com

Cryogenic Society of America Inc. (CSA)
1033 South Boulevard, Suite 13
Oak Park, IL 60302
(708) 383-6220
csa@huget.com

Gas Appliance Manufacturers Association (GAMA)
1901 North Moore Street, Suite 1100
Arlington, VA 22209
(703) 525-9565
gamaorg@aol.com

International Institute of Ammonia Refrigeration (IIAR)
1200 19th Street, NW, Suite 300
Washington, DC 20036-2401
(202) 857-1145
www.sba.com

Mechanical Contractors Association of America, Inc. (MCAA)
1385 Piccard Drive
Rockville, MD 20850
(301) 869-5800
postmaster@mcaa.com

National Association of Plumbing-Heating-Cooling Contractors (NAPHCC)
180 South Washington Street
Post Office Box 6808
Falls Church, VA 22040
(703) 237-8100
naphcc@naphcc.org

National Electrical Contractors Association (NECA)
3 Bethesda
Metro Center, 11th Floor
Bethesda, MD 20814
(301) 657-3110
www.necanet.org

North American Insulation Manufacturers Association, Inc. (NAIMA)
44 Canal Center Plaza, Suite 310
Alexandria, VA 22314
(703) 684-0084
insulation@naima.org

Refrigerating Engineers and Technicians Association (RETA)
401 North Michigan Avenue
Chicago, IL 60611
(312) 527-6763
www.reta.com

Refrigeration Service Engineers Society (RSES)
1666 Rand Road
Des Plaines, IL 60016-3552
(847) 297-6464
www.rses.org
rses@starnetinc.com

Sheet Metal & Air Conditioning Contractors' National Association (SMACNA)
4201 Lafayette Center Drive
Chantilly, VA 20151-1209
(703) 803-2980
info@smacna.org

Steel Tank Institute (STI)
570 Oakwood Road
Lake Zurich, IL 60047
(847) 438-8265

Valve Manufacturers Association of America (VMA)
1050 17th Street, Suite 7280
Washington, DC 20036
(202) 331-8105
www.vma.org

Goodheart-Willcox Welcomes Your Input

If you have comments, corrections, or suggestions regarding the textbook or its supplements, please send them to:

Managing Editor, Technology
Goodheart-Willcox Publisher
18604 West Creek Drive
Tinley Park, IL 60477–6243

1

Hand Tools

Objectives

After studying this chapter in the text, the student will be able to:

- ❑ Identify the various types of hand tools.
- ❑ Describe the advantages and disadvantages of each type of hand tool.
- ❑ Select the proper hand tool for the job.
- ❑ Demonstrate skills in using various hand tools.
- ❑ Maintain, repair, or replace damaged tools.
- ❑ Demonstrate good craftsmanship in the performance of daily tasks.

Instructional Materials

Text: Pages 13–26
Lab Manual: Pages 7–12
Instructor's Manual:
Visual Masters, Pages 29–34
Chapter 1 Quiz, Page 35

Teaching Strategy

Text

1. Review objectives.
2. Discuss and assign chapter contents. Use these visual masters as appropriate:
   HC 1-1: Wrenches
   HC 1-2: Socket Tools
   HC 1-3: Pliers
   HC 1-4: Drivers
   HC 1-5: Hammers
   HC 1-6: Drill Bits
3. Direct students to complete Test Your Knowledge questions, page 26. After completion, discuss correct answers with students.

Lab Manual

1. Provide overview of Activities. Where necessary, stress safe laboratory practices and the importance of wearing protective clothing.
2. Assign Activities to individuals or pairs of students, as appropriate.
   Activity 1-1: Identifying Tools
   Activity 1-2: Forming a Wire Loop
   Activity 1-3: Using Aviation Snips
3. Observe student performance on Activities, providing guidance or correction as needed. After reviewing the student's work, initial and date the activity in the place provided.

Evaluation

1. Administer the quiz from this Instructor's Manual.
2. Grade the quiz, and return it with appropriate comments for the student.

Answers to Test Your Knowledge, Page 26

1. mechanical
2. open-end; box; combination; flare; adjustable; Allen
3. open-end; adjustable
4. Use a box or socket because it is less likely to slip.
5. 6-point
6. true
7. electrical shock
8. lineman's; long-nose; diagonal; pump; slipjoint; locking
9. wire; metal
10. Any of the following: Torx; Allen; Posidrive; Phillips; clutchhead; Frearson; Robertson
11. ball pein

12. forward
13. false
14. d. Aviation.
15. struck

Answers to Lab Manual, Pages 7–10

Activity 1-1

1. A. open end
   B. box
   C. combination
   D. flare-nut
   E. Allen
2. Because an open-end wrench tends to slip, which could round nut edges and cause hand injuries.
3. box-end wrench
4. adjustable wrench

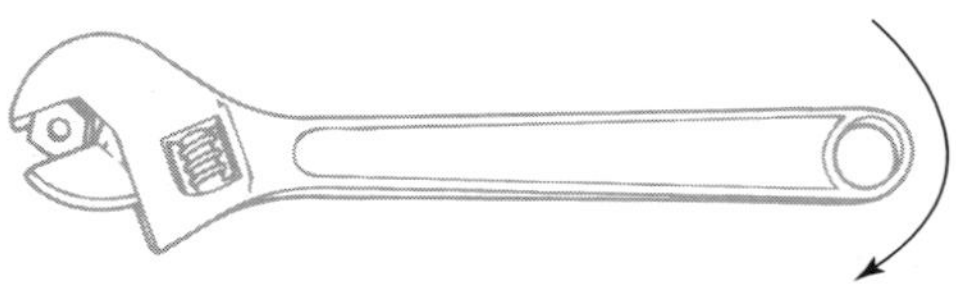

5. A. ratchet handle
   B. short socket
   C. extension bar
   D. universal joint
   E. deep-well socket
6. A. lineman's
   B. long-nose
   C. diagonal cutters (dikes)
   D. pump or groove joint
   E. slip-joint
   F. locking
7. true
8. A. Phillips
   B. clutch
   C. Torx
   D. Allen
9. A. reversible screwdriver
   B. nut driver
   C. screw-holding driver
10. A. nail or claw
    B. mallet or soft-faced
    C. engineer's or sledge
    D. ball pein
    E. setting
11. A. eye
    B. ball pein
    C. cheek
    D. neck
    E. striking face
    F. poll
12. chisels and punches; metal
13. setting hammer
14. A. twist
    B. spade-type wood bit
    C. auger-type wood bit
    D. masonry bit
    E. hole saw and kit
15. A. shank
    B. body
    C. point
    D. lips
    E. flutes
16. The flutes act as channels for the escape of metal chips from the hole being drilled.
17. wear safety goggles
18. A. straight
    B. aviation
    C. duckbill
19. They require less effort to cut metal than other types of snips.
    The blades are made of alloy steel, so they are also harder and tougher than regular snips.
    The cutting blades are serrated to prevent metal from slipping out when pressure is applied.
20. safety goggles and gloves
21. To direct the force of each hammer blow toward the center, or body, of the tool.
22. A. cold chisel
    B. pin punch
    C. center or prick punch
23. A damaged punch could slip and cause injury.
24. A grinder is used to remove the mushroomed metal and form a smooth chamfer. The cutting edge is sharpened with a hand file or a whet-stone.

Answers to Quiz

1. mechanical
2. the wrench
3. pull
4. lineman's pliers
5. b. 6-point.
6. ball pein
7. true
8. wood or metal
9. straight; left; right
10. chisels; punches

Wrenches

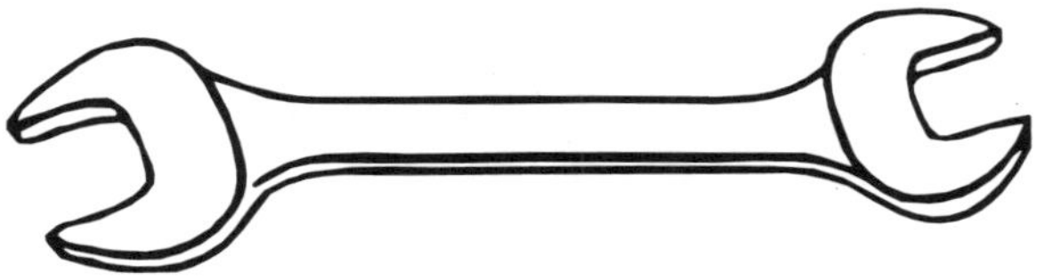

Open end

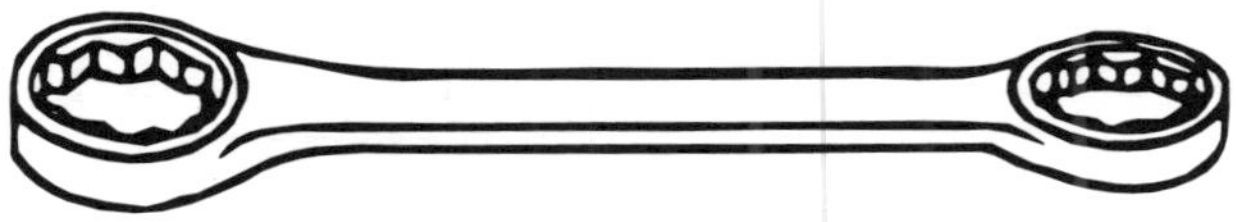

Box

Combination

Flare

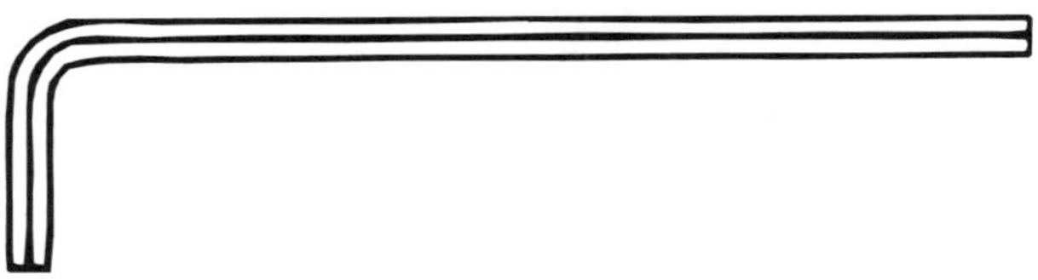

Allen

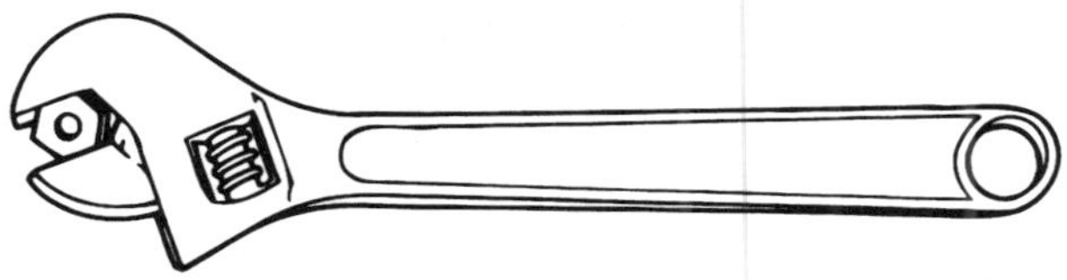

Adjustable

HC 1-1

Socket Tools

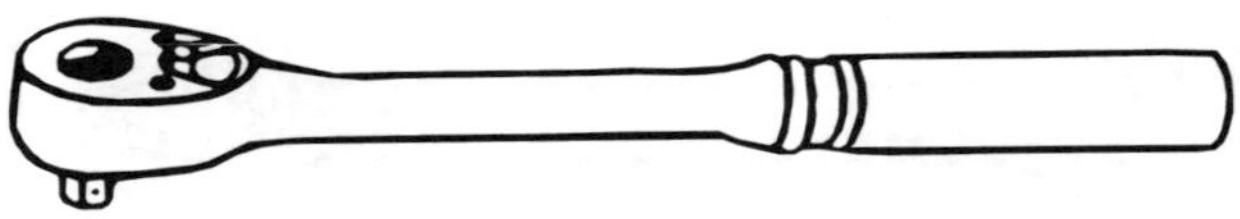

Ratchet handle

Short socket

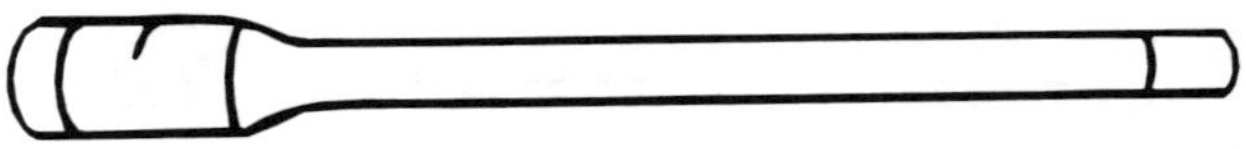

Extension bar

Universal joint

Deep socket

HC 1-2

Pliers

Lineman's

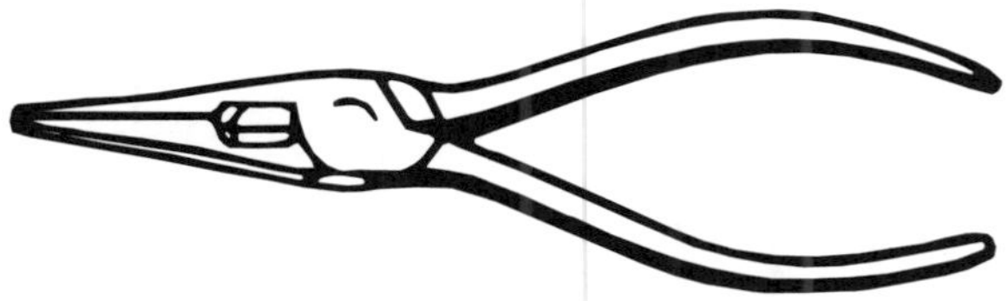

Needlenose

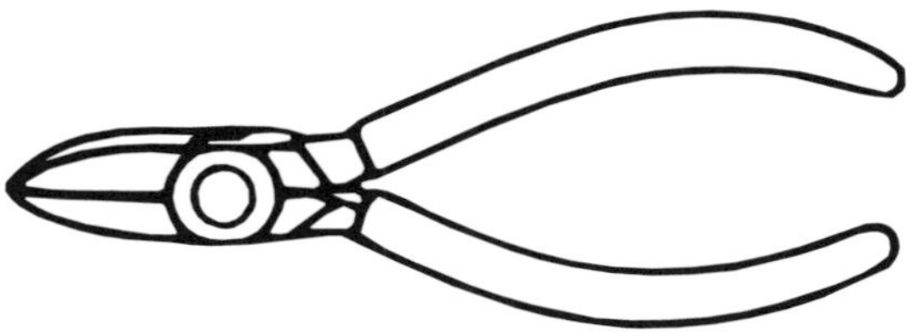

Diagonal

Pump, groove joint

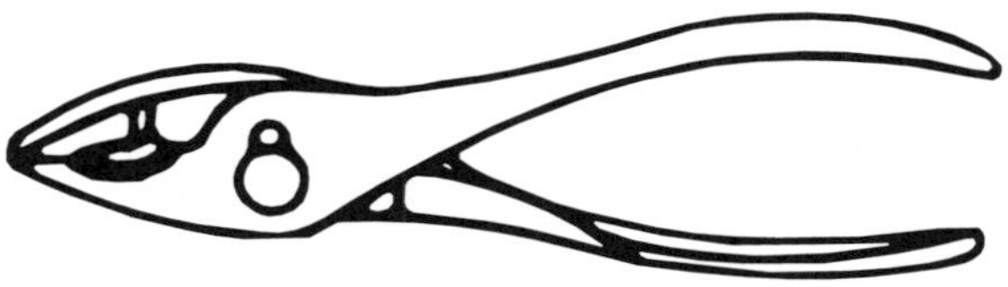

Slip joint

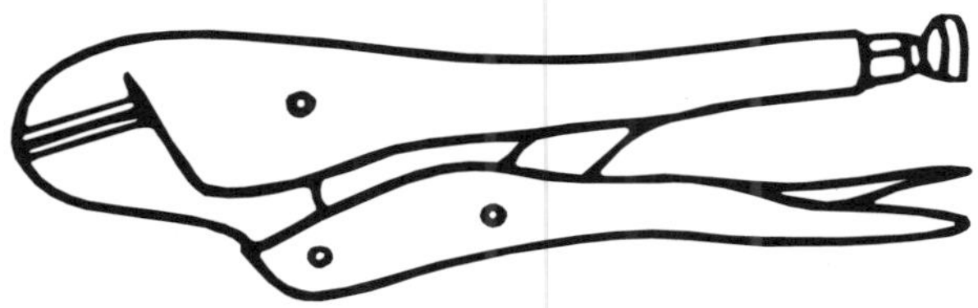

Locking

Drivers

Phillips

Torx

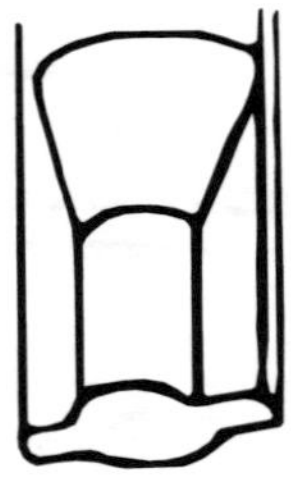

Clutch

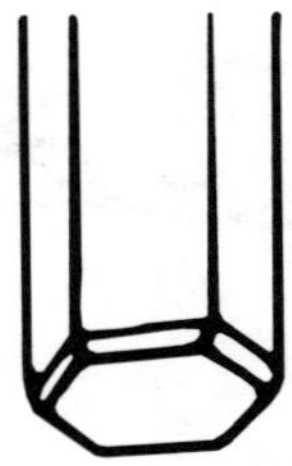

Allen

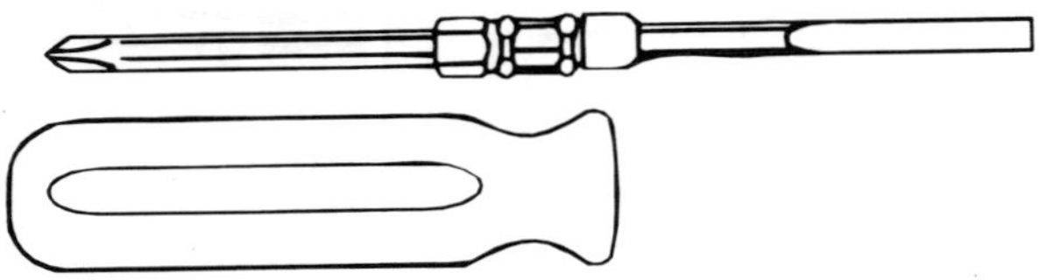

Reversible

Nut driver

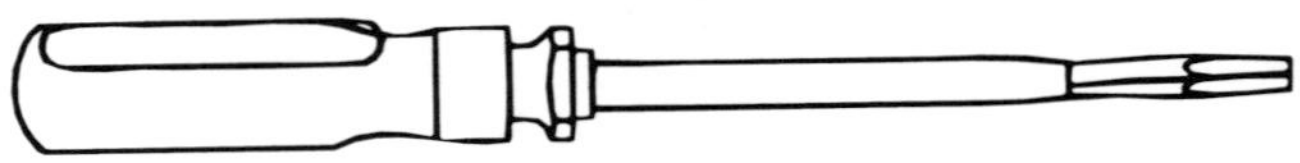

Screw-holding

HC 1-4

Hammers

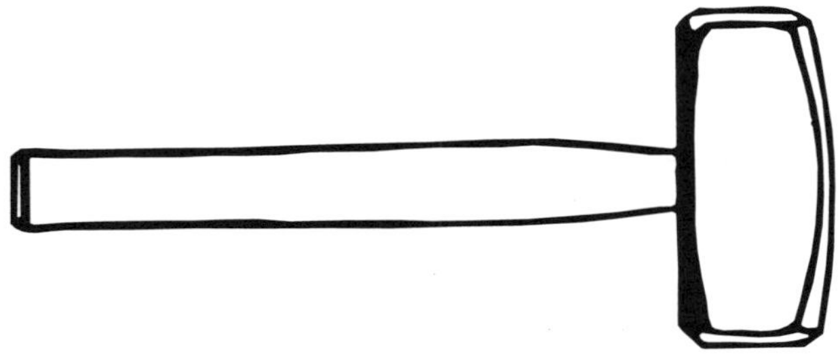

Engineer's

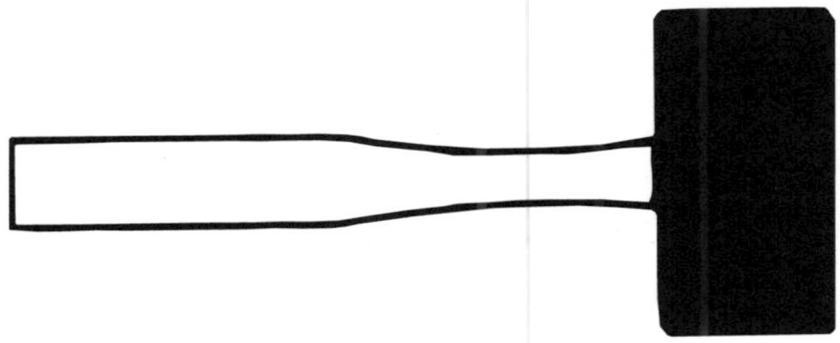

Soft-faced

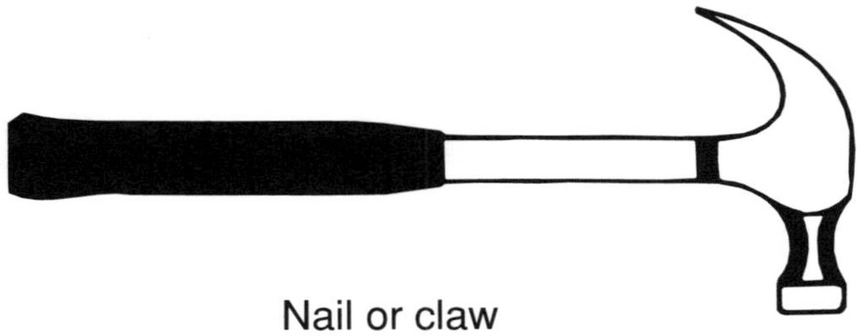

Nail or claw

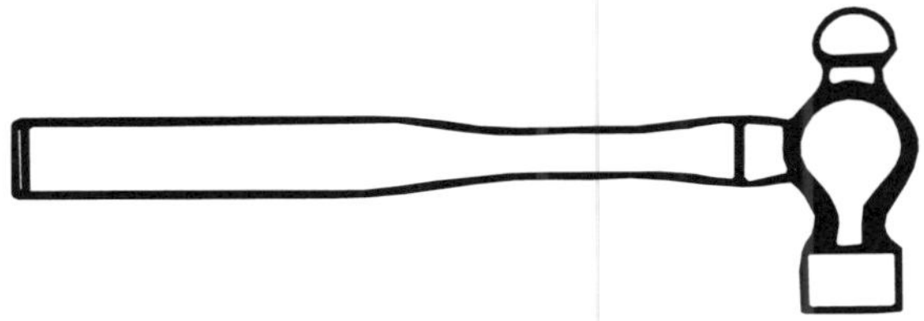

Ball pein

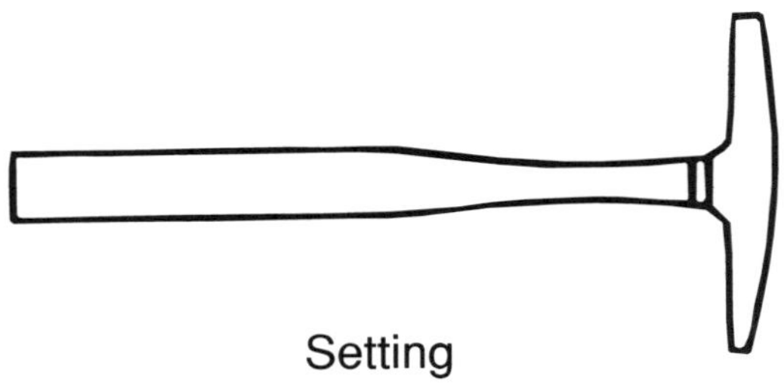

Setting

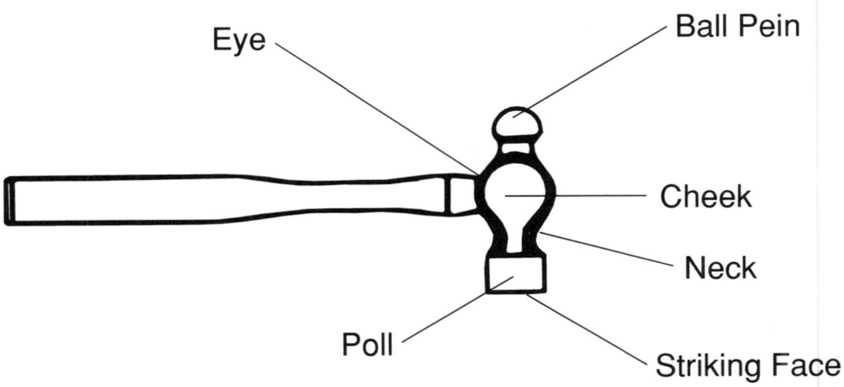

Parts of the Hammer

HC 1-5

Drill Bits

Twist

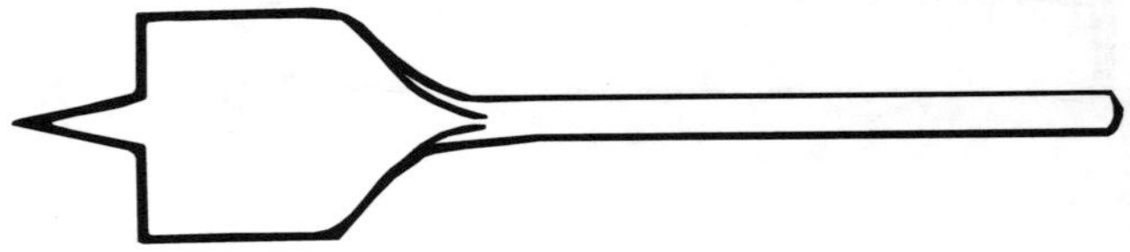

Spade

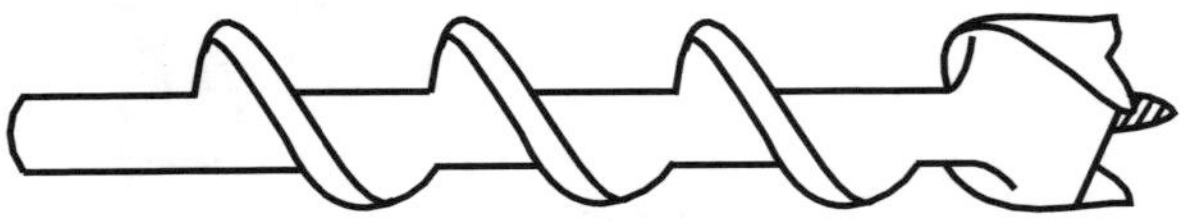

Auger-type (wood)

Concrete

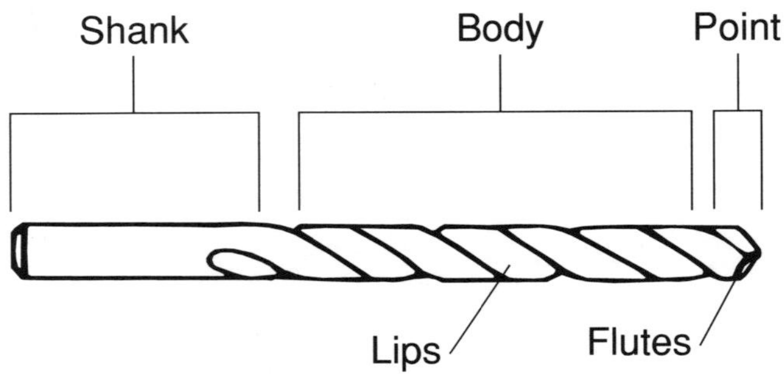

Parts of a Drill Bit

HC 1-6

Name ______________________________

Date ______________________________

Instructor ___________________________

Period______________________________

1 Hand Tools

Quiz

Instructions: Write your answers on the lines provided at the right-hand side of the page.

1. Technicians who have good tool skills can also be expected to have good _____ skills. 1. ____________________

2. What is the most widely used type of hand tool? 2. ____________________

3. Should you always pull or push on a wrench? 3. ____________________

4. What is the proper name for side-cutting pliers? 4. ____________________

5. What type of socket end is found on a nut driver? 5. ____________________
   a. 4-point.
   b. 6-point.
   c. 8-point.

6. What type of hammer is best for striking chisels and punches? 6. ____________________

7. *True or false?* Hacksaw blade teeth should point away from the handle. 7. ____________________

8. High-speed twist drill bits are used for drilling holes in _____ or _____. 8. ____________________

9. Aviation snips are available in three types of cutting patterns. Name them. 9. ____________________

10. Name two types of "struck" tools. 10. ____________________
    ____

2

Fasteners

Objectives

After studying this chapter in the text, the student will be able to:

- ❑ Define thread terminology.
- ❑ Identify standard thread forms.
- ❑ Demonstrate proper use of taps and dies.
- ❑ Identify screw types and sizes.
- ❑ Identify and correctly use bolts, nuts, and washers.
- ❑ Identify and correctly use various anchors.
- ❑ Identify and correctly use setscrews, blind rivets, and threaded rods.

Instructional Materials

Text: Pages 27–37
Lab Manual: Pages 13–18
Instructor's Manual:
Visual Masters, Pages 39-43
Chapter 2 Quiz, Page 44

Teaching Strategy

Text

1. Review objectives.
2. Discuss and assign chapter contents. Use these visual masters as appropriate:
   HC 2-1: Parts of a Screw
   HC 2-2: Plastic Anchors
   HC 2-3: Wedge Anchors
   HC 2-4: Toggle Bolts
   HC 2-5: Blind Rivets
3. Direct students to complete Test Your Knowledge questions, page 37. After completion, discuss correct answers with students.

Lab Manual

1. Provide overview of Activities. Where necessary, stress safe laboratory practices and the importance of wearing protective clothing.
2. Assign Activities to individuals or pairs of students, as appropriate.
   Activity 2-1: Identifying Fasteners
   Activity 2-2: Using Bolts to Join Metal
   Activity 2-3: Using a Tap to Cut Threads
   Activity 2-4: Using Blind Rivets to Join Metal
3. Observe student performance on Activities, providing guidance or correction as needed. After reviewing the student's work, initial and date the activity in the place provided.

Evaluation

1. Administer the quiz from this Instructor's Manual.
2. Grade the quiz, and return it with appropriate comments for the student.

Answers to Test Your Knowledge, Page 37

1. major diameter
2. root
3. Count the number of crests or roots per inch.
4. (any two) National; Unified; International; ISO Metric
5. National Fine; National Coarse; National Pipe
6. National Pipe
7. internal
8. external
9. diameter; length
10. false
11. coarse
12. a bolt extractor
13. spring; internal tooth; external tooth
14. light; solid

15. Molly; toggle
16. fan blades; pulleys
17. hacksaw
18. taps; dies

Answers to Lab Manual, Pages 13–14

Activity 2-1

1. A. flat head
   B. pan head
   C. oval head
   D. round head
   E. hex head
2. A. machine screw
   B. sheet metal screw
   C. wood screw
3. A. major diameter
   B. minor diameter
   C. thread angle
   D. root
   E. rest
   F. threads per inch
4. diameter = 4
5. 1/4-20 × 1 (A machine screw is designated first by its diameter number, second by its number of threads per inch, and finally by its length.)
6. A. flat washer
   B. lock washer
   C. hex nut
   D. bonnet or cap nut
   E. wing nut
   F. bevel washer
7. A. plastic anchor
   B. Molly bolt or hollow wall anchor
   C. masonry anchor
   D. toggle bolt
8. If the wall is *hollow,* use a Molly bolt. If it is *solid,* use a plastic screw anchor.
9. sheet metal
10. vibration

Answers to Quiz

1. fine; coarse; pipe
2. tap
3. die
4. 20
5. machine screws
6. To extend gripping area of fastener. To prevent loosening of nut or bolt.
7. true
8. solid
9. masonry; concrete
10. blind rivet

Parts of a Screw

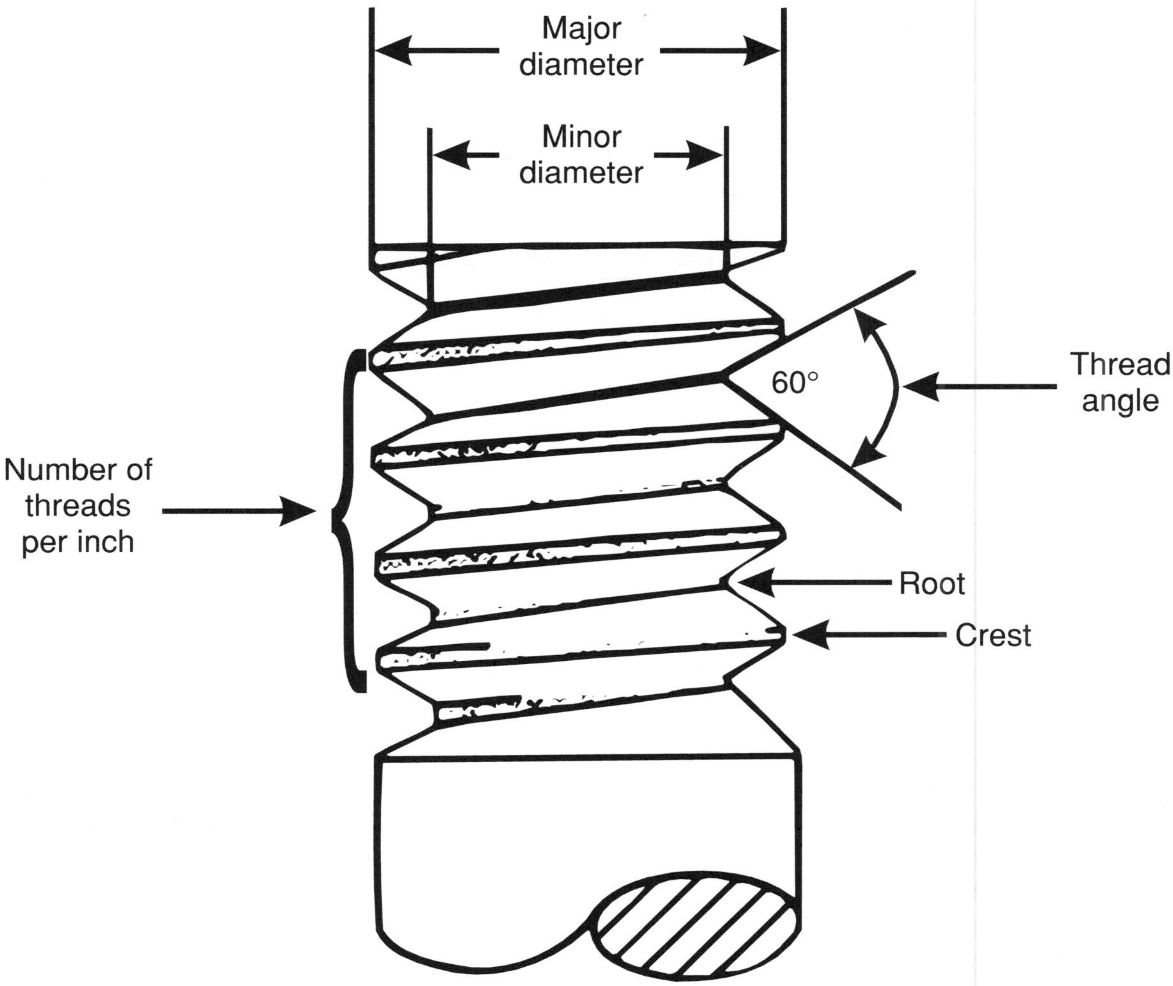

HC 2-1

Plastic Anchors

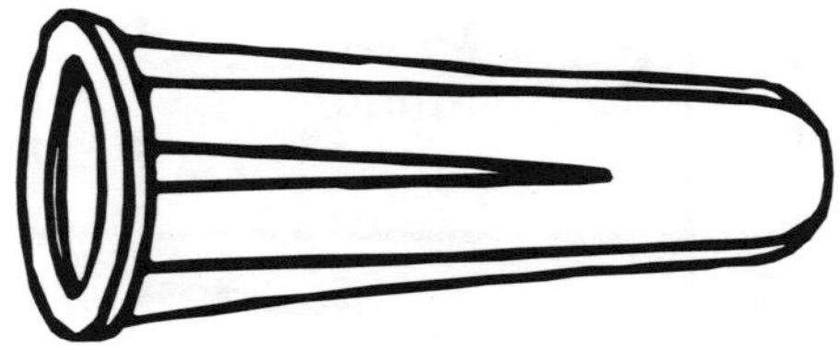

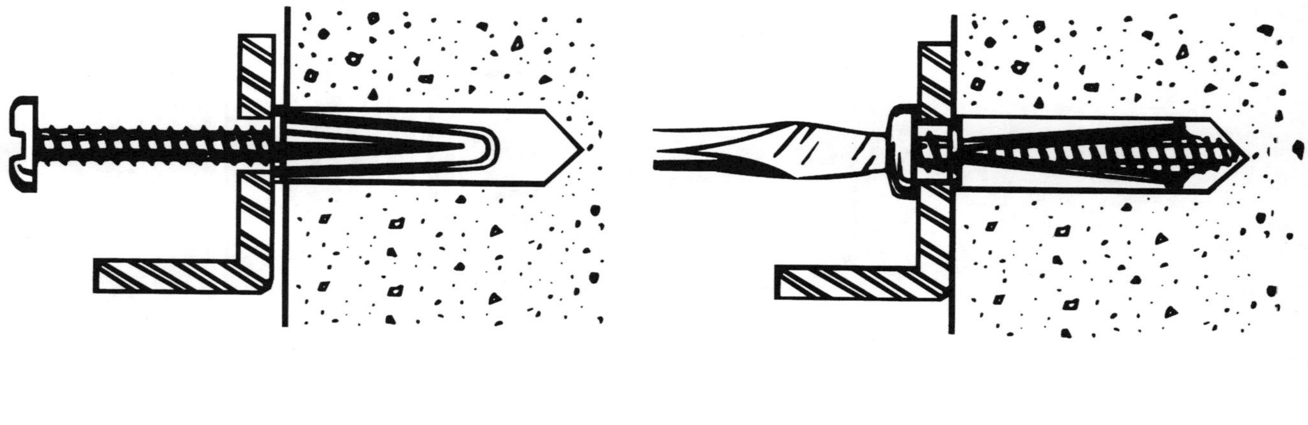

HC 2-2

Wedge Anchors

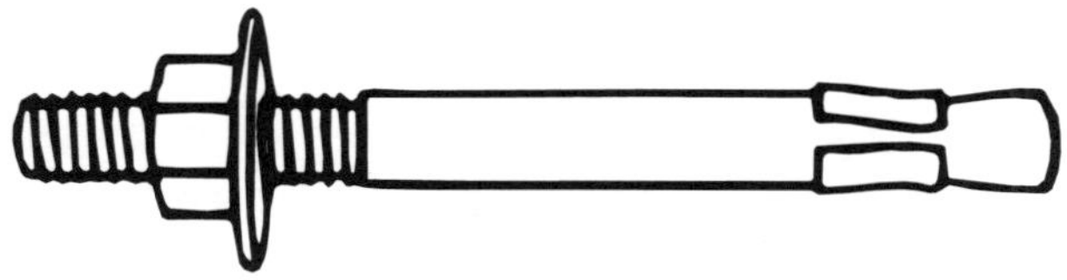

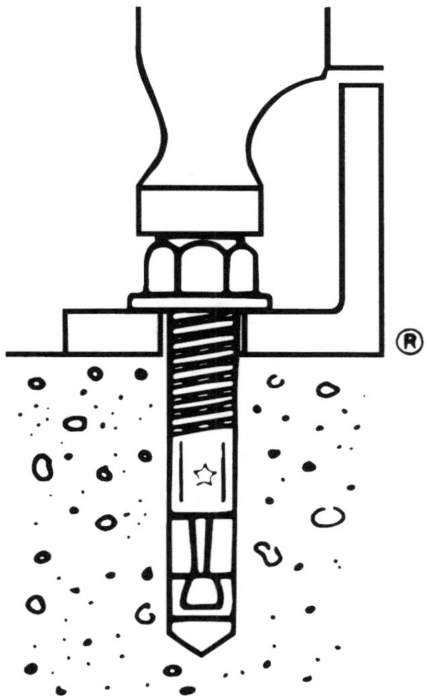

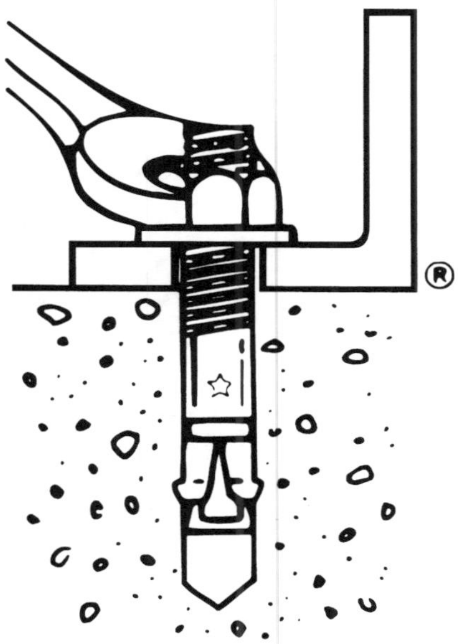

HC 2-3

Toggle Bolts

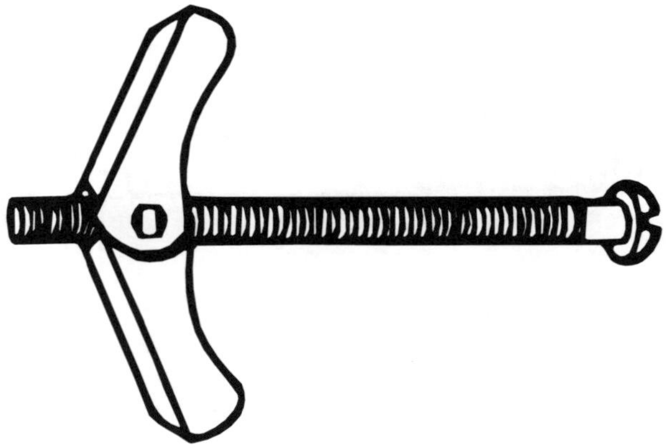

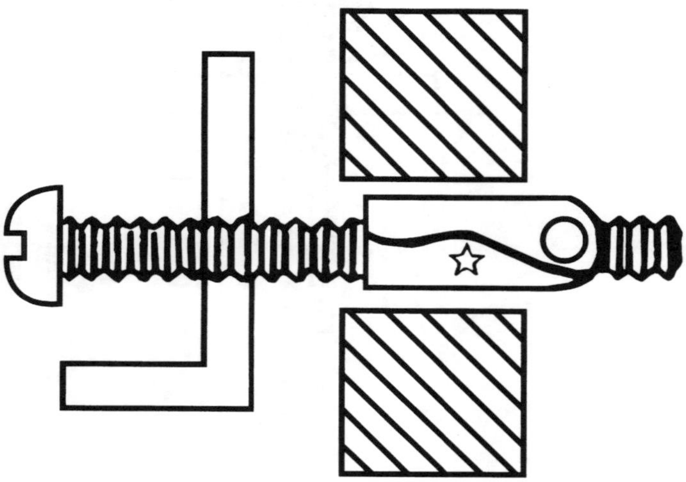

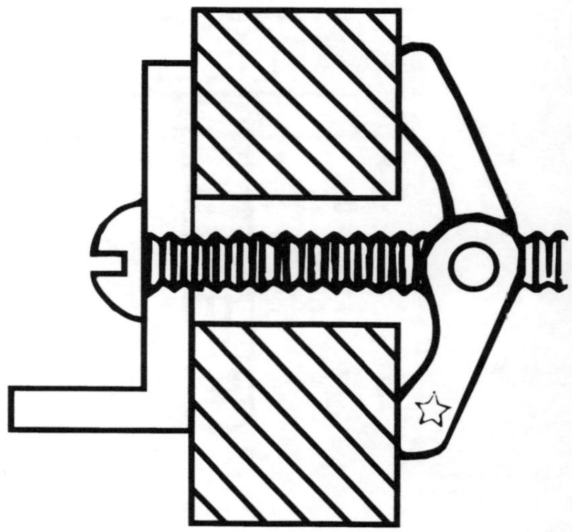

HC 2-4

Blind Rivets

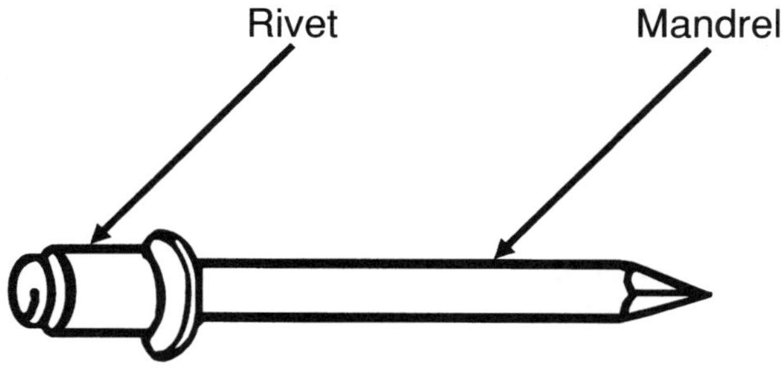

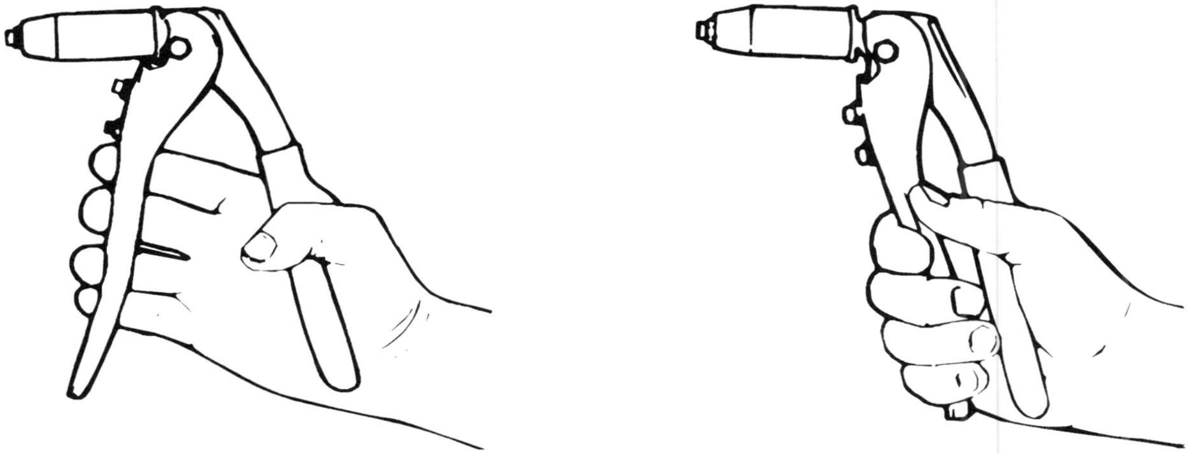

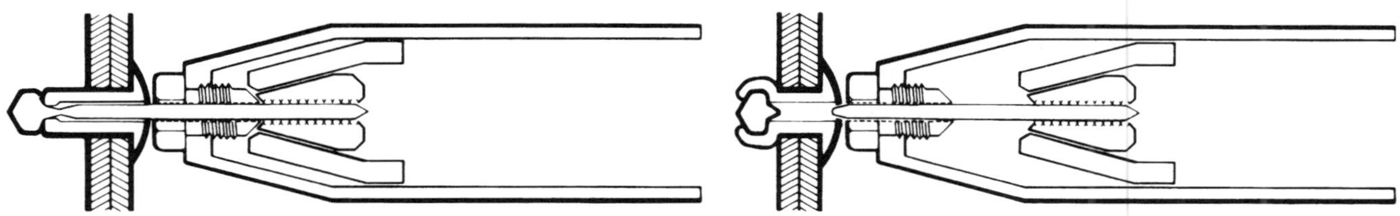

HC 2-5

Name ________________________________

Date ________________________________

Instructor ____________________________

Period________________________________

2 Fasteners

Quiz

Instructions: Write your answers on the lines provided at the right-hand side of the page.

1. Name the three common National thread forms. 1. ____________________
____
____

2. What type of tool is used to cut internal threads? 2. ____________________

3. What type of tool is used to cut external threads? 3. ____________________

4. How many threads per inch are on a machine screw designated 1/4-20 × 3/4? 4. ____________________

5. Which type of fastener fills the void between a screw and a bolt? 5. ____________________

6. Name two common uses for washers. 6. ____________________
____

7. *True or false?* A toggle bolt is used to fasten heavy objects to a hollow wall. 7. ____________________

8. Are plastic screw anchors used to fasten light loads to solid or hollow walls? 8. ____________________

9. Masonry anchors are used to fasten metal brackets to _____ or _____. 9. ____________________

10. What type of fastener is used to permanently fasten pieces of sheet metal together? 10. ____________________

3

Working with Copper Tubing

Objectives

After studying this chapter in the text, the student will be able to:

- ❑ Identify types and sizes of copper tubing.
- ❑ Select and properly use tubing tools.
- ❑ Demonstrate proper use of tube bending tools.
- ❑ Describe the steps needed to make swaged connections.
- ❑ Describe the steps needed to make flared connections.
- ❑ Identify and properly use various types of copper tubing fittings.

Instructional Materials

Text: Pages 39–50
Lab Manual: Pages 19–28
Instructor's Manual:
Visual Masters, Pages 47–50
Chapter 3 Quiz, Page 51

Teaching Strategy

Text

1. Review objectives.
2. Discuss and assign chapter contents. Use these visual masters as appropriate:
   HC 3-1: Copper Tubing
   HC 3-2: Reaming Tubing
   HC 3-3: Swaging Tubing
   HC 3-4: Flaring Tubing
3. Direct students to complete Test Your Knowledge questions, page 50. After completion, discuss correct answers with students.

Lab Manual

1. Provide overview of Activities. Where necessary, stress safe laboratory practices and the importance of wearing protective clothing.
2. Assign Activities to individuals or pairs of students, as appropriate.
   Activity 3-1: Identifying Fittings
   Activity 3-2: Identifying and Sizing Tubing and Pipe
   Activity 3-3: Cutting Soft Copper Tubing
   Activity 3-4: Reaming and Deburring Tubing
   Activity 3-5: Flaring Soft Copper Tubing
   Activity 3-6: Making Swaged Connections
   Activity 3-7: Bending Soft Copper Tubing
3. Observe student performance on Activities, providing guidance or correction as needed. After reviewing the student's work, initial and date the activity in the place provided.

Evaluation

1. Administer the quiz from this Instructor's Manual.
2. Grade the quiz, and return it with appropriate comments for the student.

Answers to Test Your Knowledge, Page 50

1. To prevent oxidation or contamination by dirt or other foreign matter.
2. nitrogen
3. hard-drawn; soft
4. c. K; L; M
5. outside; inside
6. Tubing cutter and hacksaw. The tubing cutter is preferred because it makes clean; square cuts.
7. They cause problems in assembly that could result in leakage.
8. hand; spring bender; lever bender

9. Swaging requires only one soldered joint (compared to two for a fitting); so leakage is less likely to occur.
10. sweat
11. 3
12. compression; flare
13. b. 45
14. false
15. nut

Answers to Lab Manual, Pages 19–21

Activity 3-1

1. Where the connection is permanent, such as long runs of either soft or hard copper tubing.
2. When using hard-drawn copper tubing.
3. A. 90° street elbow
   B. 45° elbow
   C. reducing coupling
   D. 90° elbow
   E. coupling
   F. cap
   G. tee
4. A. short nut
   B. long nut
   C. tee
   D. 90° elbow
   E. reducing union
5. To connect a branch to an existing line of copper tubing.
6. reducing flare nut

Activity 3-2

1. A. L
   B. K
   C. M
2. A. 5/8"
   B. 1/2"
   C. 3/8"
   D. 1/4"
3. Student response (answers will vary).

Answers to Quiz

1. false
2. a. OD
3. tubing cutter
4. lever-type tubing benders; spring benders
5. swaging
6. d. all of the above
7. 3
8. false
9. soft
10. d. All of the above.

Copper Tubing

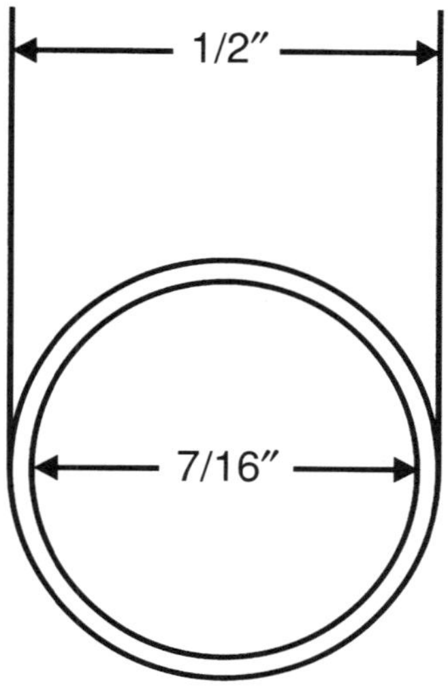

ACR Tubing:
OD Measurement

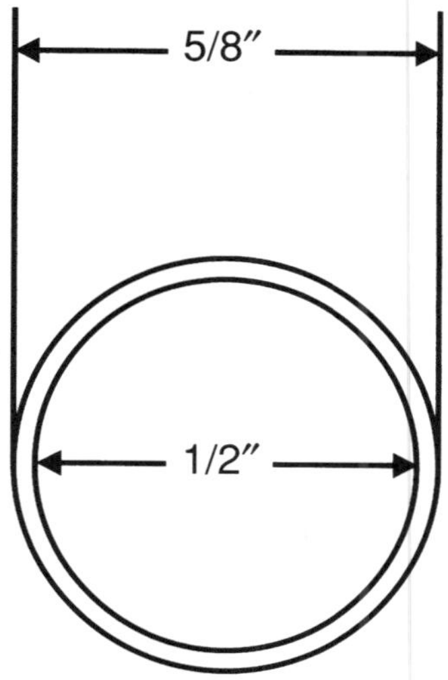

Plumber's Tubing:
Nominal ID
Measurement

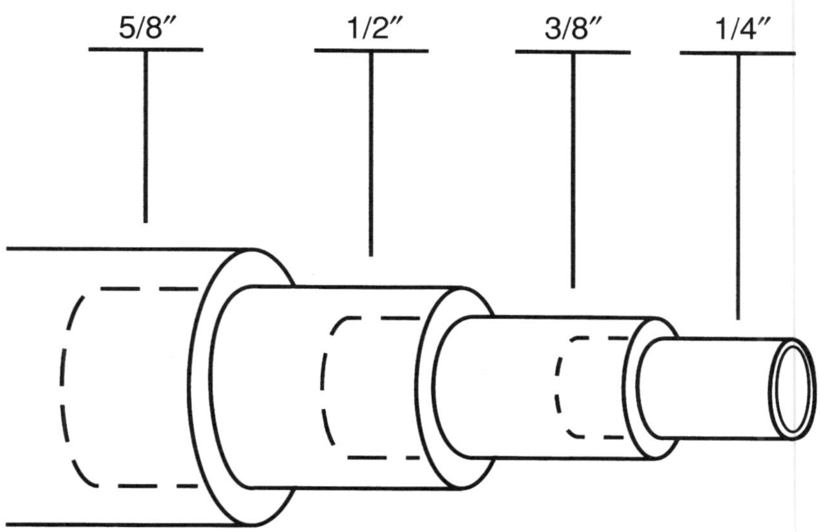

HC 3-1

Reaming Tubing

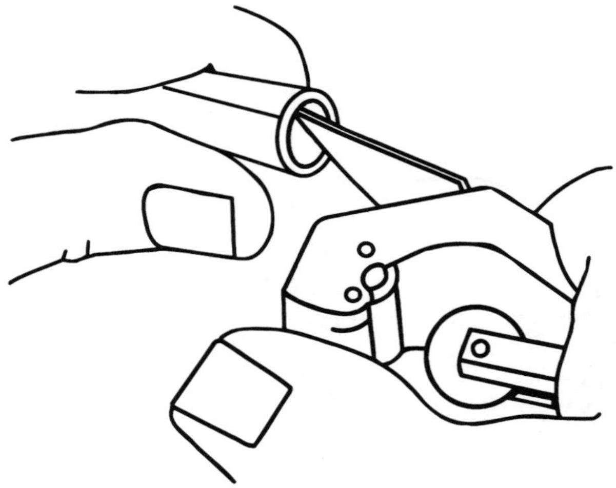

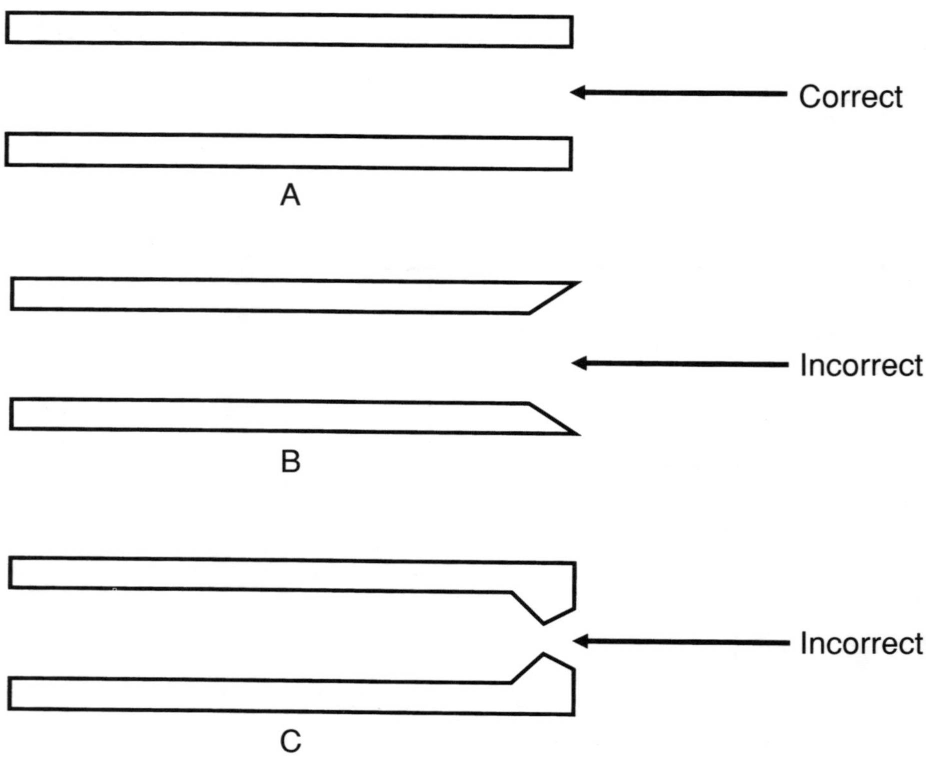

HC 3-2

Swaging Tubing

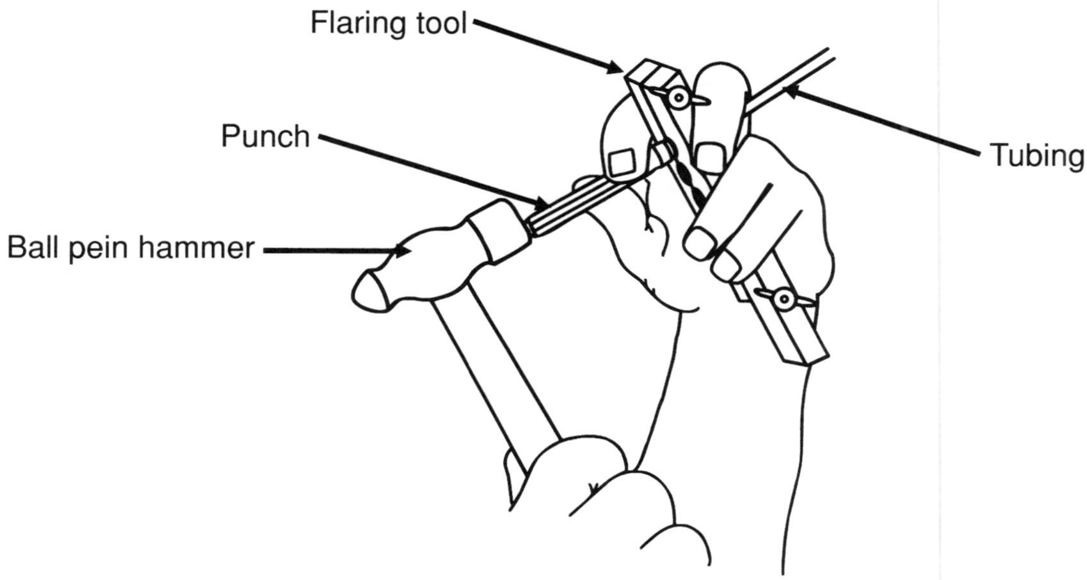

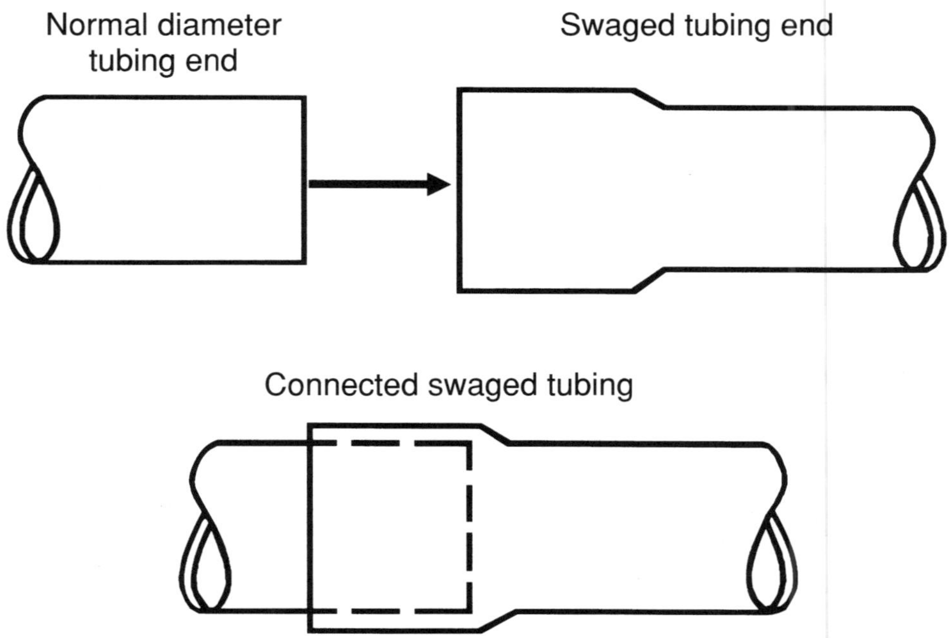

HC 3-3

Flaring Tubing

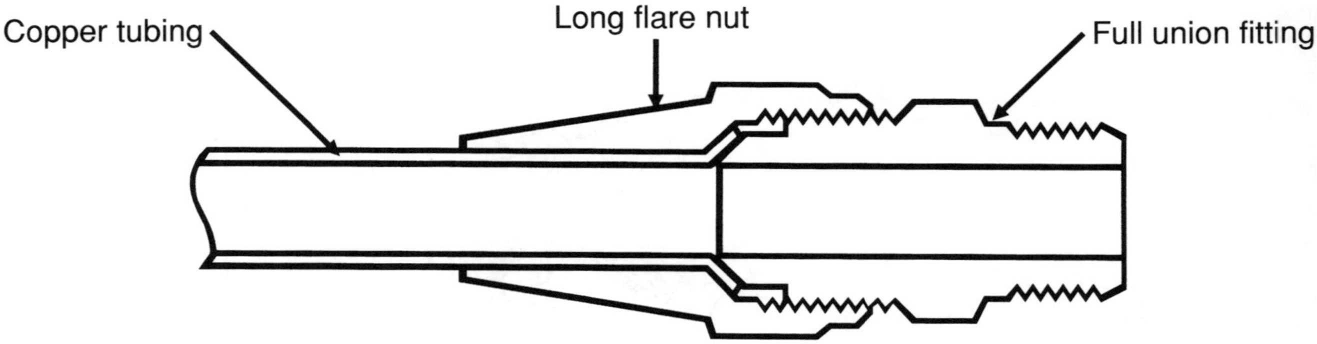

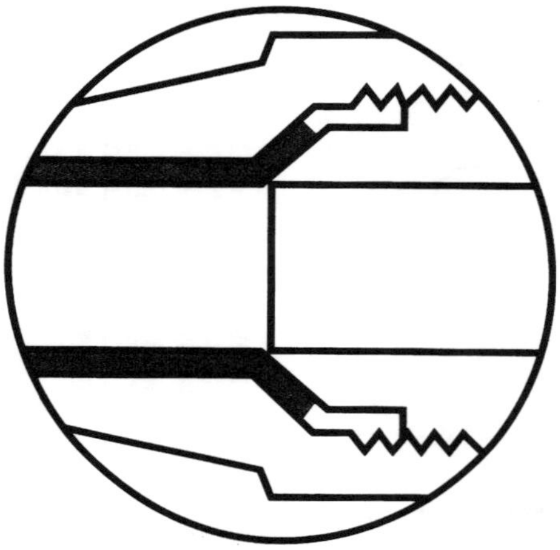
Correct

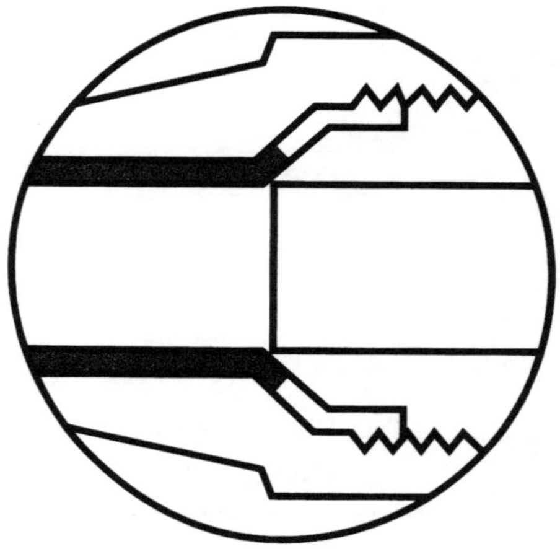
Too small

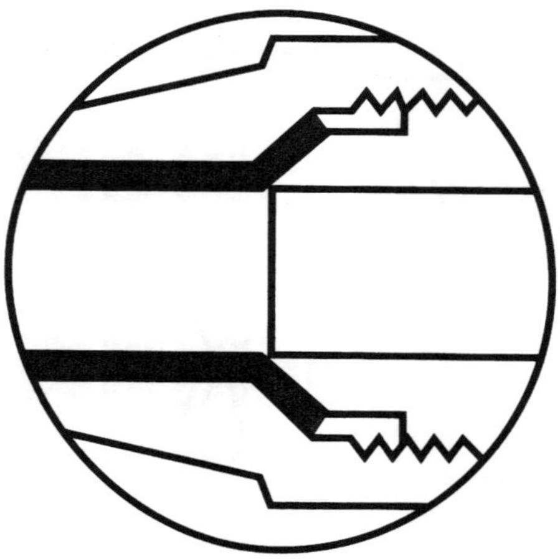
Too large

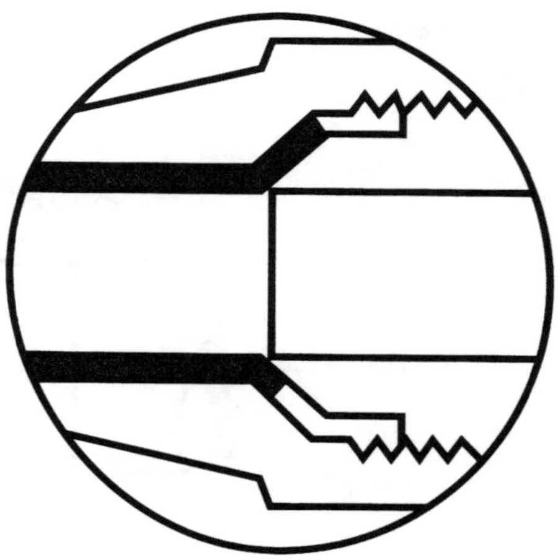
Uneven

HC 3-4

Name ______________________________

Date ______________________________

Instructor ______________________________

Period______________________________

3

Working with Copper Tubing

Quiz

Instructions: Write your answers on the lines provided at the right-hand side of the page.

1. *True or false?* The most common ACR copper tubing is Type K. 1. ____________________
2. ACR copper tubing is measured by its _____. 2. ____________________
   a. OD
   b. ID
   c. nominal diameter
3. What tool is commonly used to cut copper tubing? 3. ____________________
4. Name two common tools used to bend soft copper tubing. 4. ____________________

   ____
5. _____ is the term used to describe the process of enlarging the tubing end. 5. ____________________
6. Manufactured wrought (sweat) fittings are used where the connections are _____. 6. ____________________
   a. permanent
   b. for hard-drawn copper tubing
   c. for soft copper tubing
   d. all of the above
7. On a wrought tee fitting, the branch opening is designated with the number _____. 7. ____________________
8. *True or false?* All refrigeration flare fittings are made with a 33° angle. 8. ____________________
9. Flare fittings are mechanical fittings intended for use on _____ copper tubing. 9. ____________________
10. Wrought fittings are preferred over mechanical-type fittings because they _____. 10. ____________________
    a. are less likely to leak
    b. require less time to assemble
    c. can be used on hard or soft copper
    d. All of the above.

4

Working with Pipe

Objectives

After studying this chapter in the text, the student will be able to:

- ❑ Discuss the uses for different types of steel and plastic pipe.
- ❑ Identify various sizes of pipe.
- ❑ Recognize types of fittings and use them properly.
- ❑ Select pipe tools and describe their proper use.
- ❑ Describe how to repair leaks in piping.
- ❑ Describe the process for solvent-welding rigid plastic pipe.

Instructional Materials

Text: Pages 51–60
Lab Manual: Pages 29–34
Instructor's Manual:
Visual Masters, Pages 55–59
Chapter 4 Quiz, Page 60

Teaching Strategy

Text

1. Review objectives.
2. Discuss and assign chapter contents. Use these visual masters as appropriate:
   HC 4-1: Tubing vs. Pipe
   HC 4-2: Tools for Pipe
   HC 4-3: Steel Pipe Repair
   HC 4-4: Cutting Plastic Pipe
   HC 4-5: Joining Plastic Pipe
3. Direct students to complete Test Your Knowledge questions, page 60. After completion, discuss correct answers with students.

Lab Manual

1. Provide overview of Activities. Where necessary, stress safe laboratory practices and the importance of wearing protective clothing.
2. Assign Activities to individuals or pairs of students, as appropriate.
   Activity 4-1: Identifying Tools and Fittings Used on Pipe
   Activity 4-2: Working with Steel Pipe
   Activity 4-3: Cutting and Joining Plastic Pipe
3. Observe student performance on Activities, providing guidance or correction as needed. After reviewing the student's work, initial and date the activity in the place provided.

Evaluation

1. Administer the quiz from this Instructor's Manual.
2. Grade the quiz, and return it with appropriate comments for the student.

Answers to Test Your Knowledge, Page 60

1. black; galvanized
2. inside (or flow)
3. 1/8" larger
4. 1/16"
5. reamer
6. die
7. elbows
8. coupling
9. material cost; labor-saving
10. ABS
11. PVC
12. CPVC
13. nominal
14. solvent
15. Adapters

Answers to Lab Manual, Pages 29-31

Activity 4-1

1. A. chain vise
   B. spiral pipe reamer
   C. yoke vise
   D. hacksaw
   E. cutter
   F. pump or groove joint pliers
   G. strap-type wrench
   H. die and ratchet
   I. jaw-type wrench
2. strap-type
3. To concentrate pressure at two points and produce maximum gripping action and rotating force.
4. A. 90° elbow
   B. 45° elbow
   C. 90° street elbow
   D. reducing coupling
   E. coupling
   F. tee
   G. bushing or reducing bushing
   H. union
   I. shoulder nipple
   J. close or all-thread nipple
   K. long nipple
5. Malleable fittings can withstand more bending, pounding, and internal pressure than can ordinary cast iron fittings.
6. Pipe fittings for steel are threaded, rather than smooth like sweat-soldered tubing fittings.
   Pipe fittings are always sized according to the nominal size (ID) of the pipe being used, rather than by OD, as is ACR tubing.
7. coupling; reducing coupling
8. By installing reducing bushings (3/4" × 1/2") in each side of the valve.
9. Pipe unions
10. A. tee
    B. 45° elbow
    C. 90° elbow
    D. bushing
    E. coupling, plastic to steel pipe
    F. coupling, solder joint to plastic pipe
    G. elbow, plastic pipe to steel pipe
11. Resistant to corrosion.
    Smooth inner wall provides maximum flow rates and abrasion resistance.
    High tensile and burst strengths.
    Easy to handle, join, and install.
    Low material and labor costs.

Answers to Quiz

1. 21
2. c. nominal size
3. 2
4. reamer
5. union
6. ABS; PVC; CPVC
7. A. 1/16" to 3/32"
   B. 10° to 15°
8. true
9. d. All of the above.
10. adapter

Tubing vs Pipe

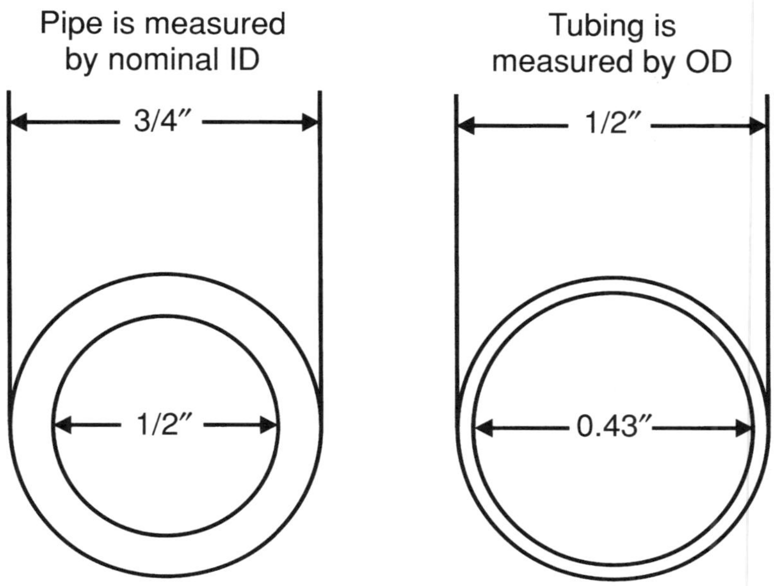

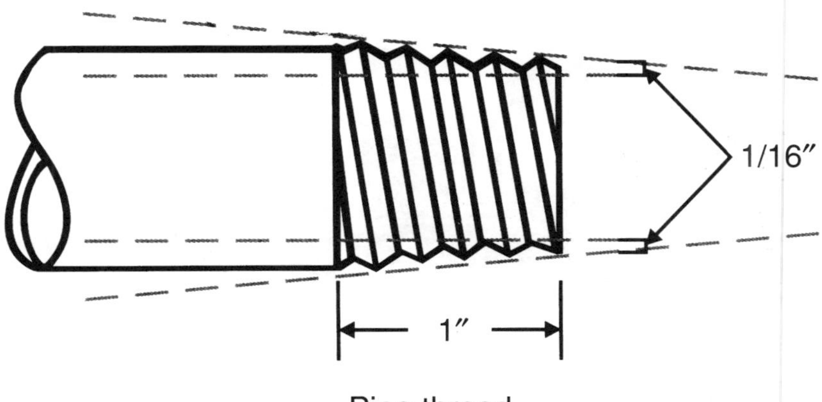

Pipe thread

HC 4-1

Tools for Pipe

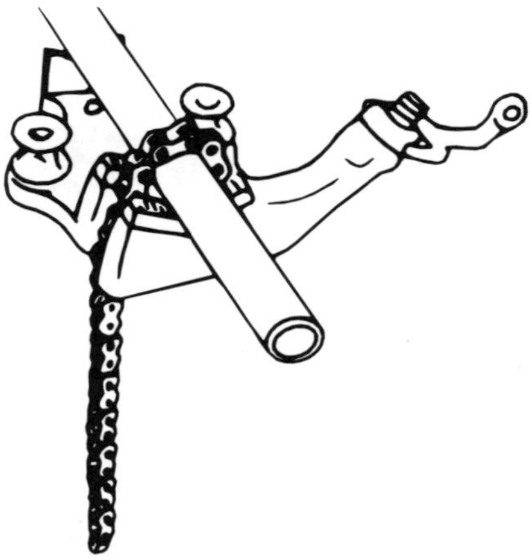

Chain vise

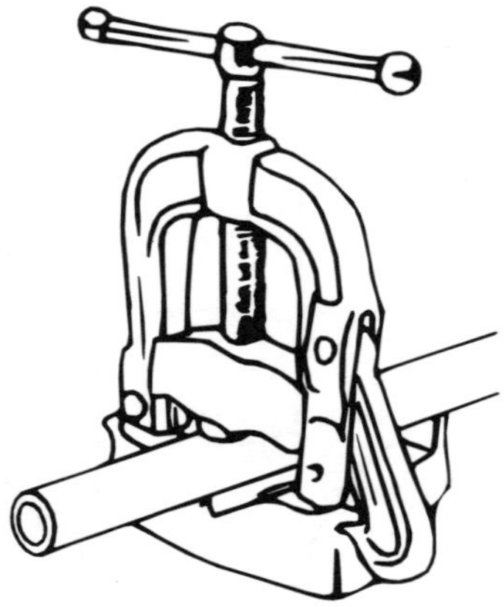

Yoke vise

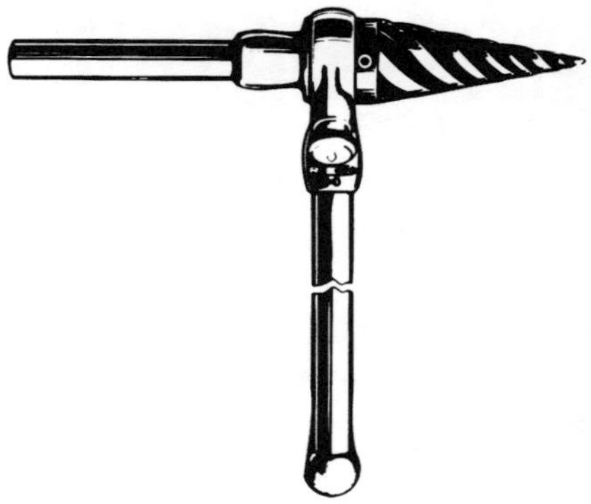

Spiral reamer

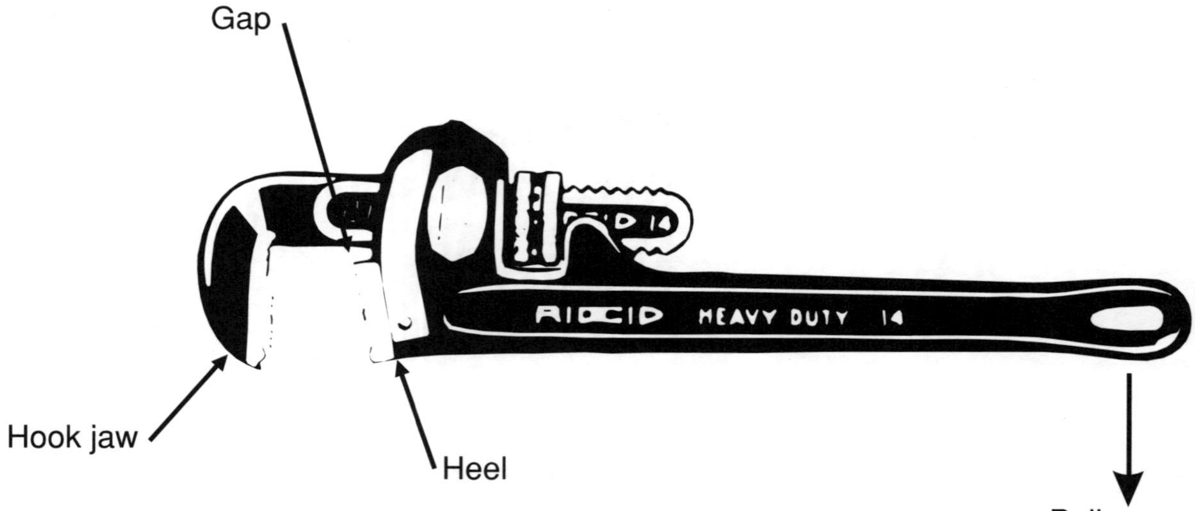

Jaw-type wrench

HC 4-2

Steel Pipe Repair

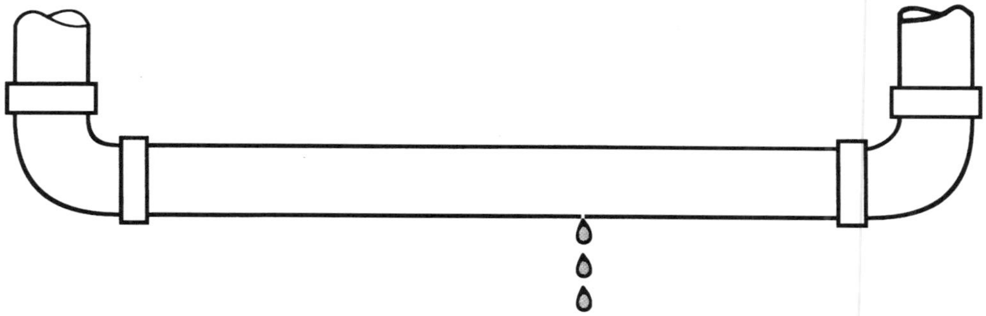

Leaking pipe

Cut out and remove defective section

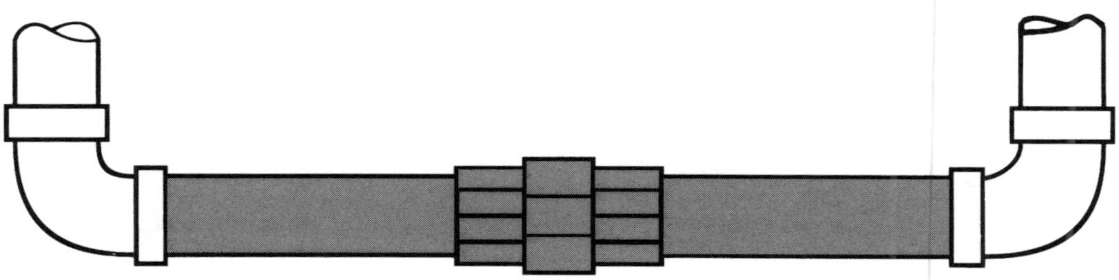

Install new pipe sections and union

HC 4-3

Cutting Plastic Pipe

Hacksaw and miter box

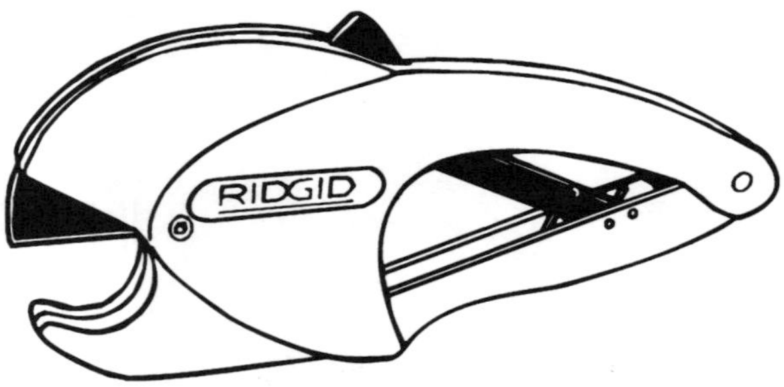

Shear-type cutter

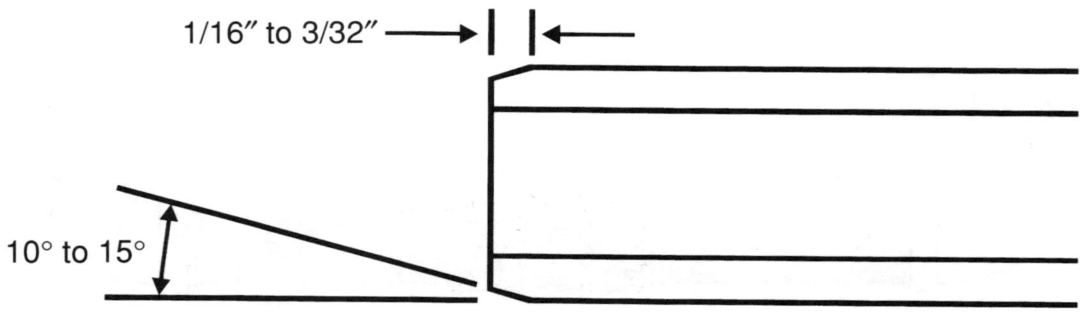

Beveling pipe

HC 4-4

Joining Plastic Pipe

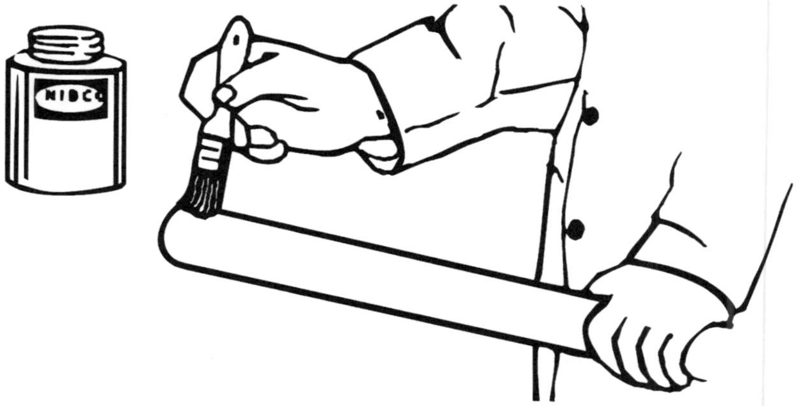

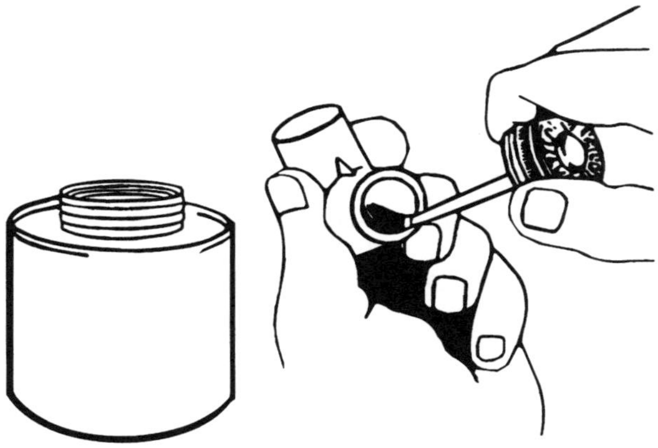

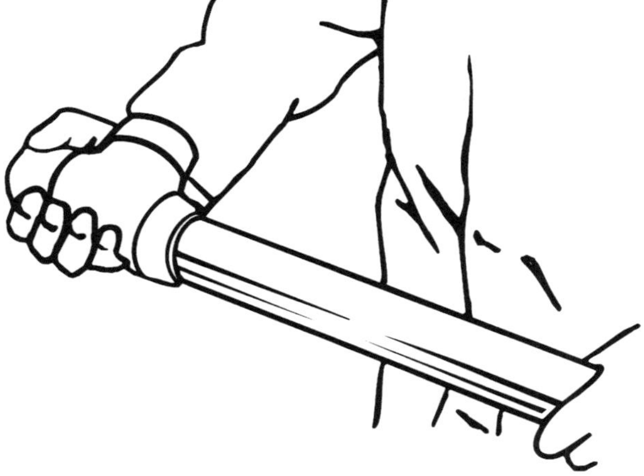

HC 4-5

Name ______________________________

Date ______________________________

Instructor ______________________________

Period______________________________

4

Working with Pipe

Quiz

Instructions: Write your answers on the lines provided at the right-hand side of the page.

1. Steel pipe is available in standard lengths of _____ feet. 1. ____________

2. Pipe is measured according to its _____. 2. ____________
   a. exact ID
   b. exact OD
   c. nominal size

3. How many 45° elbows are required to make a 90° turn? 3. ____________

4. A _____ is a tool used to remove burrs from pipe. 4. ____________

5. What fitting is installed to allow pipes to be disconnected for repair? 5. ____________

6. Name three types of plastic pipe. 6. ____________
____
____

7. On the drawing of the cut pipe end, fill in the dimensions of the bevel that should be provided to aid in proper joining. 7. a. ____________
b. ____________

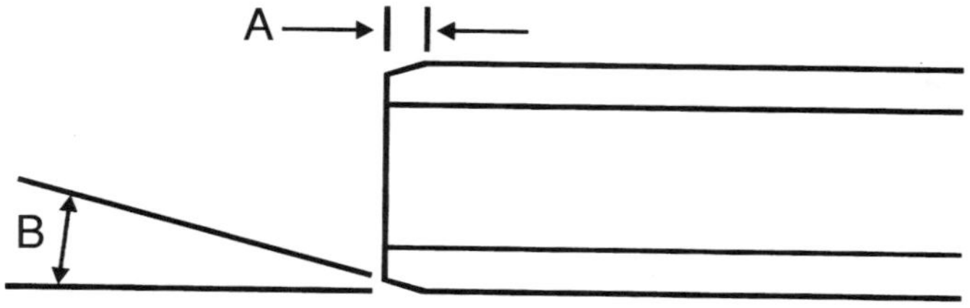

8. *True or false?* CPVC plastic pipe can be used for either hot or cold water lines. 8. ____________

9. Plastic pipe is joined by a process called: 9. ____________
   a. Gluing.
   b. Solvent welding.
   c. Cementing.
   d. All of the above.

10. An _____ is used to connect steel pipe to plastic pipe. 10. ____________

5

Soldering

Objectives

After studying this chapter in the text, the student will be able to:

- ❑ Select proper alloys and fluxes.
- ❑ Describe the use of air-fuel torches.
- ❑ Use fuel gas cylinders and torches properly and safely.
- ❑ Connect and operate pressure regulators correctly.
- ❑ Describe how to solder connections properly and safely.
- ❑ Describe air-fuel brazing.

Instructional Materials

Text: Pages 61–74
Lab Manual: Pages 35–42
Instructor's Manual:
Visual Masters, Pages 63–67
Chapter 5 Quiz, Page 68

Teaching Strategy

Text

1. Review objectives.
2. Discuss and assign chapter contents. Use these visual masters as appropriate:
   HC 5-1: Joining Metal with Solder
   HC 5-2: Applying Flux
   HC 5-3: The Air-acetylene Flame
   HC 5-4: Soldering Techniques
   HC 5-5: Heat Sink
3. Direct students to complete Test Your Knowledge questions, page 73. After completion, discuss correct answers with students.

Lab Manual

1. Provide overview of Activities. Where necessary, stress safe laboratory practices and the importance of wearing protective clothing.
2. Assign Activities to individuals or pairs of students, as appropriate.
   Activity 5-1: Identifying Air-acetylene Torch Components
   Activity 5-2: Assembling an Air-acetylene Torch Outfit
   Activity 5-3: Using the Air-acetylene Torch Outfit
   Activity 5-4: Soldering with the Air-acetylene Torch
   Activity 5-5: Brazing with the Air-acetylene Torch
3. Observe student performance on Activities, providing guidance or correction as needed. After reviewing the student's work, initial and date the activity in the place provided.

Evaluation

1. Administer the quiz from this Instructor's Manual.
2. Grade the quiz, and return it with appropriate comments for the student.

Answers to Test Your Knowledge, Page 73

1. fuel
2. approximately 5660°F (3127°C)
3. 212
4. To avoid loss of acetone and poor flame quality.
5. B
6. So the valve can be quickly closed in an emergency.
7. Acetylene is supplied by the cylinder and oxygen is obtained from the atmosphere, which produces a "soft" flame.

8. 250; 21
9. working (or flow)
10. red
11. They are left-handed.
12. 1/8; tip
13. 1982
14. 840
15. To maintain a clean surface during the soldering process.
16. tin-lead; tin-antimony; tin-silver
17. alloy or solder
18. 5 psi
19. It has a notch or groove.
20. 1/4; 1/2

Answers to Lab Manual, Pages 35–36

Activity 5-l

1. A. pressure regulator
   B. cylinder valve
   C. "B" acetylene cylinder
   D. torch
2. A. pressure adjustment
   B. cylinder gauge
   C. flow pressure gauge

Answers to Quiz

1. acetylene
2. MC
3. red
4. b. 1/2
5. To prevent connecting the hose to the oxygen supply.
6. Capillary attraction
7. b. heat sink
8. false
9. 5 psi (34 kPa)
10. 1/8" (3 mm)

Joining Metal with Solder

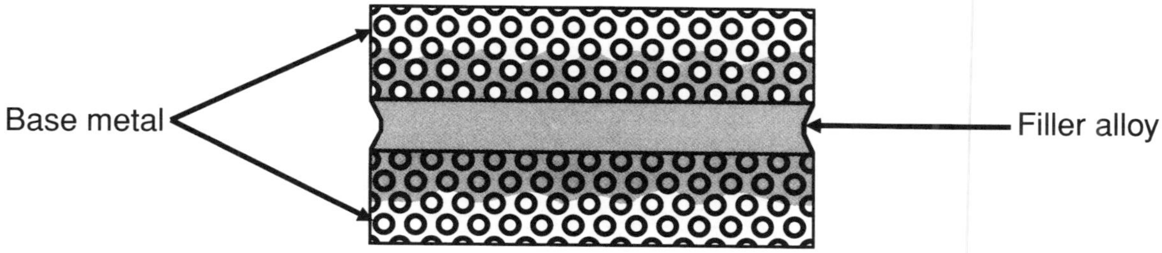

Soldering Alloys						
Solder	Composition Percent				Temperature	
	Tin (Sn)	Lead (Pb)	Antimony (Sb)	Silver (Ag)	Melts	Flows
50/50	50	50			360	420
40/60	40	60			360	460
60/40	60	40			360	375
95/5	95		5		452	464
Silver solder	96			4	430	430
Silver solder	94			6	430	535

HC 5-1

Applying Flux

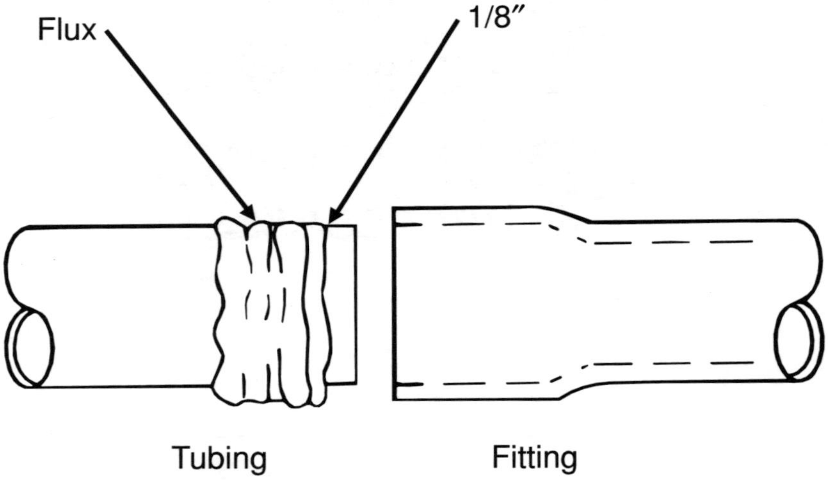

HC 5-2

The Air-acetylene Flame

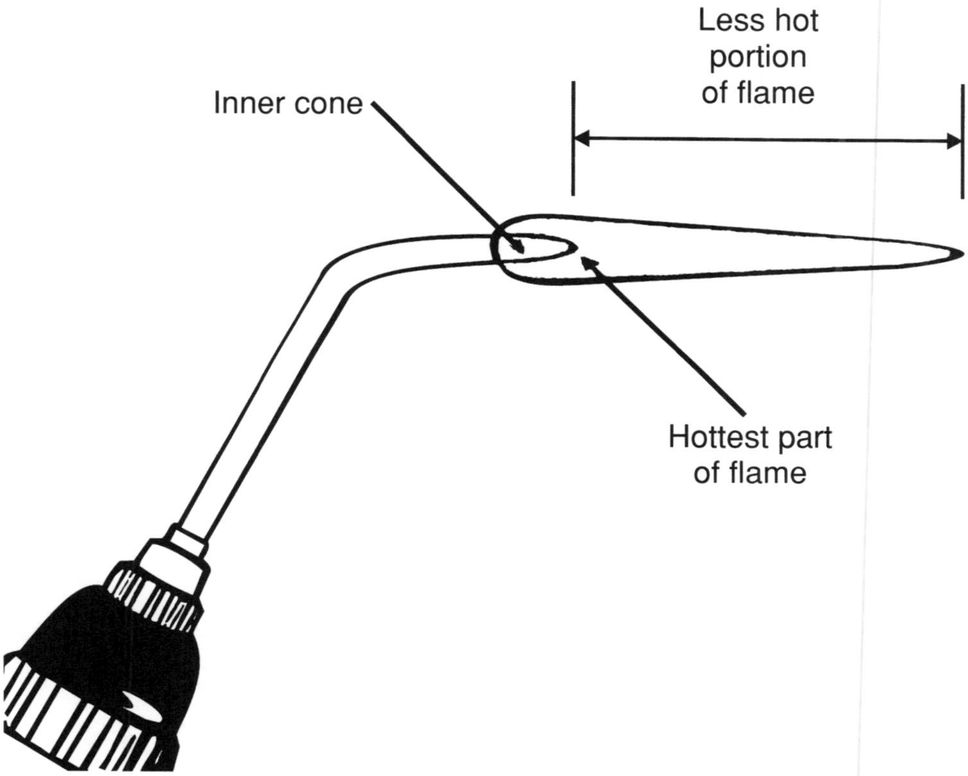

HC 5-3

Soldering Techniques

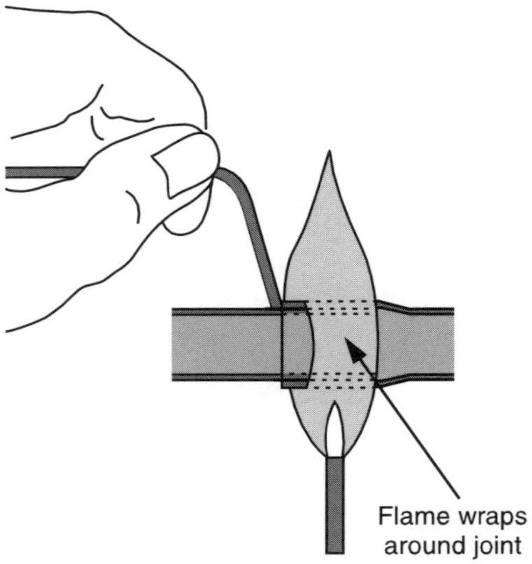

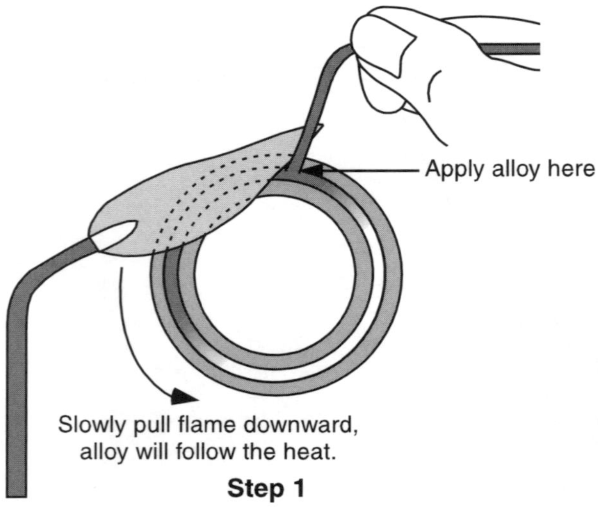

Step 1

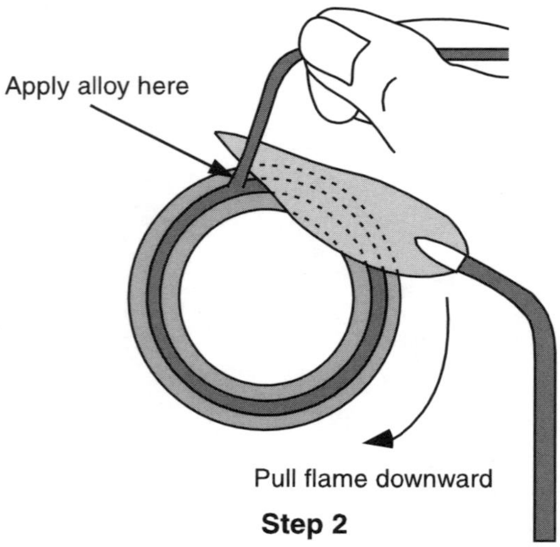

Step 2

HC 5-4

Heat Sink

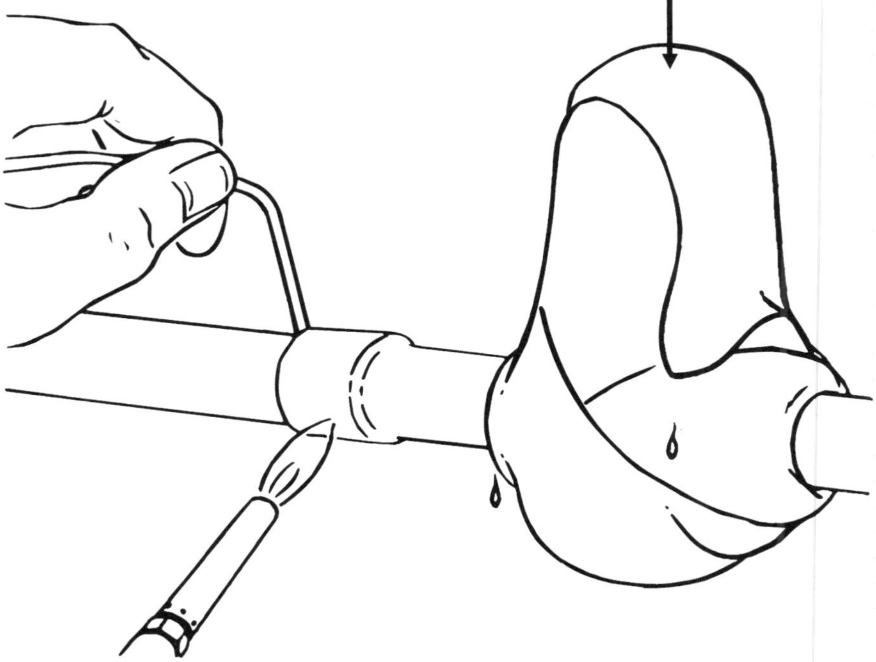

HC 5-5

Name ______________________________

Date ______________________________

Instructor __________________________

Period______________________________

5

Soldering

Quiz

Instructions: Write your answers on the lines provided at the right-hand side of the page.

1. What is the most popular fuel gas used for soldering torches? 1. ____________________

2. The smallest acetylene cylinder size is designated by the letters _____. 2. ____________________

3. What color is used to indicate an acetylene hose? 3. ____________________

4. The acetylene cylinder valve should never be opened more than _____turn. 4. ____________________
   a. 1/4
   b. 1/2
   c. one full

5. Why are left-handed threads used for acetylene connections? 5. ____________________

6. _____ is the term used to refer to the process by which the alloy "walks" itself into the fitting. 6. ____________________

7. Any heavy metal mass or device that draws heat away from the solder connection is called a _____. 7. ____________________
   a. hot-short
   b. heat sink
   c. fillet

8. *True or false?* The purpose of the flux is to clean metal prior to soldering. 8. ____________________

9. What is the normal flow pressure for the air-acetylene torch? 9. ____________________

10. The air-acetylene flame is hottest approximately _____ away from the tip of the inner cone. 10. ____________________

6

Brazing and Flame-cutting

Objectives

After studying this chapter in the text, the student will be able to:

- ❑ Describe the brazing process.
- ❑ Select appropriate brazing alloys and fluxes.
- ❑ Select and use cylinders and torches.
- ❑ Connect and operate oxyacetylene equipment for brazing metal.
- ❑ Set up, adjust, and use an oxyacetylene torch for cutting metal.

Instructional Materials

Text: Pages 75–90
Lab Manual: Pages 43–54
Instructor's Manual:
Visual Masters, Pages 71–76
Chapter 6 Quiz, Page 77

Teaching Strategy

Text

1. Review objectives.
2. Discuss and assign chapter contents. Use these visual masters as appropriate:
   HC 6-1: Brazing Alloys
   HC 6-2: Brazing Temperatures
   HC 6-3: Oxyacetylene Flame Types
   HC 6-4: Vertical Joints
   HC 6-5: Horizontal Joint
   HC 6-6: Oxyacetylene Cutting
3. Direct students to complete Test Your Knowledge questions, page 90. After completion, discuss correct answers with students.

Lab Manual

1. Provide overview of Activities. Where necessary, stress safe laboratory practices and the importance of wearing protective clothing.
2. Assign Activities to individuals or pairs of students, as appropriate.
   Activity 6-1: Matching and Identification
   Activity 6-2: Using the Oxyacetylene Torch Outfit
   Activity 6-3: Brazing with the Oxyacetylene Torch
   Activity 6-4: Welding Steel with an Oxyacetylene Torch
   Activity 6-5: Using the Oxyacetylene Cutting Torch
   Activity 6-6: Cutting with the Oxyacetylene Torch
3. Observe student performance on Activities, providing guidance or correction as needed. After reviewing the student's work, initial and date the activity in the place provided.

Evaluation

1. Administer the quiz from this Instructor's Manual.
2. Grade the quiz, and return it with appropriate comments for the student.

Answers to Test Your Knowledge, Page 90

1. 840
2. 1981; 1083
3. Any three of these:
   The joint is stronger than the base metals.
   The joint will tolerate vibration without cracking.
   The joint is leakproof under high pressure.
   The brazing material is noncorrosive.

4. oxyacetylene; air-acetylene
5. AWS BCuP-2
6. copper
7. flux
8. aircraft
9. 15 169
10. To backseat the valve and prevent leakage at the stem.
11. c. Neutral.
12. both are 5 psi (34 kPa)
13. preheat
14. 1625
15. bright orange
16. The jet of oxygen burns and melts the metal to make a cut.
17. Depending upon the size of the cutting tip, 20 to 35 psi (138 to 241 kPa).

Answers to Lab Manual, Pages 43–44

Activity 6-1

1. D
2. C
3. B
4. A
5. A. working pressure adjustment knob
   B. cylinder pressure gauge
   C. working (flow) pressure gauge
   D. cylinder connection
   E. hose connection
6. A. carburizing
   B. neutral
   C. oxidizing

Answers to Quiz

1. a. 840°F (450°C)
2. true
3. flux
4. b. Neutral.
5. oxyacetylene
6. vertical-down; vertical-up; horizontal
7. oxyacetylene torch
8. For quick shutoff in case of emergency.
9. To avoid a backfire into the mixer.
10. bright orange

Brazing Alloys

Chemical Composition of Brazing Alloys							
AWS Number	Copper (Cu)%	Phosphorus (P)%	Silver (Ag)%	Zinc (Zn)%	Tin (Sn)%	Temperature	
						Melts	Flows
BCuP-2	92.88	7.12				1310	1485
BCuP-3	88.75	6.25	5			1190	1485
BCuP-4	87.00	7.00	6			1190	1480
BCuP-5	80.00	5.00	15			1190	1500
BAg-5	38.00		30	25		1250	1370
BAg-7	22.00		56	17	5	1145	1200
BAg-20	38.00		30	32		1250	1410

HC 6-1

Brazing Temperatures

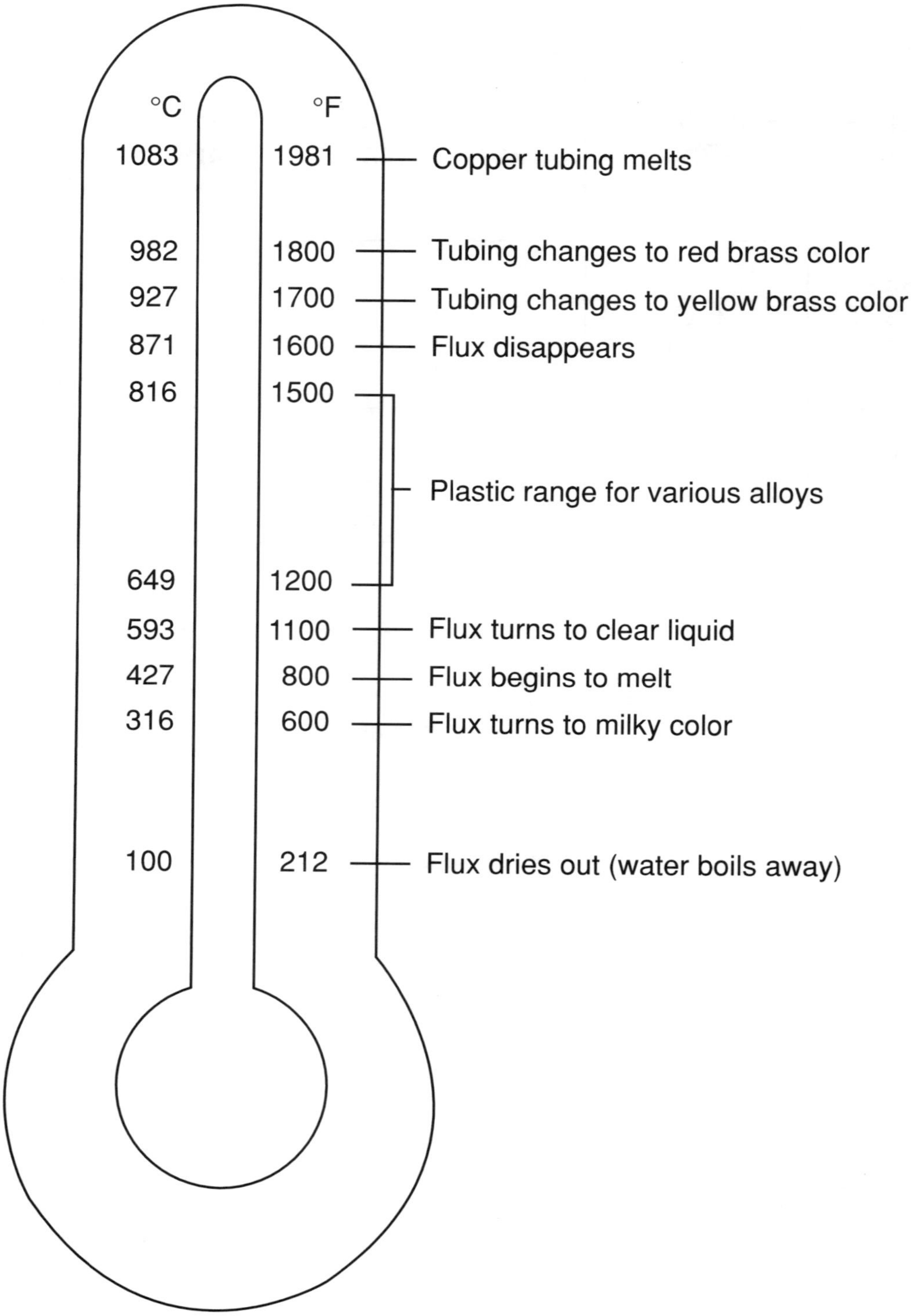

HC 6-2

Oxyacetylene Flame Types

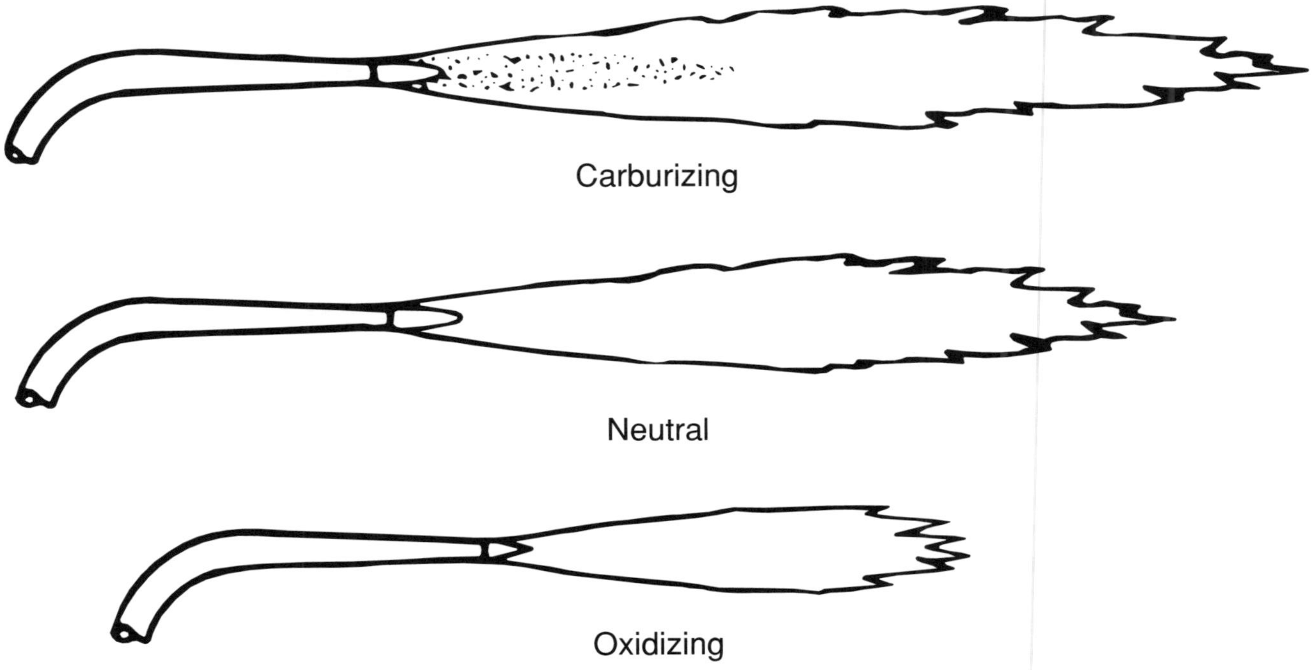

HC 6-3

Vertical Joints

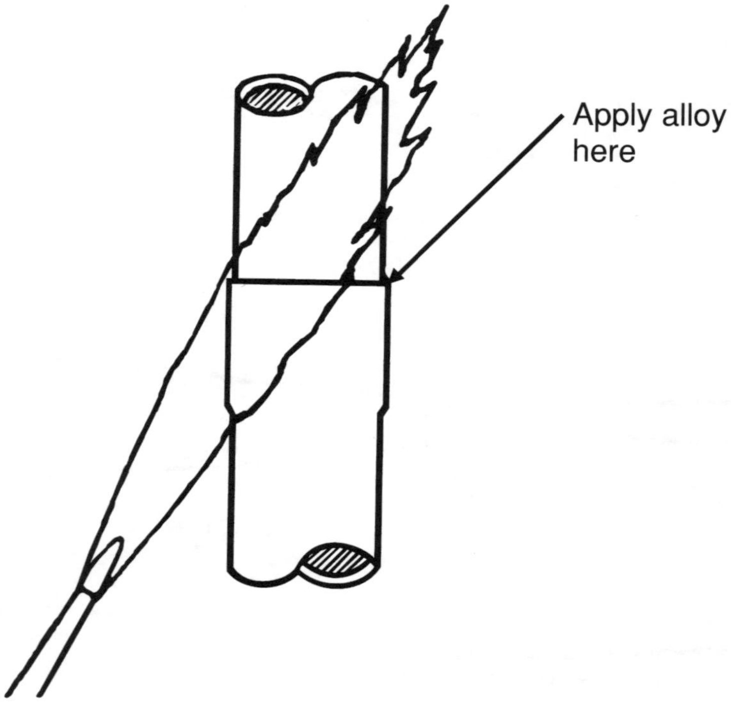

Vertical-down

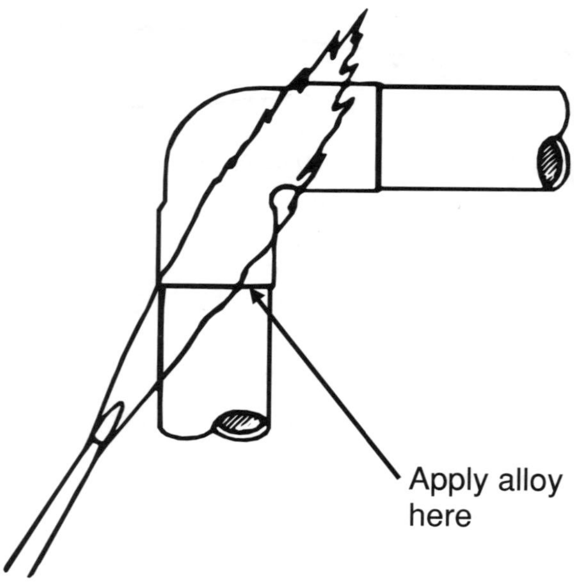

Vertical-up

HC 6-4

Horizontal Joint

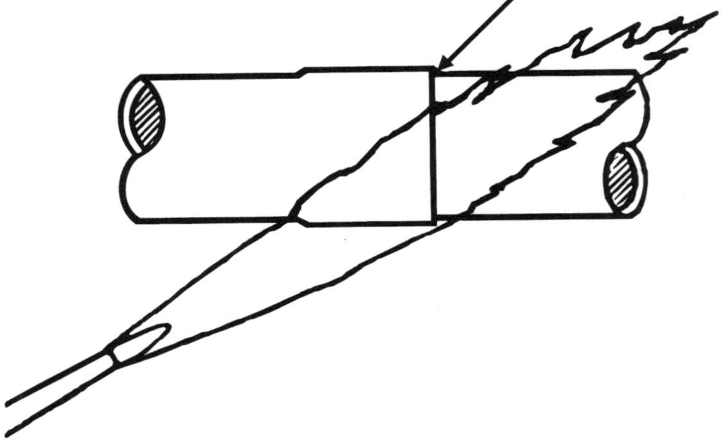

HC 6-5

Oxyacetylene Cutting

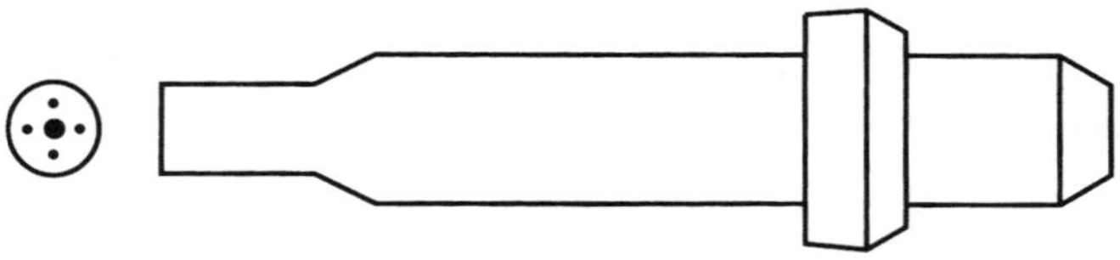

000,00
Light preheat

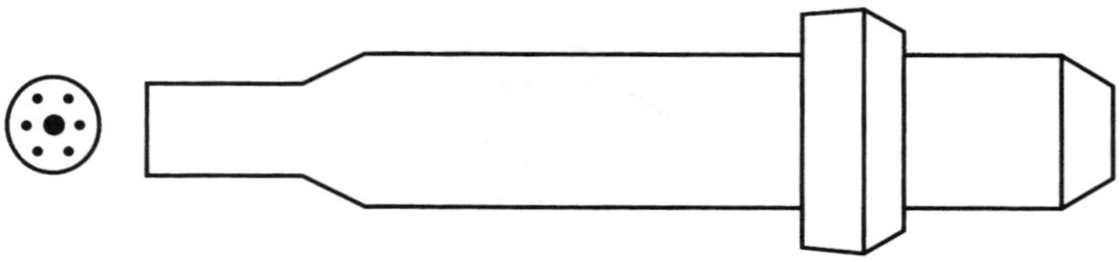

0, 1, 2
Medium preheat, general hand cutting

Cutting Tip Pressure Settings			
Metal Thickness	**Tip Size**	**Oxygen (psig)**	**Acetylene (psig)**
1/8″	000	20–25	5
1/4″	00	20–25	5
3/8″	0	25–30	5
1/2″	0	30–35	5
3/4″	1	30–40	5
1″	2	35–50	6

HC 6-6

Name ______________________________

Date ______________________________

Instructor ___________________________

Period______________________________

6

Brazing and Flame-cutting

Quiz

Instructions: Write your answers on the lines provided at the right-hand side of the page.

1. Brazing is the joining of metals at temperatures above: 1. ____________________
   a. 840°F (450°C).
   b. 1200°F (650°C).
   c. 300°F (167°C).

2. *True or false?* A nonferrous metal does not contain any iron. 2. ____________________

3. A _____ is required when using silver-bearing brazing alloys to join copper to steel. 3. ____________________

4. What type of flame is used for brazing with the oxyacetylene torch? 4. ____________________
   a. Oxidizing.
   b. Neutral.
   c. Carburizing.

5. Brazing procedures should be performed with the _____ torch. 5. ____________________

6. Name three types of brazing connections (as identified by direction of alloy flow). 6. ____________________
   ____
   ____

7. What is the preferred type of brazing torch when a heat sink is involved? 7. ____________________

8. Why is the acetylene cylinder valve wrench always left on the valve? 8. ____________________

9. When shutting down the oxyacetylene torch, why is the oxygen turned off first? 9. ____________________

10. When using the cutting attachment, the base metal is preheated to a _____ color before the oxygen cutting lever is pressed. 10. ____________________

7

Mathematics for Technicians

Objectives

After studying this chapter in the text, the student will be able to:

- ❑ Read the place values of numbers.
- ❑ Work with fractions and decimals.
- ❑ Make linear, two-, and three-dimensional measurements.
- ❑ Convert Fahrenheit temperature readings to Celsius and vice versa.
- ❑ Use percentages to calculate profit margin on parts.

Instructional Materials

Text: Pages 91–102
Lab Manual: Pages 55–64
Instructor's Manual:
Visual Masters, Pages 81–85
Chapter 7 Quiz, Page 86

Teaching Strategy

Text

1. Review objectives.
2. Discuss and assign chapter contents. Use these visual masters as appropriate:
   HC 7-1: Reading a Ruler
   HC 7-2: Working with Circles
   HC 7-3: Cubic Volume
   HC 7-4: Volume of a Cylinder
   HC 7-5: Converting Temperatures
3. Direct students to complete Test Your Knowledge questions, page 102. After completion, discuss correct answers with students.

Lab Manual

1. Provide overview of Activities. Where necessary, stress safe laboratory practices and the importance of wearing protective clothing.
2. Assign Activities to individuals or pairs of students, as appropriate.
   Activity 7-1: Linear and Angular Measurements
   Activity 7-2: Calculating Surface Area
   Activity 7-3: Calculating Volume
   Activity 7-4: Measuring and Converting Temperatures
   Activity 7-5: Time Sheet Calculation
   Activity 7-6: Completing a Work Order/Invoice
3. Observe student performance on Activities, providing guidance or correction as needed. After reviewing the student's work, initial and date the activity in the place provided.

Evaluation

1. Administer the quiz from this Instructor's Manual.
2. Grade the quiz, and return it with appropriate comments for the student.

Answers to Test Your Knowledge, Page 102

1. b. Denominator.
2. true
3. parts (or fractions) of a whole
4. six and four hundred thirty-five thousandths
5. U.S. Conventional
6. 12; 100
7. 16
8. true
9. area
10. "Five squared" or "five to the second power." It means that the factor (5) is multiplied by itself (5×5).
11. radius
12. height
13. c. 3.1416

14. °F = (1.8 × °C) + 32
15. $55.26

Answers to Lab Manual, Pages 55–63

Activity 7-1

1. A. 2 1/16"
   B. 2 7/8"
   C. 3 11/16"
   D. 4 3/16"
2. A. 53 mm
   B. 72 mm
   C. 90 mm
   D. 102 mm
3. A. 1/16"
   B. 3/8"
   C. 3/4"
   D. 1"
   E. 1 1/2"
4. Student response (answers will vary).
5. A. 45°
   B. 90°
   C. 135°
   D. 180°
   E. 225°
   F. 270°
   G. 315°
6. A. 45°
   B. 90°
   C. 45°
   D. 45°
   E. 90°
   F. 90°
   G. 2
   H. 4
7. A. 1 1/2"
   B. 3"

Activity 7-2

1.–9.Student response (answers will vary).
10. A= L × W; 18,819 sq. in. or 130.69 sq. ft.
11. $A = \pi r^2$; 50.26 sq. in.

Activity 7-3

1. V = L × W × H
2. 4.95 cu. ft.
3. 8 (7.5)
4. $V = \pi r^2 H$; 50.24 cu. in.
5. 209.33 cu. ft./min. (361,728 cu. in./min.)

Activity 7-4

1. (°F – 32) ÷ 1.8
2. (1.8 × °C) + 32

3.–4.Student response (answers will vary).

Activity 7-5

1. 57 1/2
2. 17 1/2
3. Straight time: $340.00
   Overtime: $223.13
   Total pay: $563.13

Activity 7-6

BLAIR'S
Air Conditioning & Heating, Inc.
2260 Bay Shore Blvd.
St. Petersburg, Fl. 33990
(813) 439-1200

JOB INVOICE
09482

☑ SERVICE ☐ INSTALL	☐ PICK UP ☐ DELIVERY	REPAIR IN ☑ HOME ☐ SHOP	CUSTOMER ORDER NO. 33456	DATE OF ORDER 7/5

NAME: *Mrs. E. F. Jones* — PHONE: *948-2130* — DATE PROMISED: *7/5*
ADDRESS: *409 Fifth Street* — MECHANIC: *YOUR NAME*
CITY: *Tarpon Springs, Fl.* — DATE OF ORIG. INSTAL: ?
MAKE: *Trane* — MODEL: *B4J928* — SERIAL NO.: *9-23478*
NATURE OF SERVICE REQUEST: *AC Not Cooling*
☐ ESTIMATE ☐ WARRANTY ☐ CONTRACT ☑ CASH ☐ CHARGE ☐ C.O.D.

QUAN.	PART NO.	DESCRIPTION	PRICE	AMOUNT
6	#12	wire connectors	.31	1 88
1	#428	filter-drier	43.13	43 13
5	lbs	R-22	3.75	18 75

SERVICE PERFORMED: *Located and repaired leak on flare nut. Charged system to capacity. Repaired defective wiring connections.*

HOURS	LABOR	AMOUNT
2 hrs	MECHANICS @	87 50
	HELPERS @	
	DATE COMPLETED 7 / 5 / — TOTAL LABOR	87 50

TOTAL MATERIAL	63 76
TOTAL LABOR	87 50
TAX	3 83
TOTAL	155 09

I hereby accept above performed service, and charges, as being satisfactory and acknowledge that equipment has been left in good condition

Customer's Signature ____________________

THANK YOU!

Answers to Quiz

1. bottom
2. c. mixed number
3. and; point
4. b. 18.462
5. $12.75 per hour
6. square
7. U.S. Conventional; SI Metric (inches and centimeters are also acceptable)
8. 27
9. 360
10. false

Reading a Ruler

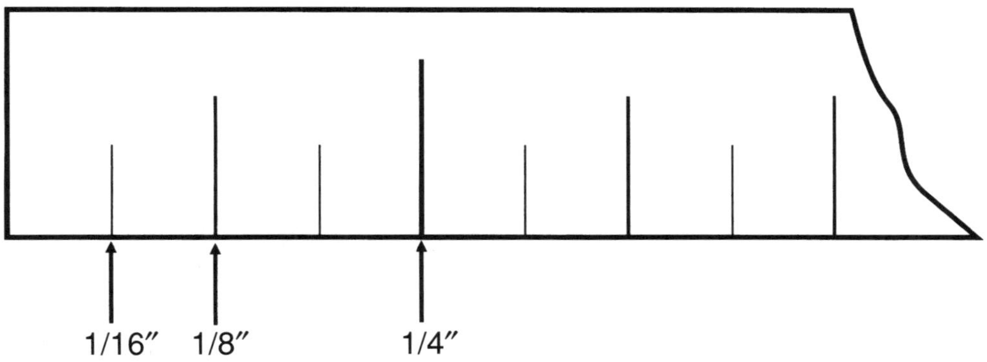

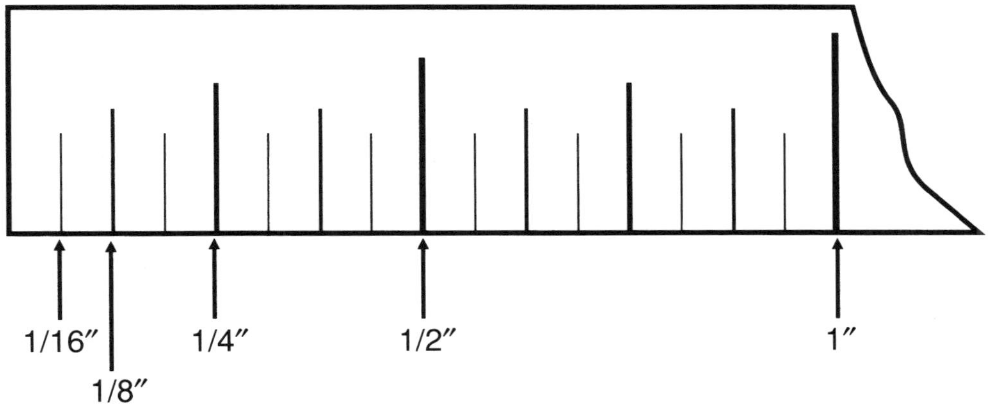

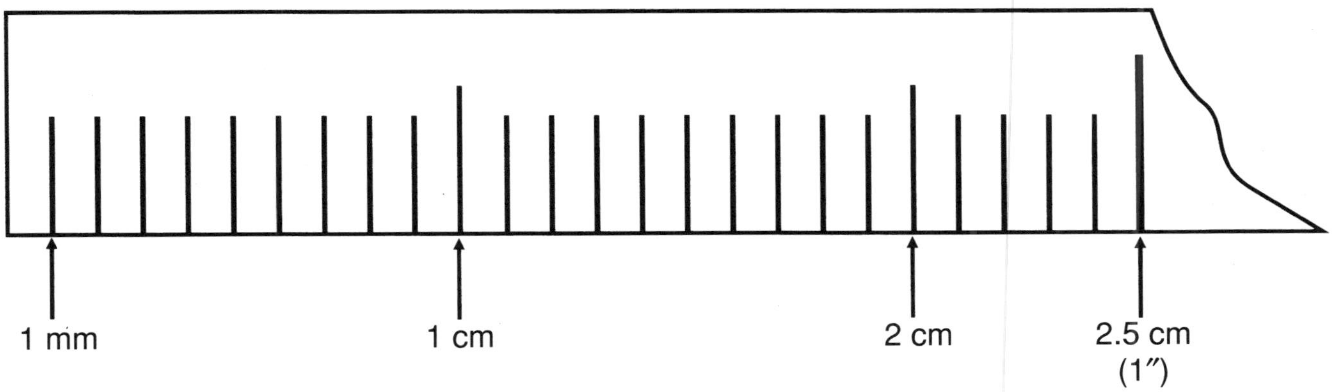

HC 7-1

Working with Circles

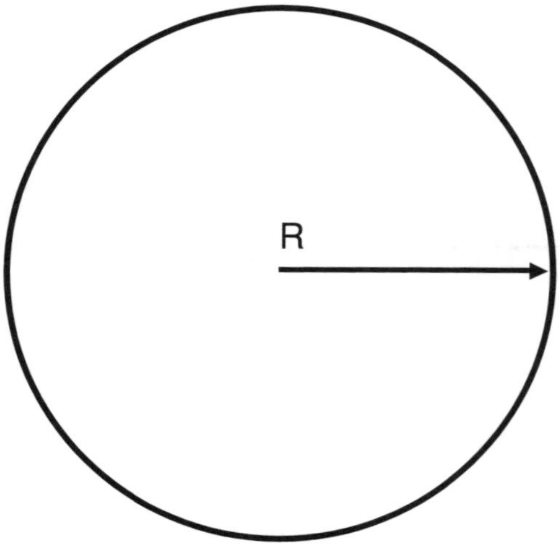

R = Radius

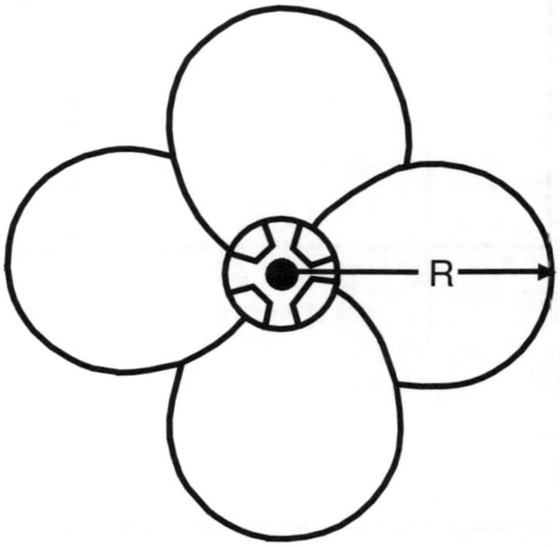

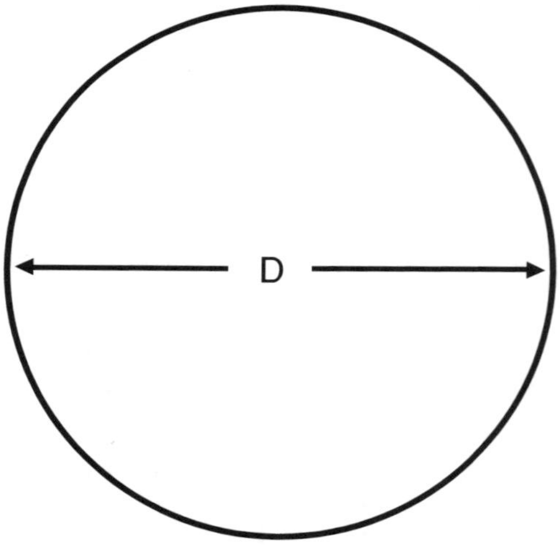

D = Diameter

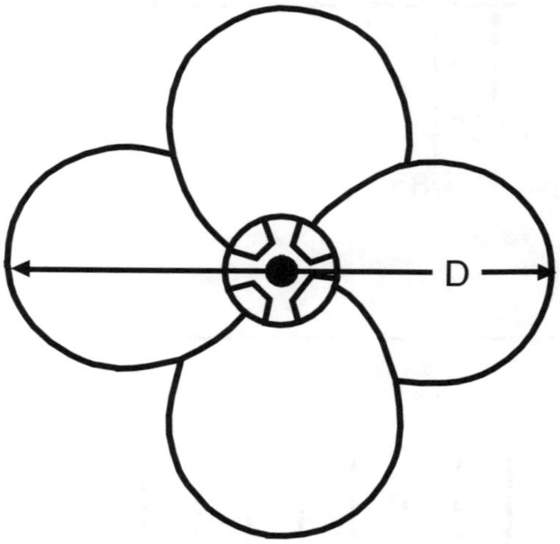

HC 7-2

Cubic Volume

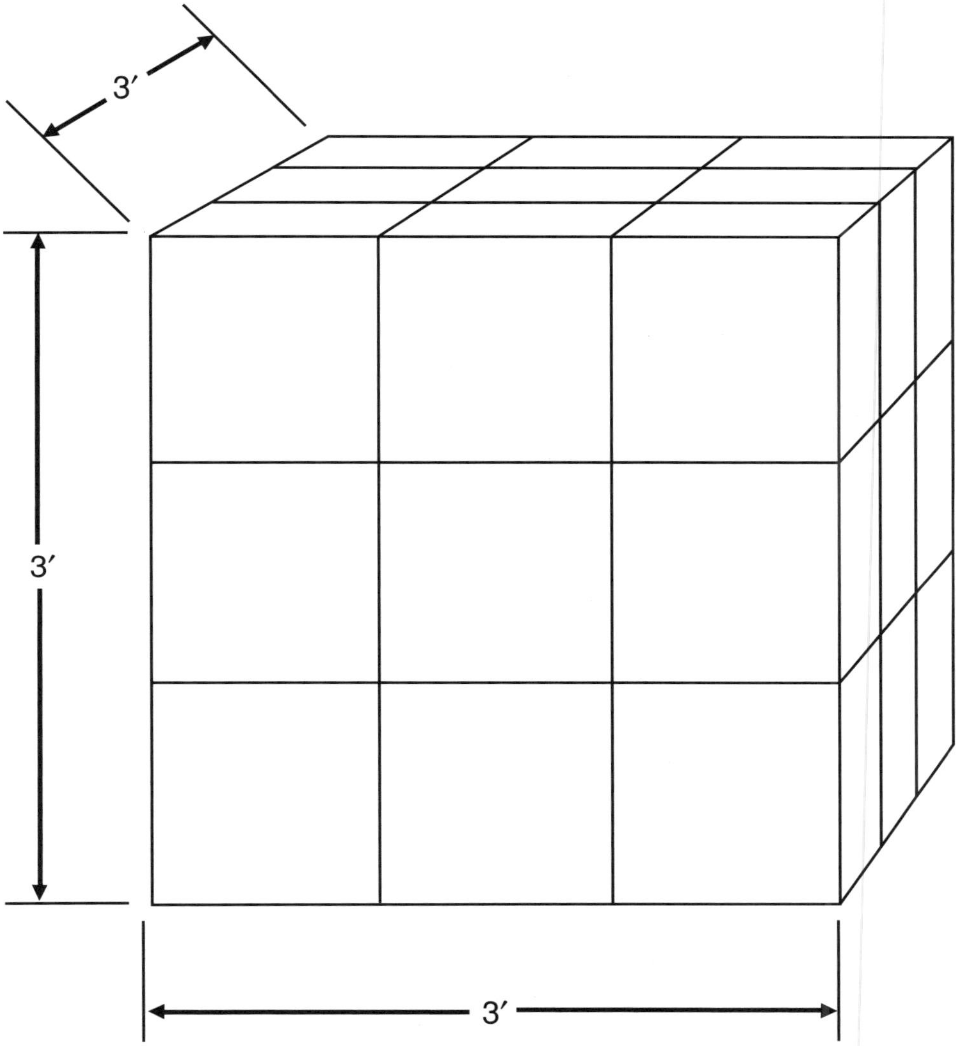

3 ft. × 3 ft. × 3 ft. = 27 cu. ft.

HC 7-3

Volume of a Cylinder

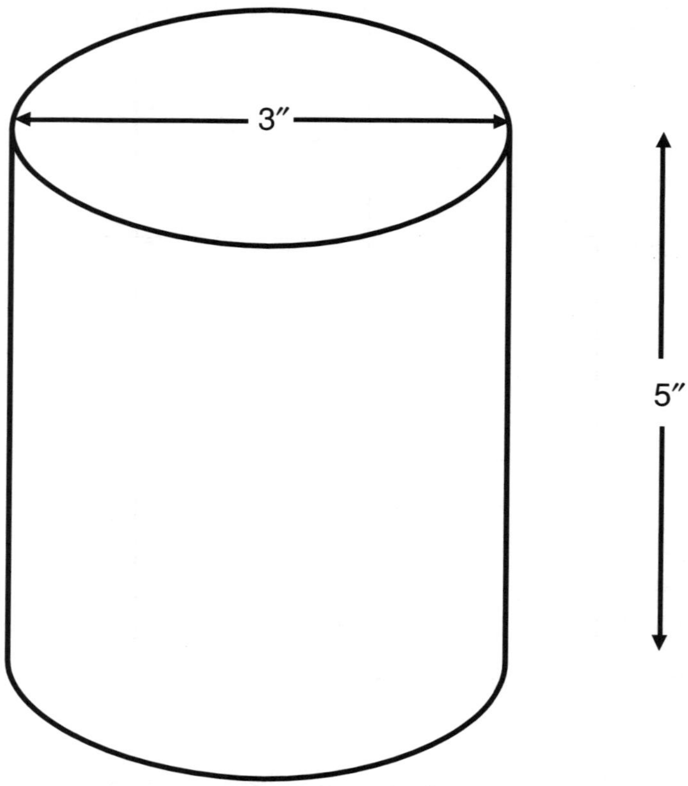

$V = \pi r^2 H$

HC 7-4

Converting Temperatures

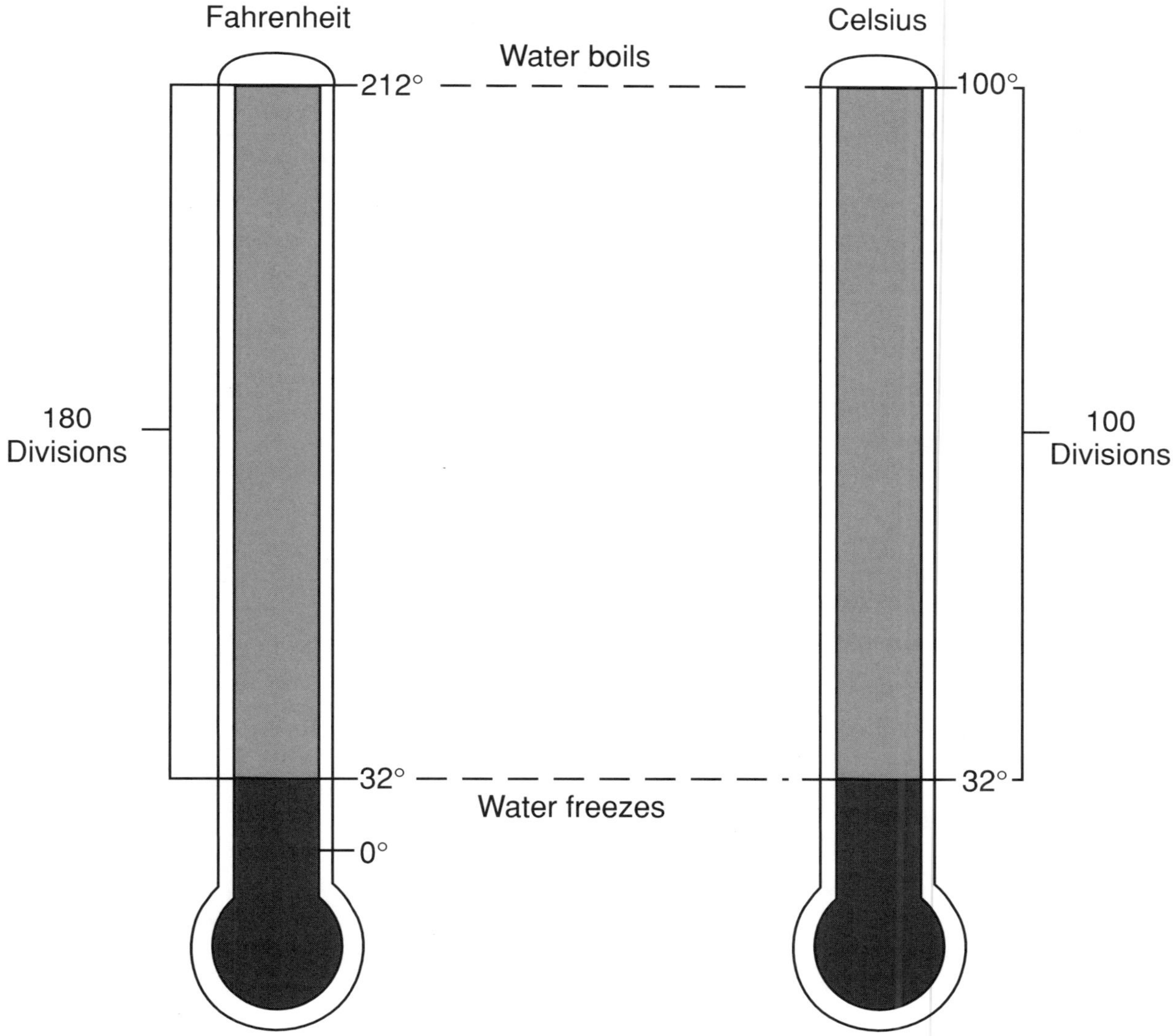

°F = (1.8 × °C) + 32
°C = (°F – 32) ÷ 1.8

HC 7-5

Name ______________________________

Date ______________________________

Instructor ______________________________

Period______________________________

7 Mathematics for Technicians

Quiz

Instructions: Write your answers on the lines provided at the right-hand side of the page.

1. In a fraction, is the top or bottom number the denominator? 1. ______________

2. The number 6 5/8 is called a(n) _____. 2. ______________
   a. proper fraction
   b. improper fraction
   c. mixed number

3. Name two words used to indicate a decimal. 3. ______________

4. Eighteen and four-hundred sixty-two thousandths is written: 4. ______________
   a. 18.0462
   b. 18.462
   c. 18.4062

5. If a technician's regular pay is $8.50 per hour, how much per hour would the technician be paid for overtime? 5. ______________

6. Is a measurement of two dimensions known as square or cubic measure? 6. ______________

7. Name two types of divisions that can be found on rulers and used to measure length. 7. ______________
   ____

8. What is the volume, in cubic feet, of the object shown? 8. ______________

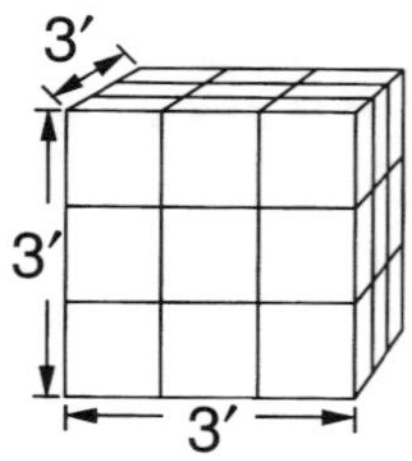

9. How many degrees are in a complete circle? 9. ______________

10. *True or false?* The Celsius thermometer is the most common type used in the United States. 10. ______________

8

Basic Thermodynamic Principles

Objectives

After studying this chapter in the text, the student will be able to:

- ❑ Identify the ways heat is transferred.
- ❑ Describe how a material undergoes a change of state.
- ❑ Define and distinguish sensible heat, specific heat, and latent heat.
- ❑ Explain the significance of the British thermal unit.
- ❑ Name the five latent heats.
- ❑ Define saturated conditions, superheat, and subcooling.
- ❑ Explain and calculate a ton of refrigeration.

Instructional Materials

Text: Pages 103–112
Lab Manual: Pages 65–68
Instructor's Manual:
Visual Masters, Pages 89–94
Chapter 8 Quiz, Page 95

Teaching Strategy

Text

1. Review objectives.
2. Discuss and assign chapter contents. Use these visual masters as appropriate:
   HC 8-1: Thermodynamic Laws
   HC 8-2: Motion of Molecules
   HC 8-3: British Thermal Unit
   HC 8-4: Latent Heats
   HC 8-5: Saturation Point
   HC 8-6: Superheating and Subcooling
3. Direct students to complete Test Your Knowledge questions, page 112. After completion, discuss correct answers with students.

Lab Manual

1. Provide overview of Activities. Where necessary, stress safe laboratory practices and the importance of wearing protective clothing.
2. Assign Activities to individuals or pairs of students, as appropriate.
   Activity 8-1: Forced Convection
   Activity 8-2: Computing Btu Requirements
   Activity 8-3: Identifying Latent Heat
   Activity 8-4: Calculating Refrigeration Effect
   Activity 8-5: Subcooling and Superheating
3. Observe student performance on Activities, providing guidance or correction as needed. After reviewing the student's work, initial and date the activity in the place provided.

Evaluation

1. Administer the quiz from this Instructor's Manual.
2. Grade the quiz, and return it with appropriate comments for the student.

Answers to Test Your Knowledge, Page 112

1. Lack of heat.
2. Energy cannot be created or destroyed but can be converted from one form to another.
3. hot; cold
4. radiation; conduction; convection
5. Dark colors absorb heat and light colors tend to reflect heat.
6. solid; liquid; gas
7. Heat that causes a change in temperature.
8. heat; pound; Fahrenheit

9. The amount of heat required to raise the temperature of one pound of a given substance one degree Fahrenheit.
10. Heat that causes a change in state.
11. fusion (freezing); melting; vaporization (evaporation); condensation; sublimation
12. Dry ice, which goes directly from a solid state to a gaseous state.
13. 0.095
14. A saturated substance is one that has absorbed or lost as much heat energy as it possibly can without changing state.
15. true
16. Any heat added to a gas.
17. supercooled
18. 12,000 Btu/hr.
19. 1-1/2 tons
20. ice

Answers to Lab Manual, Pages 65–68

Activity 8-1

1. A. blower
   B. furnace area
   C. evaporator
   D. supply airflow
   E. register
   F. return air grille
   G. return airflow
   H. filter
2.

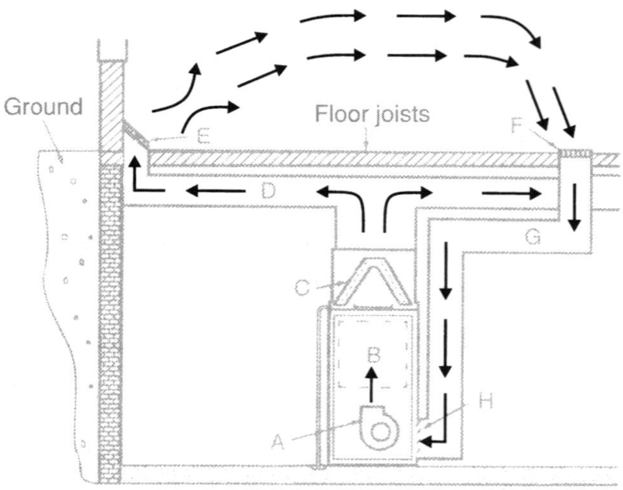

Activity 8-2

1. A. 35.06
   B. 144
   C. 180
   D. 970.4
   E. 3.68
   F. 1333.14
2. A. 2180.85
   B. 830.13
   C. 568,000

Activity 8-3

1. A. Freezing (fusion)
   B. Melting
   C. Vaporization (evaporation)
   D. Condensation
   E. Sublimation

Activity 8-4

1. Ice Melting Equivalent; ton
2. 288,000; one-ton
3. A. 1 ton
   B. 1 ton
   C. 1 ton
   D. 3 tons
   E. 1 1/2 tons
   F. 1/2 ton

Activity 8-5

1. liquid; superheated vapor
2. saturation point; subcooled
3. A. 10 degrees Fahrenheit
   B. 13 degrees Fahrenheit

Answers to Quiz

1. A. 1 ton
   B. 3 tons
   C. 2 tons
2. true
3. solid; liquid; gas (or vapor)
4. Heat that causes a change in temperature.
5. Heat that causes a change in state.
6. saturation point
7. A. added
   B. removed
   C. added
   D. removed
8. water; Fahrenheit
9. superheated
10. subcooled

Thermodynamic Laws

First Law

Energy cannot be created or destroyed, but can be *converted* from one to another.

Forms of energy: Heat
Electrical
Light (solar)
Mechanical (motor)
Chemical (battery)
Nuclear (atomic)

Second Law

Heat always travels from hot to cold; the greater the temperature difference, the faster the heat transfer will take place.

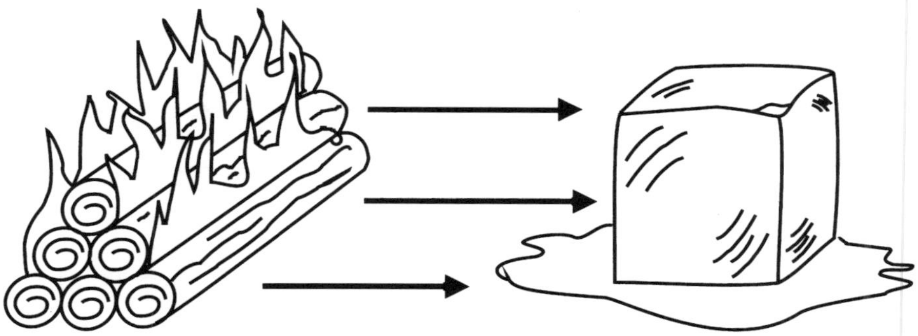

HC 8-1

Motion of Molecules

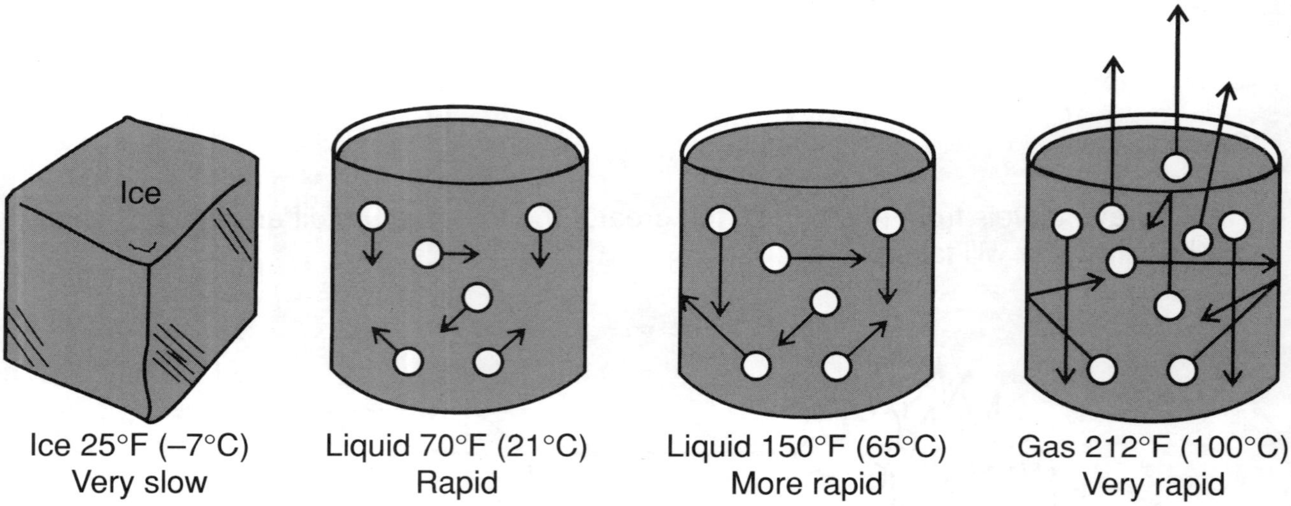

HC 8-2

British Thermal Unit

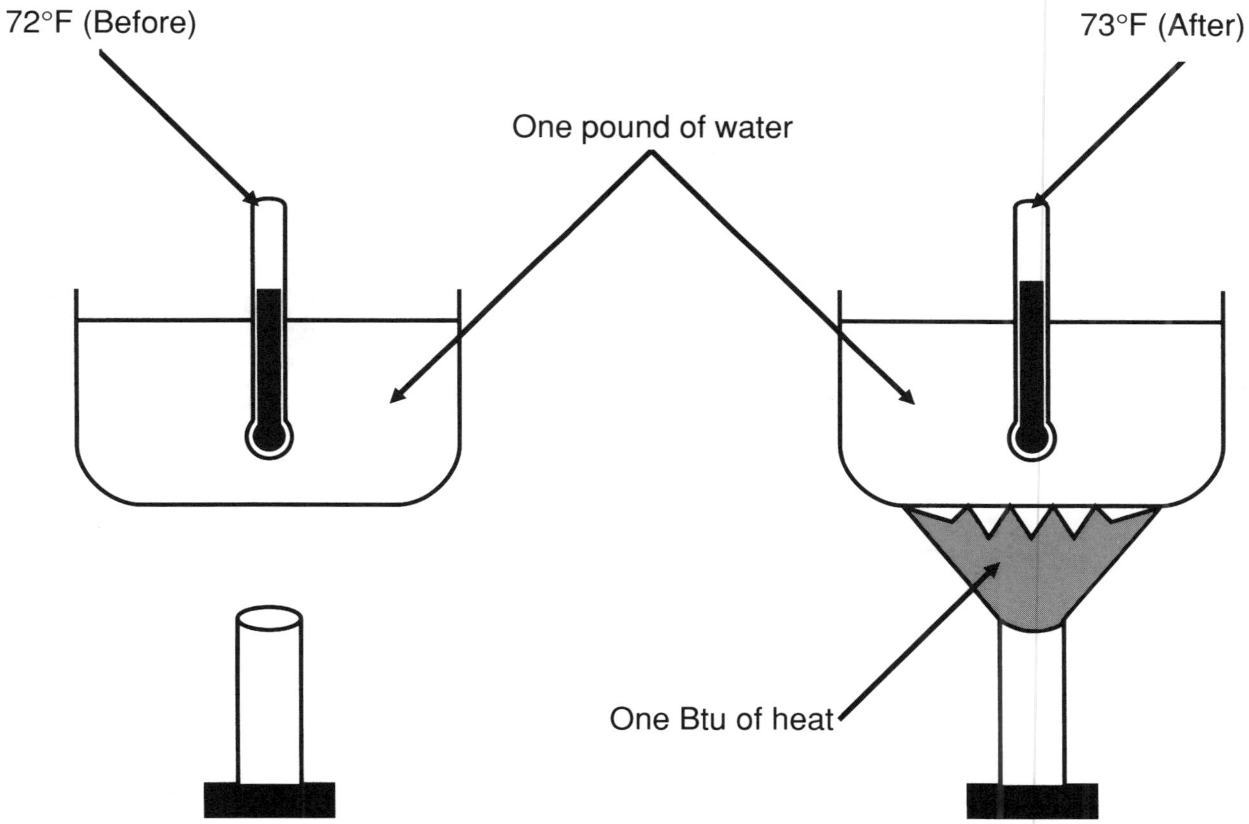

HC 8-3

Latent Heats

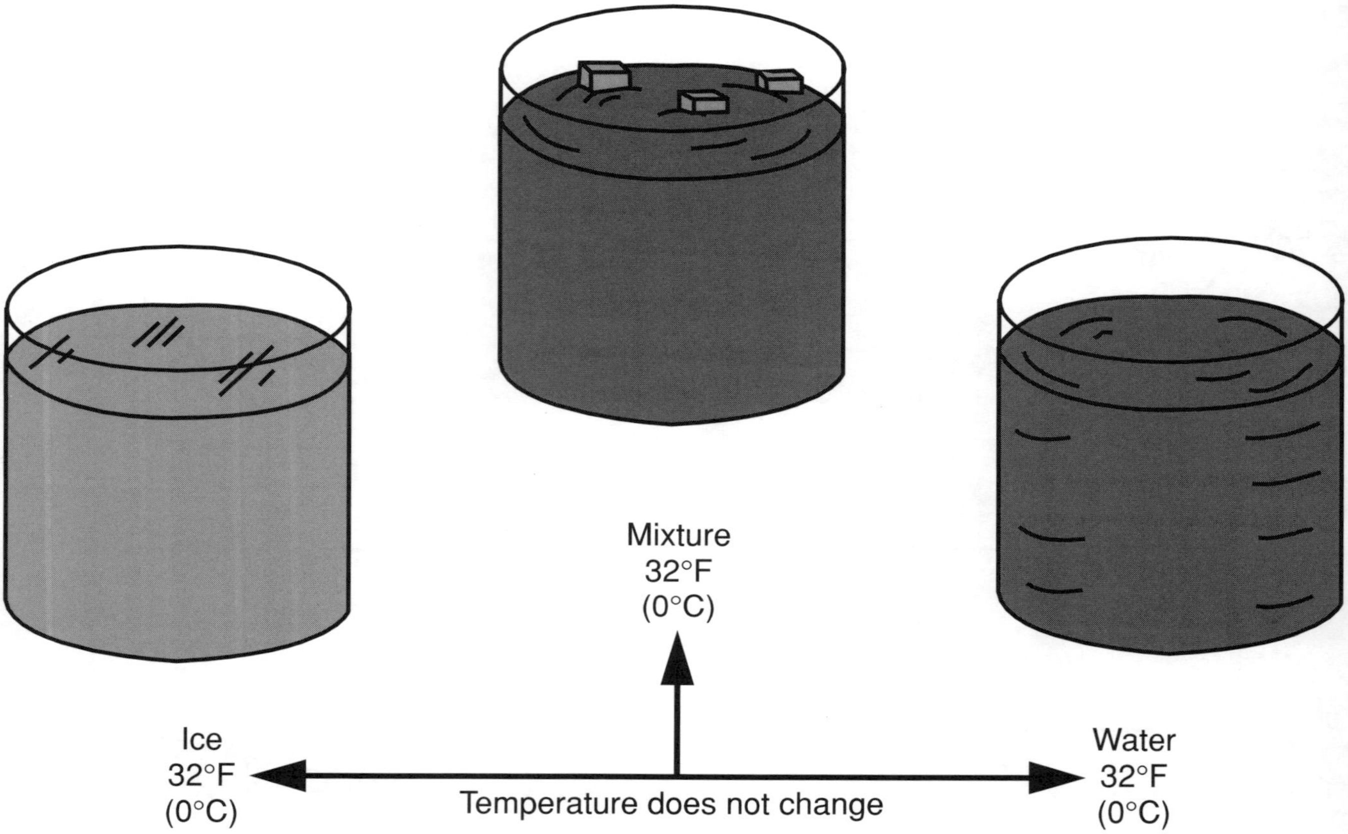

HC 8-4

Saturation Point

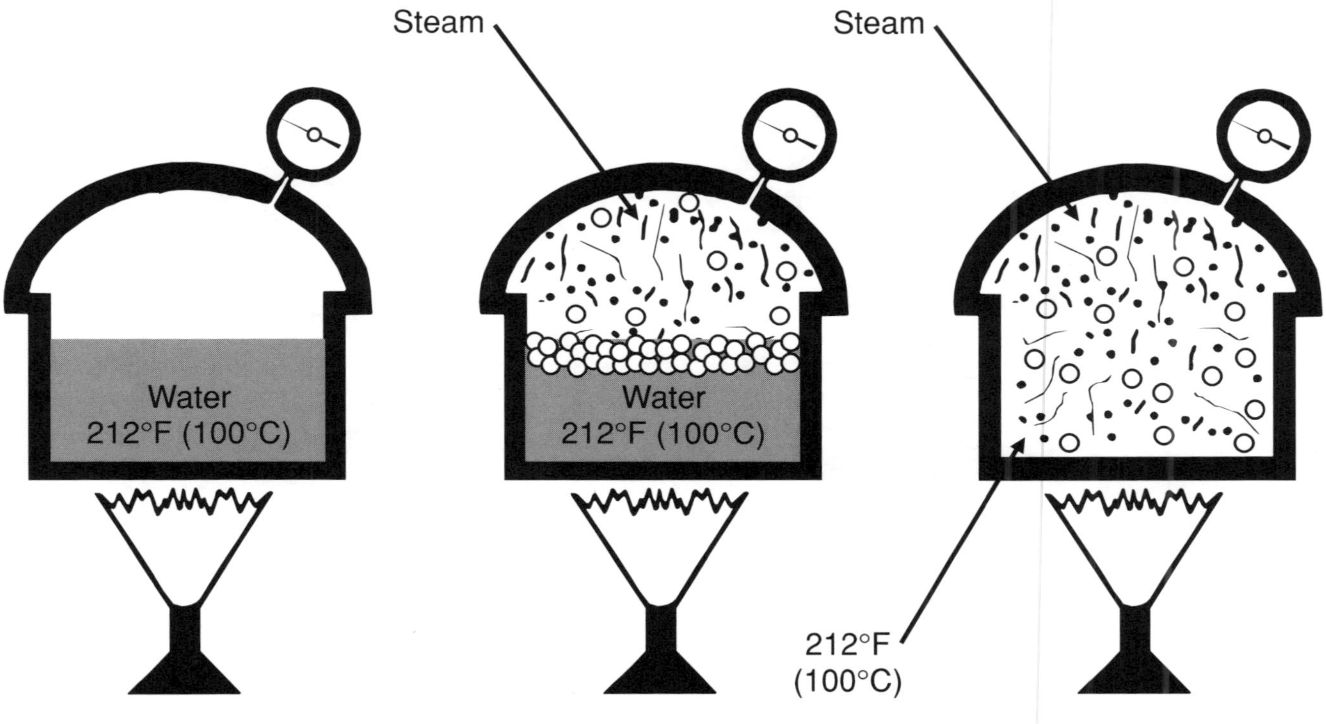

Atmospheric Pressure

HC 8-5

Superheating and Subcooling

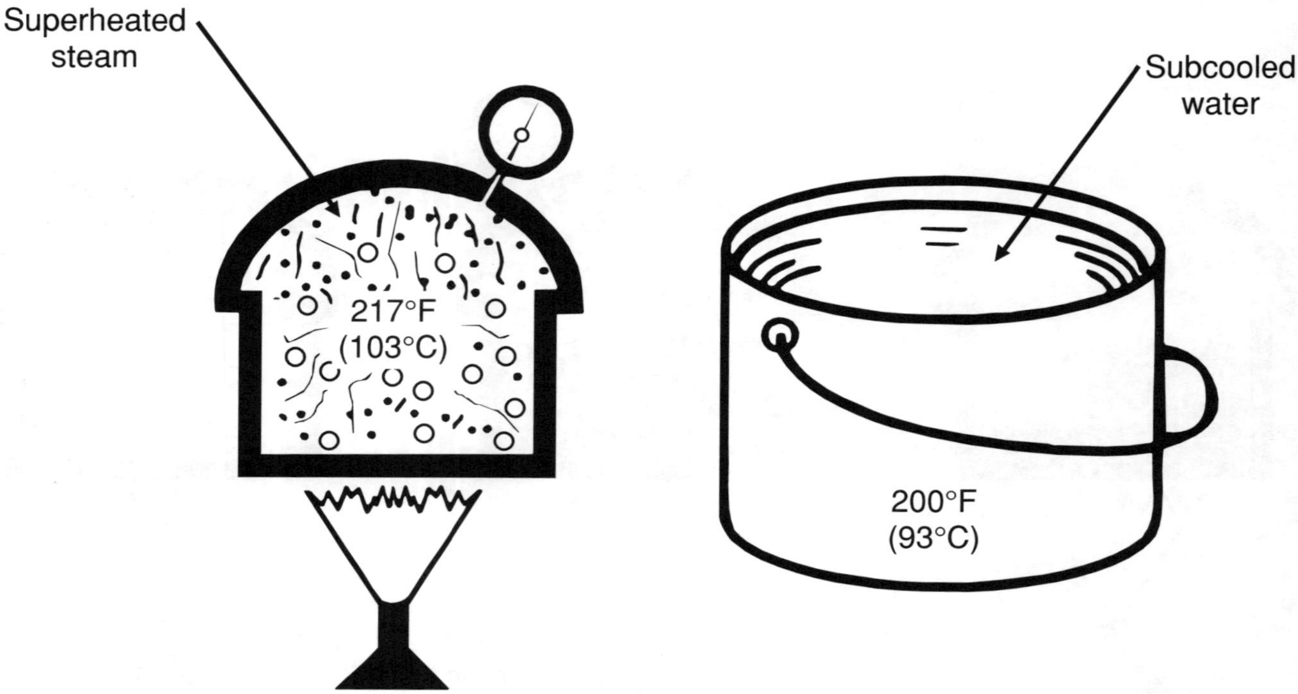

HC 8-6

Name ______________________________

Date ______________________________

Instructor ______________________________

Period______________________________

8 Basic Thermodynamic Principles

Quiz

Instructions: Write your answers on the lines provided at the right-hand side of the page.

1. Each of the following is equivalent to how many "tons" of refrigeration?
   a. 288,000 Btu/24 hrs.
   b. 36,000 Btu/hr.
   c. 400 Btu/min.

   1. a. ______________ b. ______________ c. ______________

2. *True or false?* Heat always travels from the hotter to the colder substance.

   2. ______________

3. Name the three physical states of matter.

   3. ______________ ______________ ______________

4. What is sensible heat?

   4. ______________

5. What is latent heat?

   5. ______________

6. What is the HVAC term for *boiling point*?

   6. ______________

7. Note whether heat is being added or removed in each of the following:
   a. Latent heat of melting.
   b. Latent heat of freezing.
   c. Latent heat of evaporation.
   d. Latent heat of condensation.

   7. a. ______________ b. ______________ c. ______________ d. ______________

8. A British Thermal Unit is the amount of heat required to raise the temperature of one pound of _____ one degree _____.

   8. ______________ ______________

9. A substance with a temperature above its saturation point is said to be _____.

   9. ______________

10. A substance with a temperature below its saturation point is said to be _____.

    10. ______________

9

Temperature and Pressure

Objectives

After studying this chapter in the text, the student will be able to:

- ❑ Identify four types of thermometer scales.
- ❑ Discuss the importance of Fahrenheit and Celsius absolute temperatures.
- ❑ Describe atmospheric pressure and how pressures are measured and read.
- ❑ Describe vacuum measurement and the use of the compound gauge.
- ❑ Explain and apply Boyle's Law and Charles' Law.

Instructional Materials

Text: Pages 113–124
Lab Manual: Pages 69–74
Instructor's Manual:
  Visual Masters, Pages 99–101
  Chapter 9 Quiz, Page 102

Teaching Strategy

Text

1. Review objectives.
2. Discuss and assign chapter contents. Use these visual masters as appropriate:
   HC 9-1: Temperature Scales
   HC 9-2: Temperature and Pressure
   HC 9-3: Gas Laws
3. Direct students to complete Test Your Knowledge questions, page 124. After completion, discuss correct answers with students.

Lab Manual

1. Provide overview of Activities. Where necessary, stress safe laboratory practices and the importance of wearing protective clothing.
2. Assign Activities to individuals or pairs of students, as appropriate.
   Activity 9-1: Temperature Scales
   Activity 9-2: Temperature-Pressure Relationships
3. Observe student performance on Activities, providing guidance or correction as needed. After reviewing the student's work, initial and date the activity in the place provided.

Evaluation

1. Administer the quiz from this Instructor's Manual.
2. Grade the quiz, and return it with appropriate comments for the student.

Answers to Test Your Knowledge, Page 124

1. – 460°F; –273°C; 0°R; 0°K
2. cryogenic
3. The temperature conditions surrounding an object.
4. The pressure exerted upon it.
5. false
6. 14.7
7. 0
8. 0
9. atmospheric; vacuum
10. inversely; temperature
11. Po × Vo = Pn × Vn
12. 85 psig
13. The absolute pressure of a gas will vary in direct proportion to the absolute temperature, provided the volume remains constant; and the volume of a gas will vary in direct proportion to the absolute temperature, provided the pressure remains constant.
14. pressure; volume
15. Pressures in a situation with a mixture of gases.

Answers to Lab Manual, Pages 69–72

Activity 9-1

1. A. 180
   B. 212
   C. 32
   D. 0
2. A. 100
   B. 100
   C. 0
3. A. 672
   B. –273
   C. 32
   D. 0

Activity 9-2

1. A. Charles' Law
   B. Dalton's Law
   C. Boyle's Law
2. A. 2
   B. 1
   C. psig
   D. 0
   E. 0
   F. 2
   G. 30 in. Hg
3. A. 0.6 psig
   B. 32°F (0°C)
   C. –45°F (–43°C)
   D. 19.9 psig
   E. 14.1 in. Hg
   F. 15.7 psig
   G. 121.4 psig
   H. 15.8 psig
   I. 10.1 psig
   J. –50°F (–46°C)
4. A. 19
   B. 6
   C. –5
   D. 22
   E. 122
   F. 10
   G. –13
   H. not shown
   I. not shown
   J. 129

Answers to Quiz

1. Fahrenheit; Celsius; Rankine; Kelvin
2. false
3. d. All of the above.
4. atmospheric
5. Charles' Law
6. Charles' Law
7. Dalton's Law
8. boiling point
9. c. 30 in. Hg
10. saturation charts

Temperature Scales

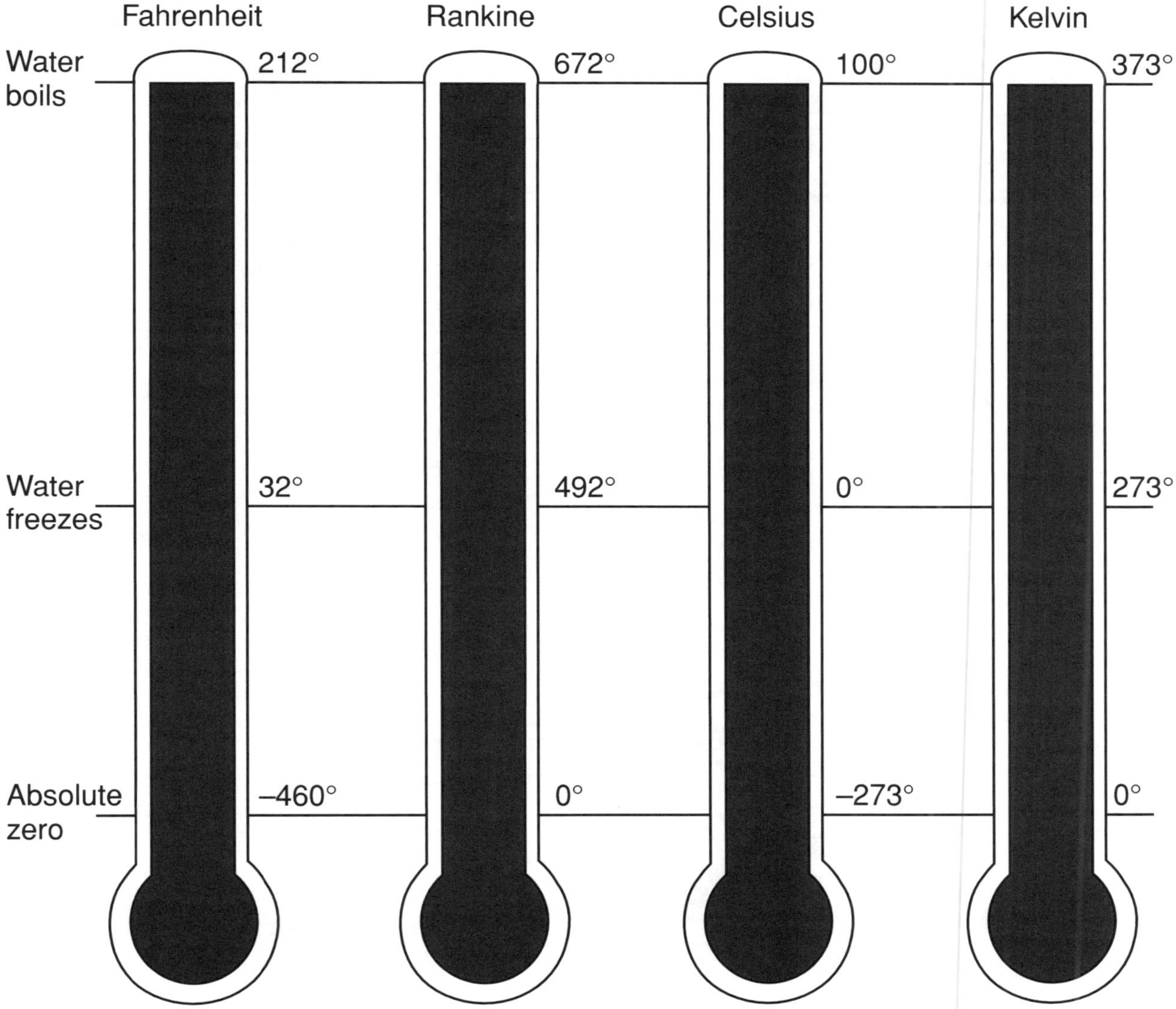

HC 9-1

Temperature and Pressure

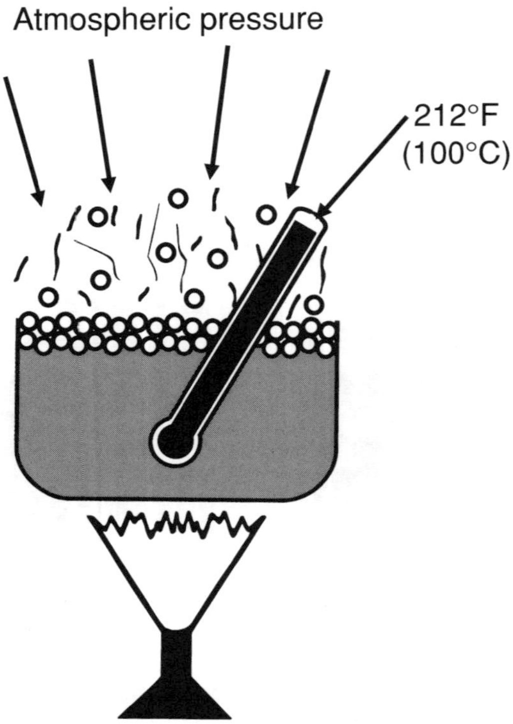

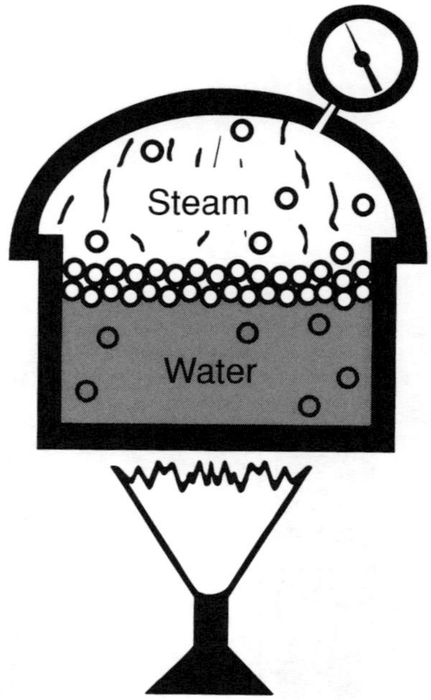

Pressure Above Atmospheric	Boiling Point
5 psi	228°F (109°C)
10 psi	244°F (118°C)
15 psi	250°F (121°C)
35 psi	281°F (138°C)
85 psi	328°F (164°C)

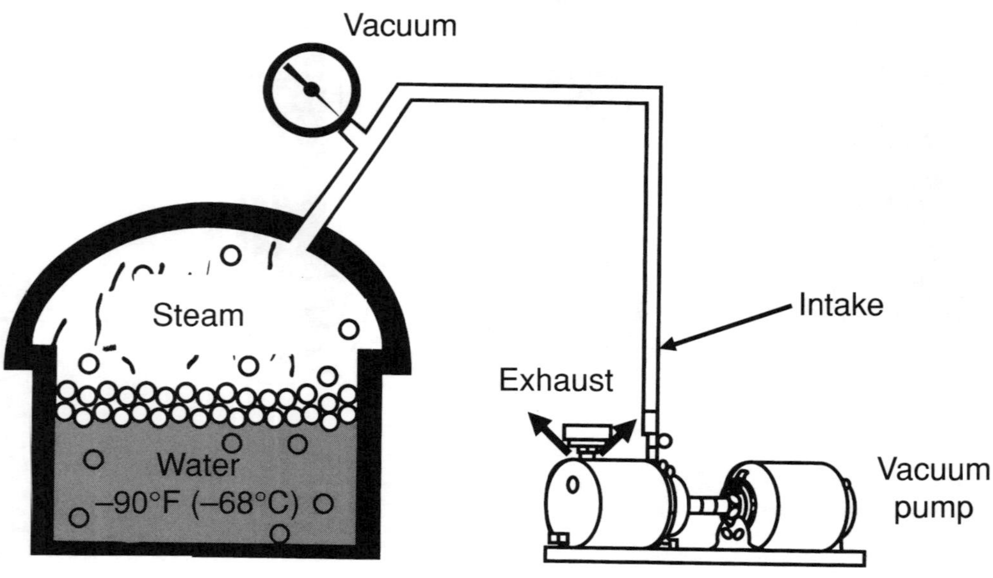

HC 9-2

Gas Laws

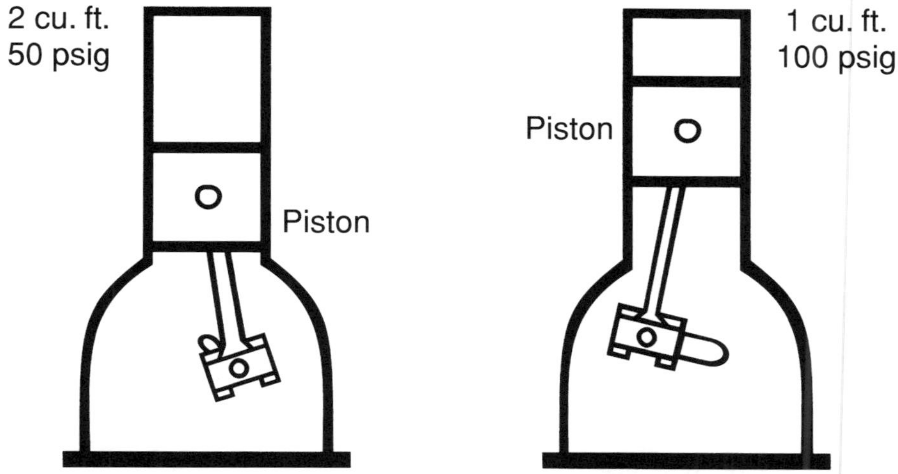

Boyle's Law

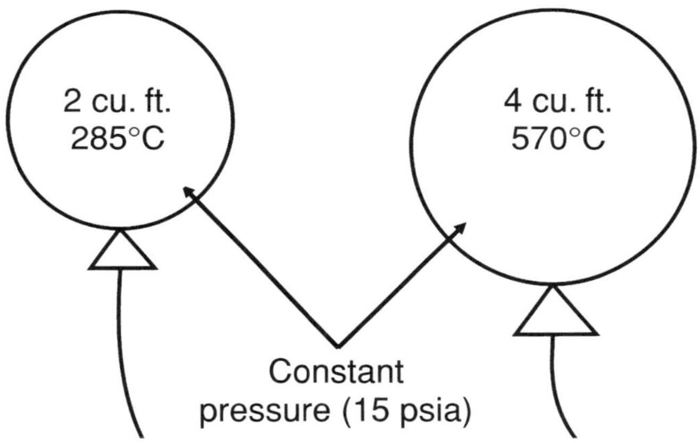

Charles' Law

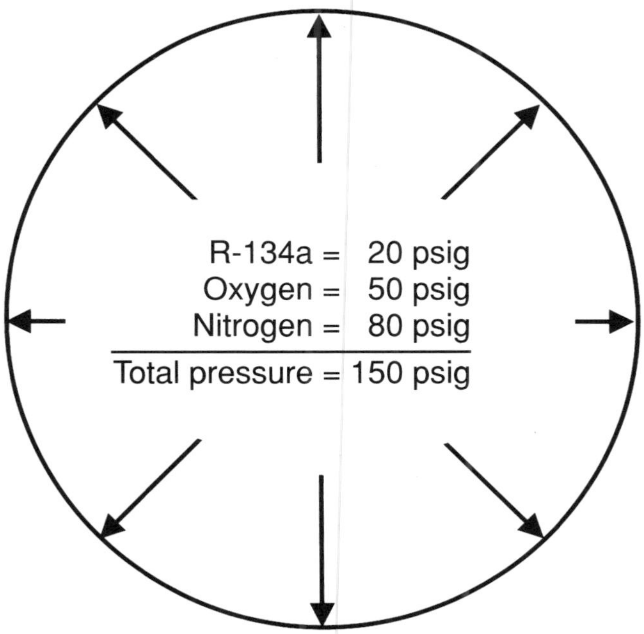

Dalton's Law

HC 9-3

Name ______________________________

Date ______________________________

Instructor ______________________________

Period______________________________

9

Temperature and Pressure

Quiz

Instructions: Write your answers on the lines provided at the right-hand side of the page.

1. Name four different temperature scales. 1. ______________________________
____
____
____

2. *True or false?* "Ambient" refers to temperatures between –250°F and absolute zero. 2. ______________________________

3. Atmospheric pressure is _____. 3. ______________________________
   a. 14.7 psia
   b. 0 psig
   c. 0 in. Hg
   d. All of the above.

4. Vacuum refers to pressures that are lower than _____ pressure. 4. ______________________________

5. Which of the Gas Laws can be summarized in these words: "At a constant volume, if the absolute temperature rises, the absolute pressure also rises." 5. ______________________________

6. Which of the Gas Laws describes the normal operation of a refrigeration system? 6. ______________________________

7. Which of the Gas Laws becomes applicable when atmospheric air enters the refrigeration system? 7. ______________________________

8. When the pressure on a liquid is changed, the _____ also changes. 8. ______________________________

9. On the compound gauge of a gauge manifold set, absolute zero would be shown as _____. 9. ______________________________
   a. 15 psig
   b. 0 in. Hg
   c. 30 in. Hg
   d. None of the above.

10. Charts or tables that show the boiling point of a substance at a specific pressure-temperature relationship are called _____. 10. ______________________________

10

Basic Refrigeration Cycle

Objectives

After studying this chapter in the text, the student will be able to:

- ❑ Draw a diagram of the basic refrigeration cycle.
- ❑ Describe the condition of the refrigerant in each component.
- ❑ Explain the purpose of each component of the refrigeration system.
- ❑ Identify refrigeration system component variations.

Instructional Materials

Text: Pages 125–146
Lab Manual: Pages 75–82
Instructor's Manual:
Visual Masters, Pages 106–109
Chapter 10 Quiz, Page 110

Teaching Strategy

Text

1. Review objectives.
2. Discuss and assign chapter contents. Use these visual masters as appropriate:
   HC 10-1: Basic Refrigeration System
   HC 10-2: Evaporator Operation
   HC 10-3: Compressor Operation
   HC 10-4: Condenser Operation
3. Direct students to complete Test Your Knowledge questions, pages 144–145. After completion, discuss correct answers with students.

Lab Manual

1. Provide overview of Activities. Where necessary, stress safe laboratory practices and the importance of wearing protective clothing.
2. Assign Activities to individuals or pairs of students, as appropriate.
   Activity 10-1: Refrigeration Cycle
   Activity 10-2: Identification
   Activity 10-3: Temperature Difference
   Activity 10-4: Compressor Operation
   Activity 10-5: Installing Tubing Insulation
   Activity 10-6: Removing the Compressor Head and Valve Plate
3. Observe student performance on Activities, providing guidance or correction as needed. After reviewing the student's work, initial and date the activity in the place provided.

Evaluation

1. Administer the quiz from this Instructor's Manual.
2. Grade the quiz, and return it with appropriate comments for the student.

Answers to Test Your Knowledge, Pages 144-145

1. evaporator; suction line; compressor; hot gas discharge line; condenser; liquid line; refrigerant control
2. To absorb heat from where it is unwanted (such as a room or the interior of a refrigerator).
3. low-temperature, low-pressure, saturated
4. 20 degrees Fahrenheit (11 degrees Celsius)
5. copper tubing
6. low-temperature, low-pressure, superheated gas
7. The refrigerant control and the compressor valve plate.
8. It converts a low-temperature, low-pressure gas to a high-temperature, high-pressure gas by means of compression.

9. high-temperature, high-pressure, highly superheated gas.
10. true
11. To dispose of the heat absorbed by the refrigerant as it passed through the evaporator.
12. hot
13. cold
14. c. High-temperature, high-pressure, highly superheated gas.
15. b. Medium-temperature, high-pressure liquid.
16. It meters the correct amount of liquid refrigerant into the evaporator.
17. temperature
18. shell; shelf; wall; plate; finned-tube with forced convection
19. d. Osmotic.
20. finned forced-convection; wire static; wall static

Answers to Lab Manual, Pages 75–80

Activity 10-1

1. A. evaporator
   B. suction line
   C. compressor
   D. hot gas discharge line
   E. condenser
   F. liquid line
   G. refrigerant control

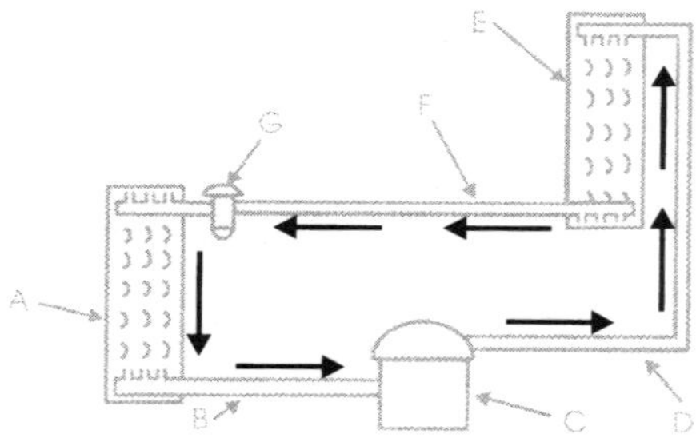

2. A. compressor
   B. hot gas discharge line
   C. condenser
   D. liquid line
   E. refrigerant control
   F. evaporator
   G. suction line

Activity 10-2

1. A. freezer section
   B. evaporator
   C. cold plate evaporator
   D. refrigeration section
   E. section 1

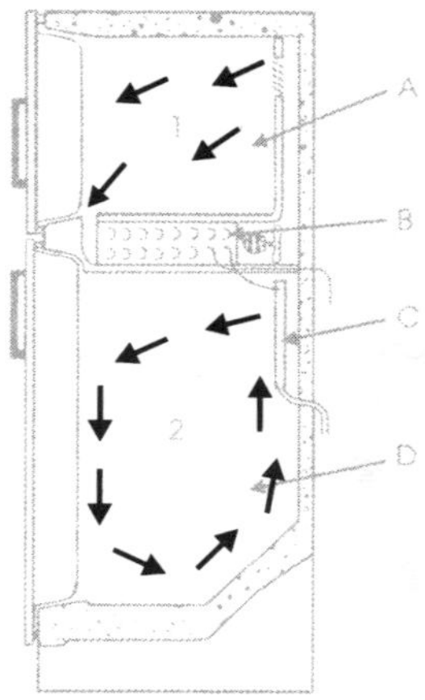

2. A. spray nozzles
   B. air intake
   C. catch basin
   D. water pump
   E. water makeup
   F. condenser

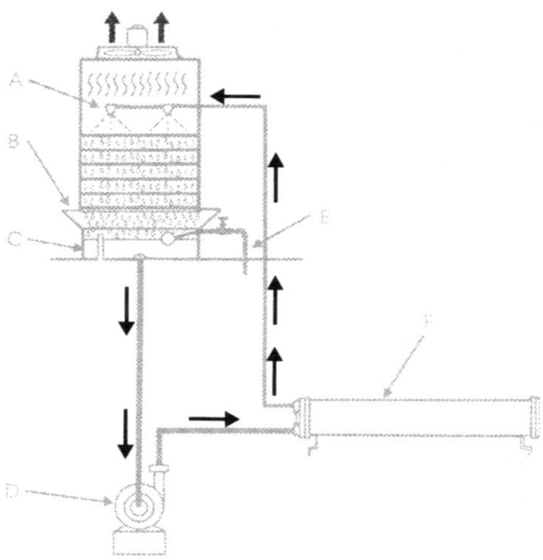

3. A. open-type compressor
   B. semihermetic compressor
   C. hermetic-type compressor

Activity 10-3

1. refrigerant
2. 10
3. 10
4. 12 psig (82.8 kPa)
5. 30°F (–1°C)
6. 10°F (–12°C)
7. 0 psig (0 kPa)
8. 10
9. warmer
10. 20 degrees Fahrenheit (11 degrees Celsius)

Activity 10-4

1. 115°F (46°C)
2. 95°F (35°C)
3. 195.9 psig (1352 kPa)
4. 110°F (43°C)
5. 155°F (68°C)
6. 296.8 psig (2048 kPa)
7. 80°F (27°C)
8. 214.4 psig (1479 kPa)
9. 143.6 psig (991 kPa)
10. The discharge pressure goes up as well.

Answers to Quiz

1. low temperature, low pressure, saturated
2. high temperature, high pressure, saturated
3. b. 20 degrees Fahrenheit.
4. sucking action; compressing action
5. 30
6. refrigerant control; compressor
7. open-type; hermetic; semihermetic
8. true
9. high
10. To cool condensing water for reuse.

Basic Refrigeration System

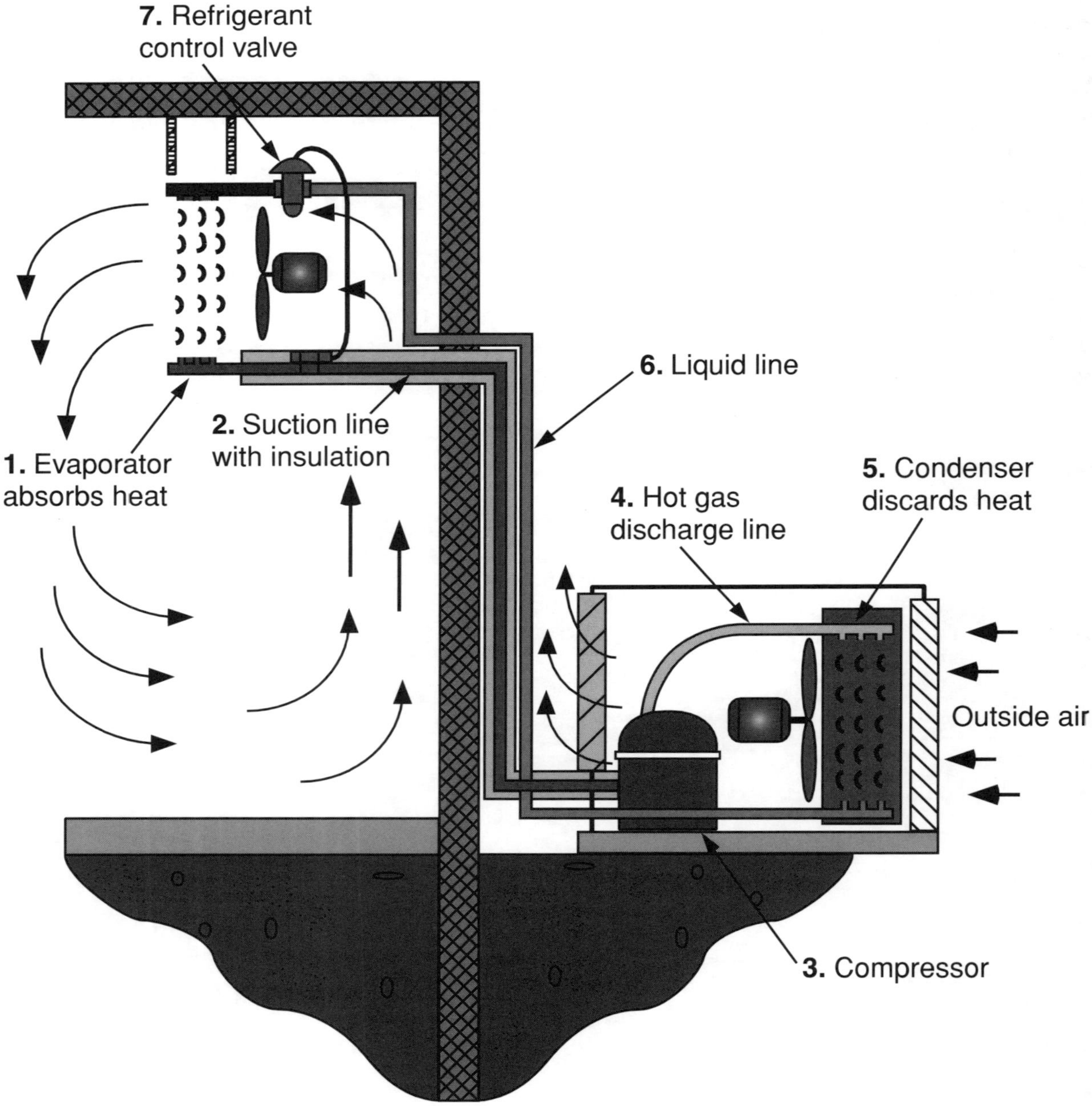

HC 10-1

Evaporator Operation

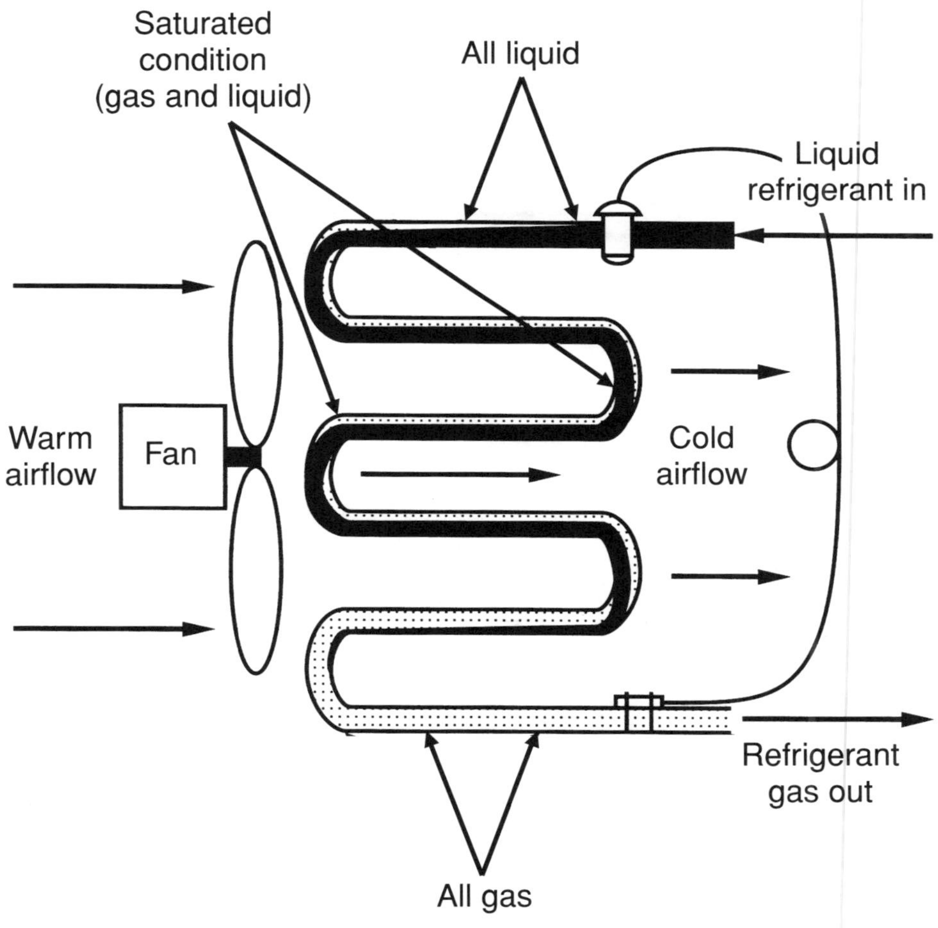

HC 10-2

Compressor Operation

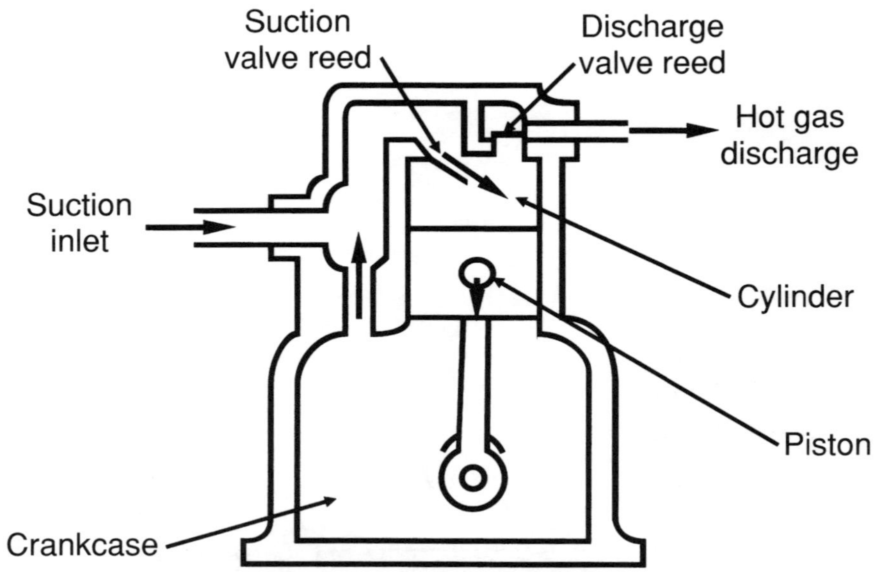

Intake

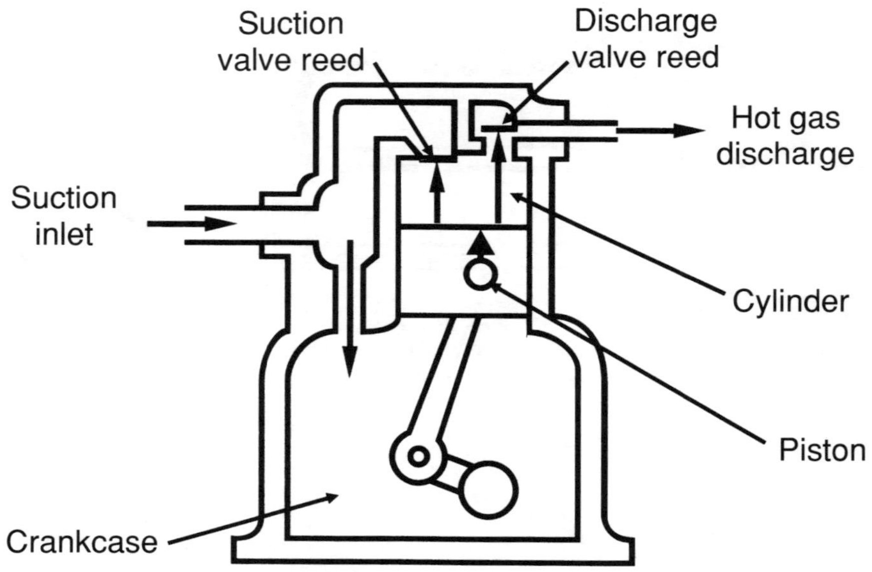

Discharge

HC 10-3

Condenser Operation

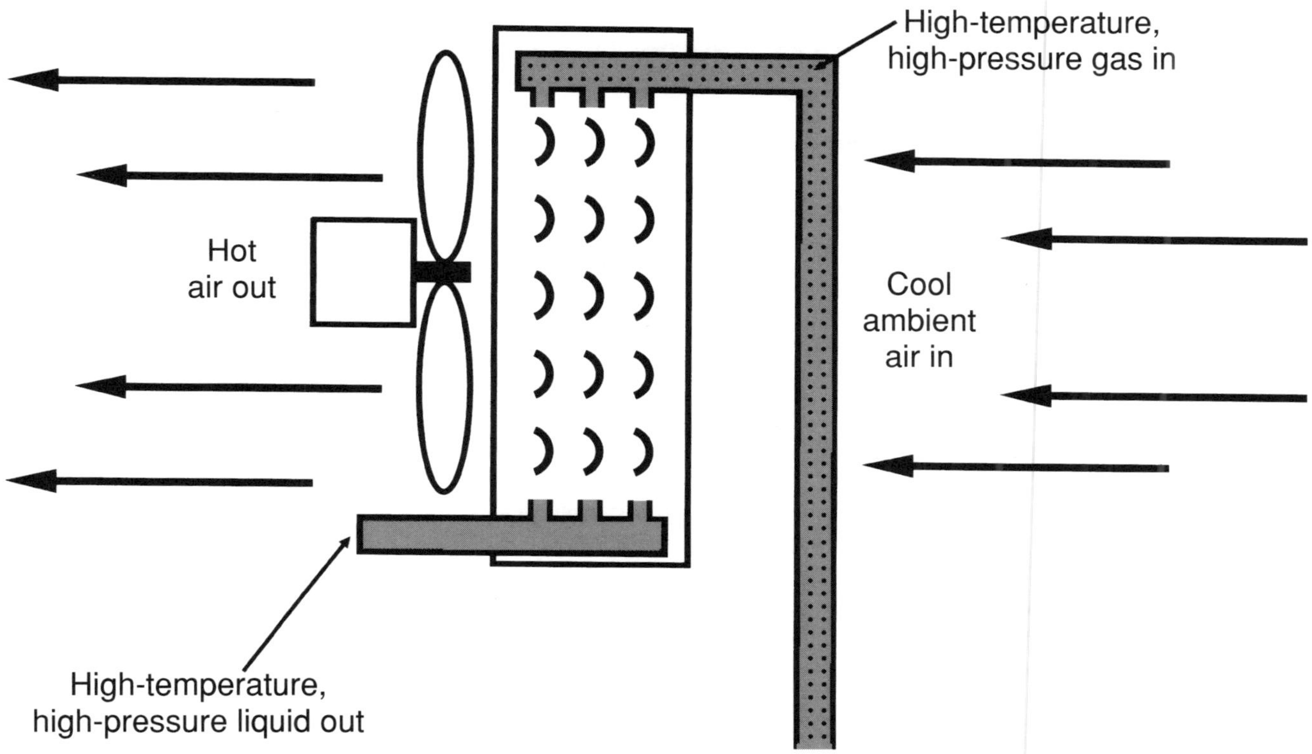

HC 10-4

Name ____________________

Date ____________________

Instructor ____________________

Period ____________________

10

Basic Refrigeration Cycle

Quiz

Instructions: Write your answers on the lines provided at the right-hand side of the page.

1. What is the condition of the refrigerant inside the evaporator? 1. ____________

2. What is the condition of the refrigerant inside the condenser? 2. ____________

3. When the system is running, what is the temperature difference between the air flowing across the evaporator and the refrigerant inside the evaporator? 3. ____________
   a 10 degrees Fahrenheit.
   b. 20 degrees Fahrenheit.
   c. 30 degrees Fahrenheit.
   d. There is no difference.

4. What two functions must a refrigeration system compressor perform? 4. ____________ ____________

5. The saturation temperature-pressure inside the condenser is about _____ degrees Fahrenheit higher than the ambient air temperature entering the condenser. 5. ____________

6. Name the two division points between the high-pressure and low-pressure sides of a refrigeration system. 6. ____________ ____________

7. What are the three basic styles of reciprocating compressors? 7. ____________ ____________ ____________

8. *True or false?* The suction line is insulated to keep heat in the air from warming the suction gas. 8. ____________

9. Does a dirty condenser cause high or low discharge pressure? 9. ____________

10. What is the purpose of a water cooling tower? 10. ____________

11

Other System Components

Objectives

After studying this chapter in the text, the student will be able to:

- ❑ Identify condensing and evaporator units.
- ❑ Describe the condition of refrigerant in various accessory components.
- ❑ Describe the purpose of system accessory components.
- ❑ Identify component variations.
- ❑ Name accessory components and describe the purpose of each.
- ❑ Install and use a gauge manifold.
- ❑ Discriminate between components in domestic and commercial systems.

Instructional Materials

Text: Pages 147–172
Lab Manual: Pages 83–94
Instructor's Manual:
Visual Masters, Pages 113–118
Chapter 11 Quiz, Page 119

Teaching Strategy

Text

1. Review objectives.
2. Discuss and assign chapter contents. Use these visual masters as appropriate:
   HC 11-1: Split System
   HC 11-2: Service Valves
   HC 11-3: Valve Positions
   HC 11-4: Evacuating a System
   HC 11-5: Charging a System
   HC 11-6: Suction Accumulator
3. Direct students to complete Test Your Knowledge questions, page 171. After completion, discuss correct answers with students.

Lab Manual

1. Provide overview of Activities. Where necessary, stress safe laboratory practices and the importance of wearing protective clothing.
2. Assign Activities to individuals or pairs of students, as appropriate.
   Activity 11-1: Refrigeration System Components
   Activity 11-2: Recalibrating a Gauge Manifold Set
   Activity 11-3: Installing a Gauge Manifold to Check Low-side Pressure
   Activity 11-4: Installing a Gauge Manifold to Check High-side Pressure
   Activity 11-5: Using the Gauge Manifold to Read System Pressures
   Activity 11-6: Obtaining a High-pressure Reading at the Liquid Receiver Service Valve
   Activity 11-7: System Pumpdown
   Activity 11-8: Replacing a Filter-drier
3. Observe student performance on Activities, providing guidance or correction as needed. After reviewing the student's work, initial and date the activity in the place provided.

Evaluation

1. Administer the quiz from this Instructor's Manual.
2. Grade the quiz, and return it with appropriate comments for the student.

Answers to Test Your Knowledge, Page 171

1. suction line and liquid line
2. in the bottom
3. To hold excess refrigerant and make repairs less expensive.

4. 80
5. suction service valve (SSV); discharge service valve (DSV); liquid receiver service valve (LRSV)
6. DSV; LRSV
7. the service port
8. None. All openings are open.
9. The opening where the suction gas enters the valve.
10. half union; 1/8″ MPT × 1/4″ MFT
11. clamp-on type; Schrader (core) type
12. center (yellow)
13. closed
14. In the liquid line close to the liquid receiver.
15. To filter out impurities and absorb moisture.
16. A probable leak in the system.
17. Immediately after the filter-drier.
18. To indicate safe or unsafe moisture levels.
19. To subcool liquid refrigerant and superheat suction gas.
20. evaporator; suction line; suction accumulator; SSV; compressor; DSV; hot gas discharge line; condenser; liquid receiver; LRSV; filter-drier; sight glass; liquid line; heat exchanger; refrigerant control

Answers to Lab Manual, Page 83–92

Activity 11-1

A. evaporator
B. refrigerant control valve
C. liquid line
D. sight glass
E. filter-drier
F. liquid receiver service valve
G. liquid receiver
H. condenser
I. hot gas discharge line
J. discharge service valve
K. compressor
L. suction service valve
M. suction accumulator
N. suction line
O. heat exchanger

Activities 11-3 to 11-7

Student response (answers will vary).

Answers to Quiz

1. suction line; liquid line
2. 80
3. suction service valve; discharge service valve; liquid receiver service valve
4. true
5. b. Backseated.
6. A. inlet/outlet
   B. service port
   C. to compressor or receiver
7. normally closed
8. gas; moisture
9. b. liquid line
10. false

Split System

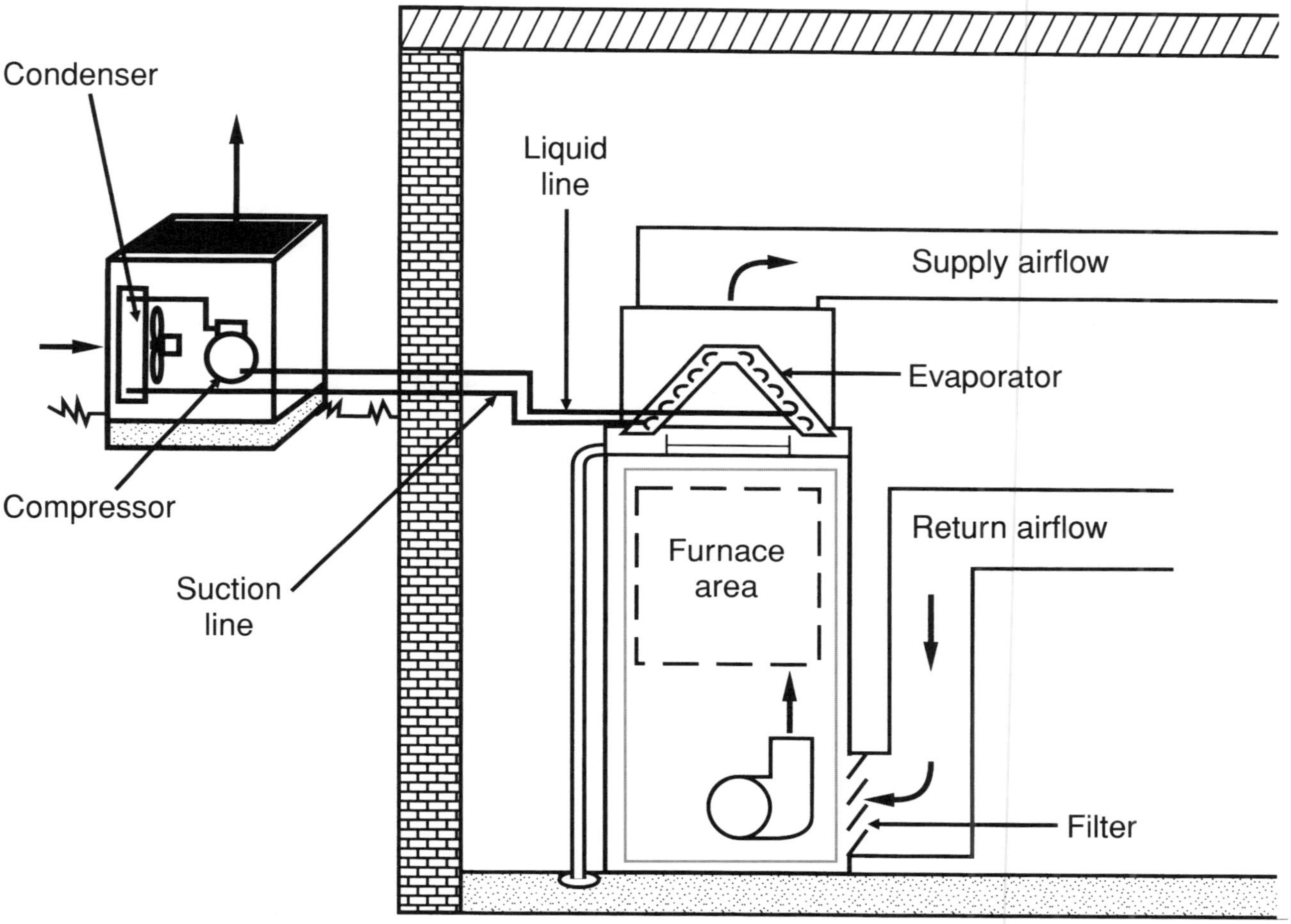

HC 11-1

Service Valves

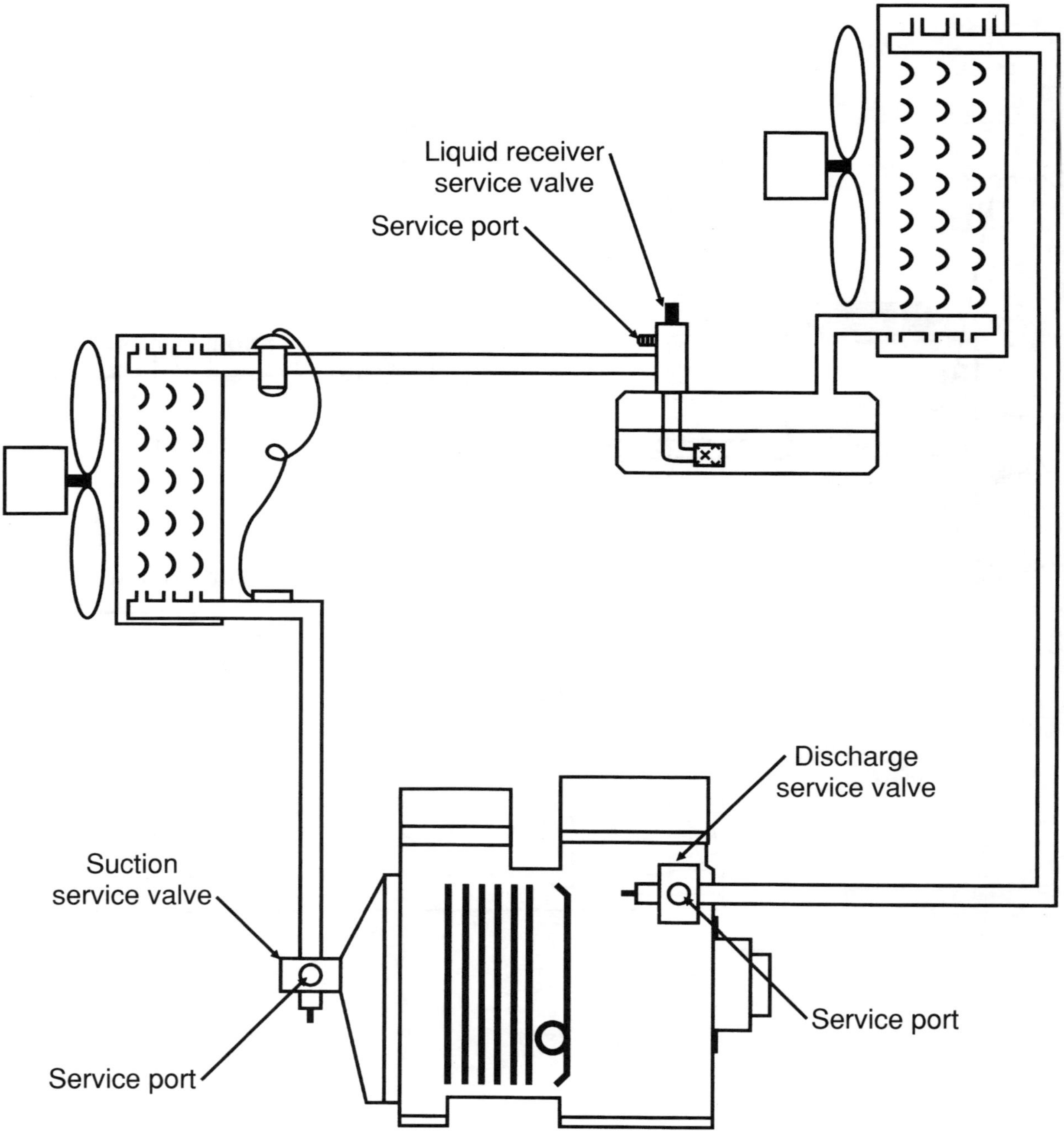

HC 11-2

Valve Positions

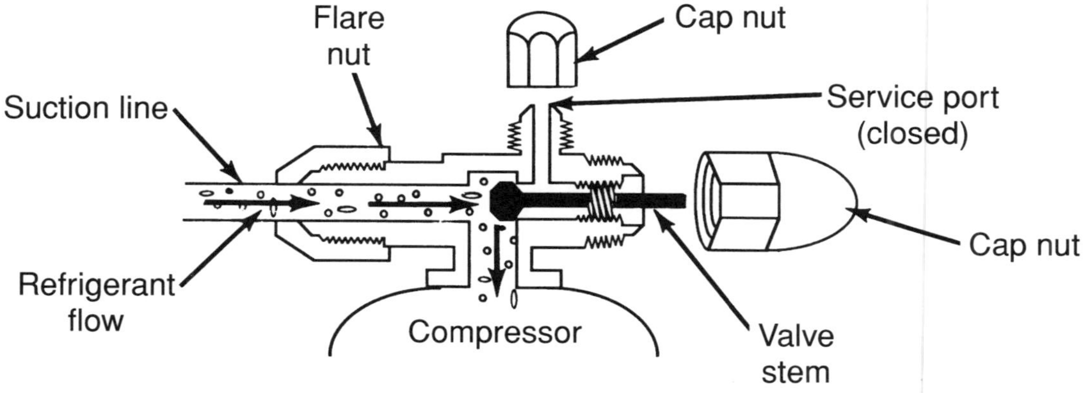

Backseated

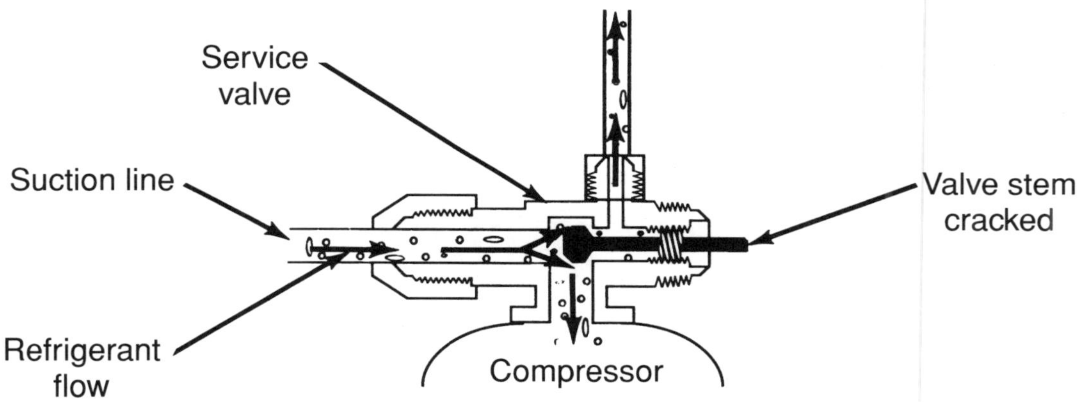

Cracked

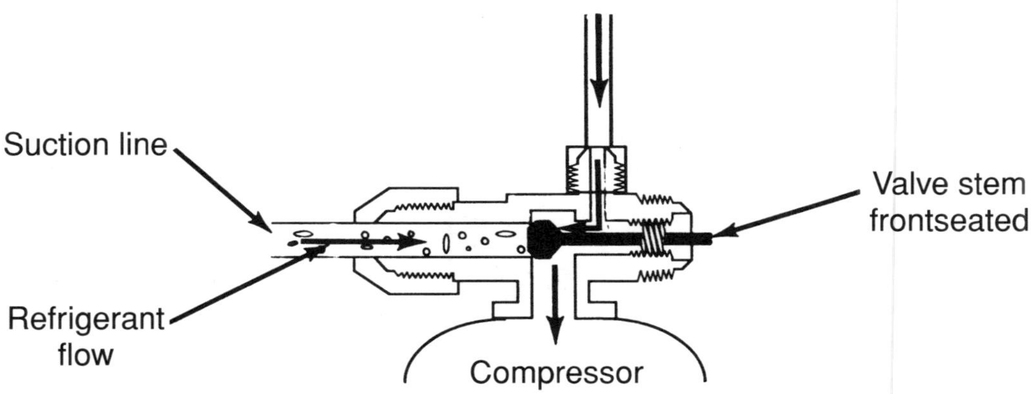

Frontseated

HC 11-3

Evacuating a System

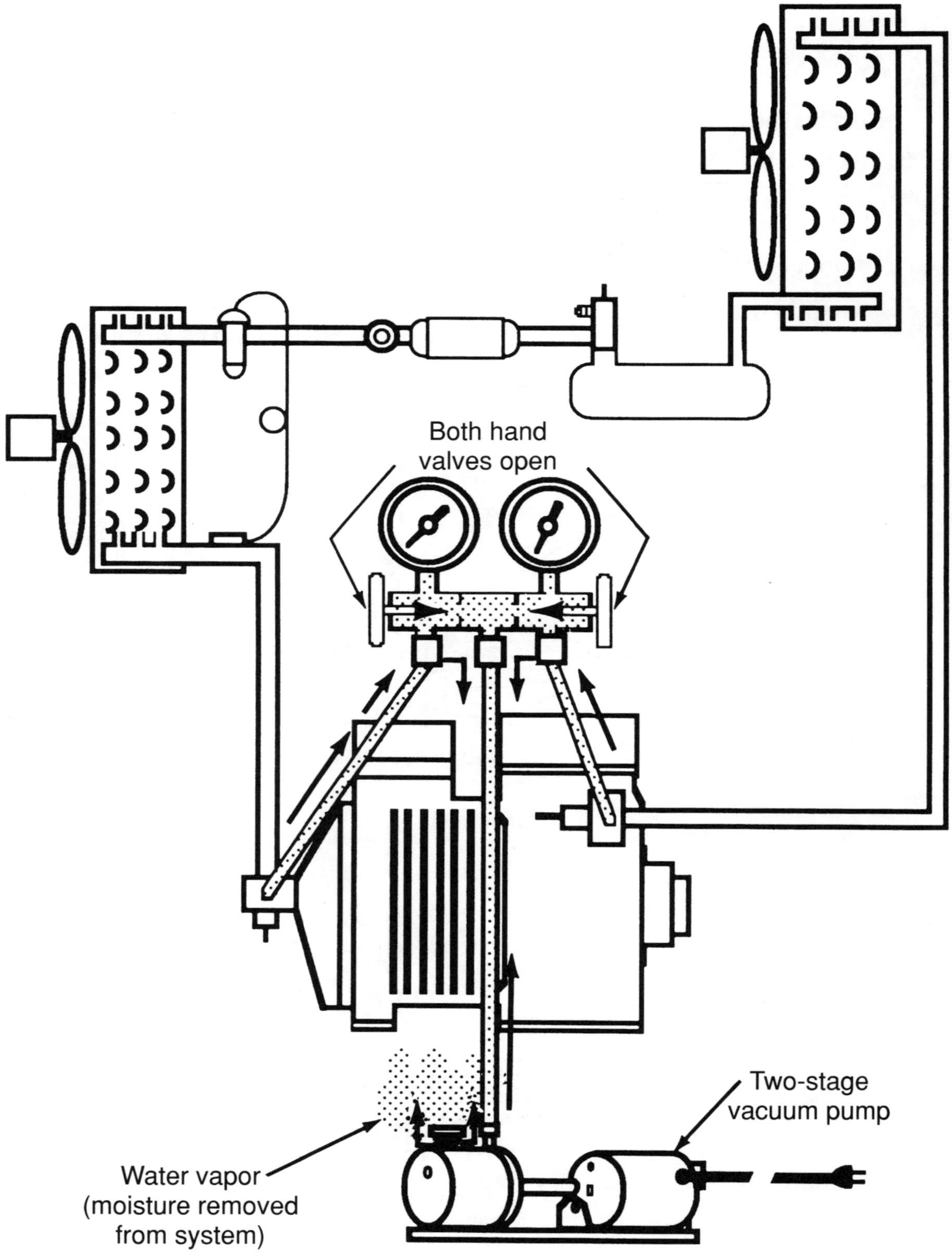

HC 11-4

Charging a System

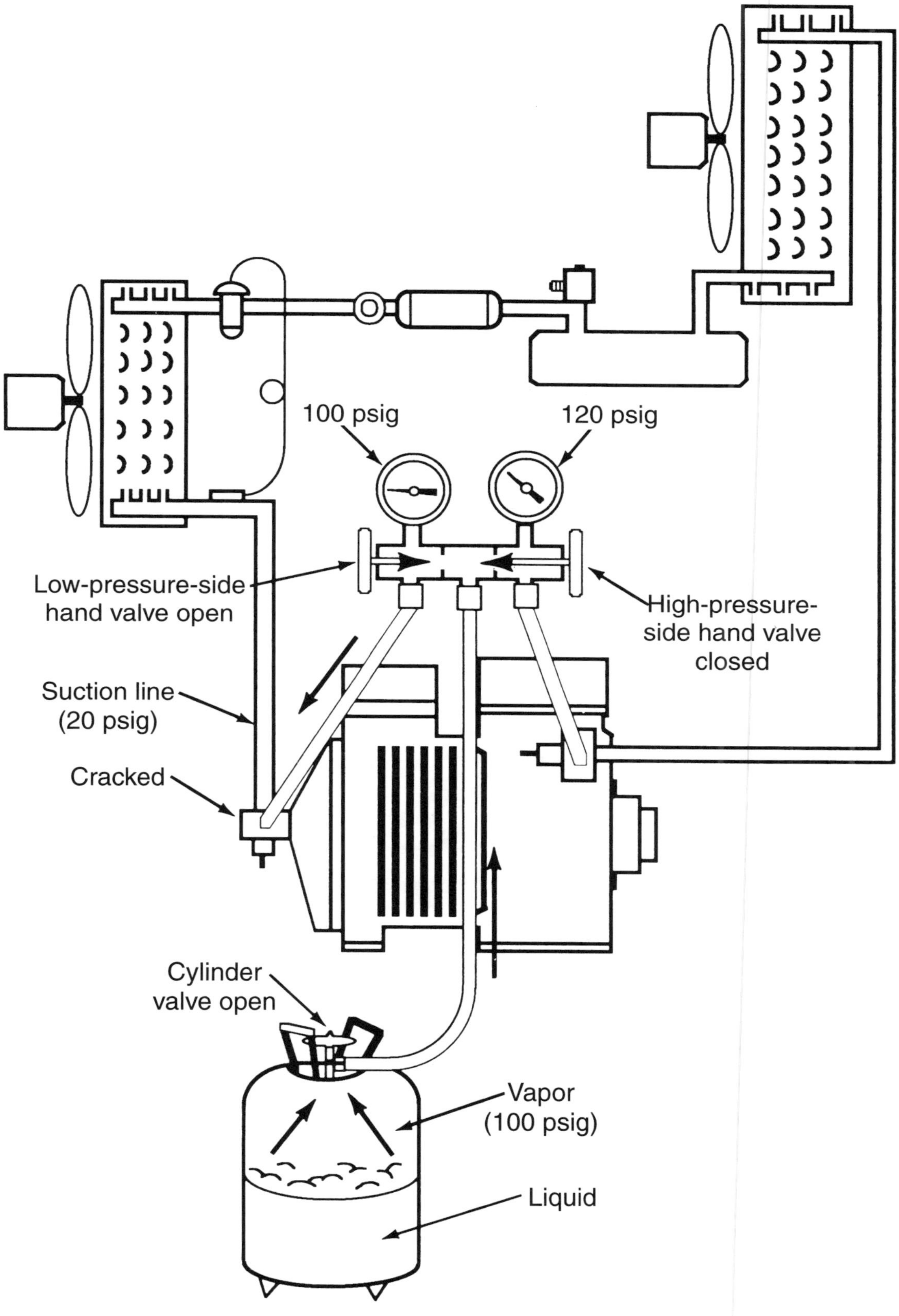

HC 11-5

Suction Accumulator

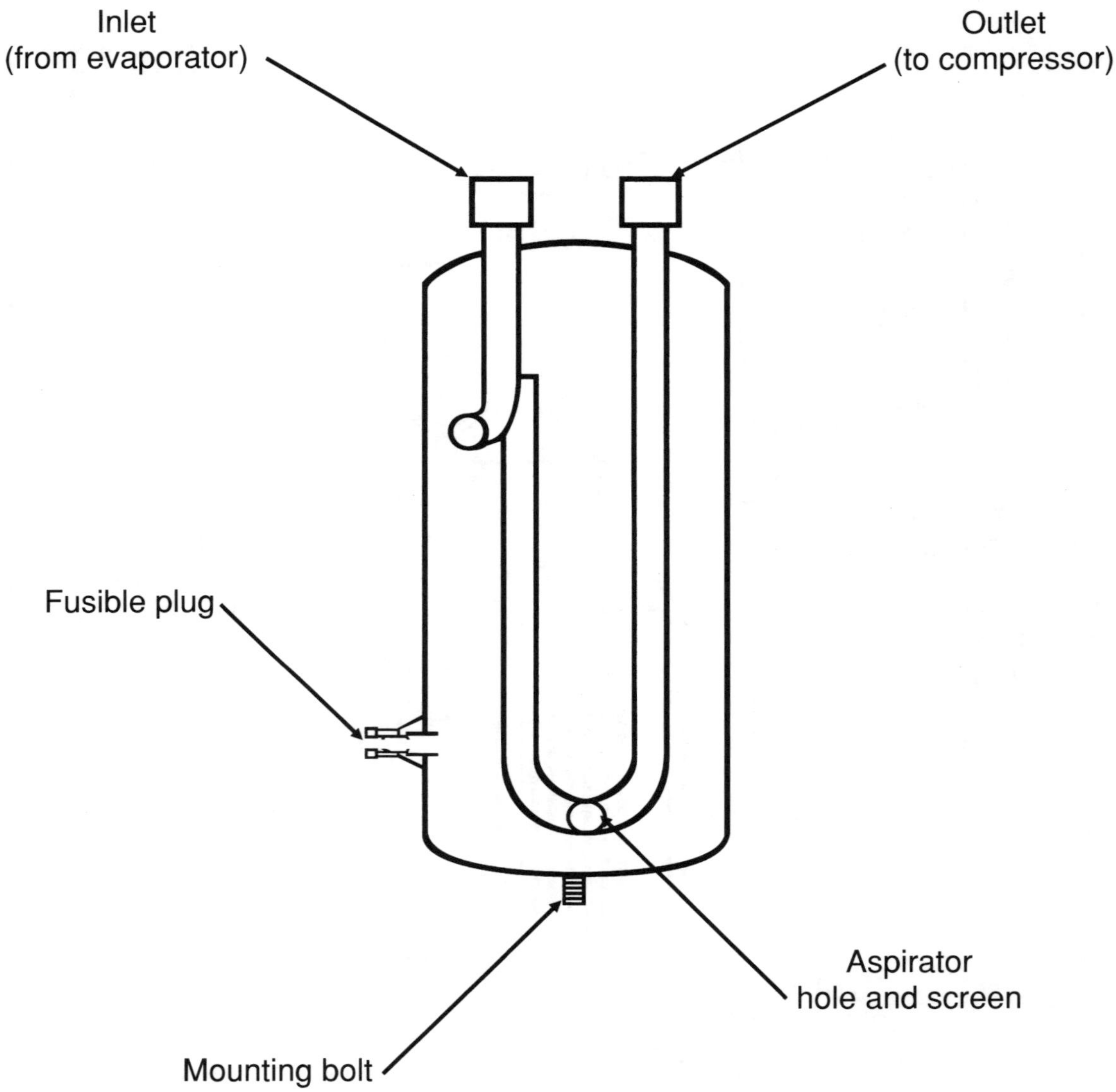

HC 11-6

Name ______________________________

Date ______________________________

Instructor ______________________________

Period______________________________

11

Other System Components

Quiz

Instructions: Write your answers on the lines provided at the right-hand side of the page.

1. Name the two copper lines that connect the evaporating unit with the condensing unit. 1. ______________________________ ______________________________

2. The liquid receiver should never be more than _____% full. 2. ______________________________

3. Name three common service valves. 3. ______________________________ ______________________________ ______________________________

4. *True or false?* Never frontseat the discharge service valve when the compressor is running. 4. ______________________________

5. What is the normal position for service valves? 5. ______________________________
   a. Frontseated.
   b. Backseated.
   c. Cracked open.

6. Identify the three openings of a typical service valve, as shown in the drawing. 6. a. ______________________________ b. ______________________________ c. ______________________________

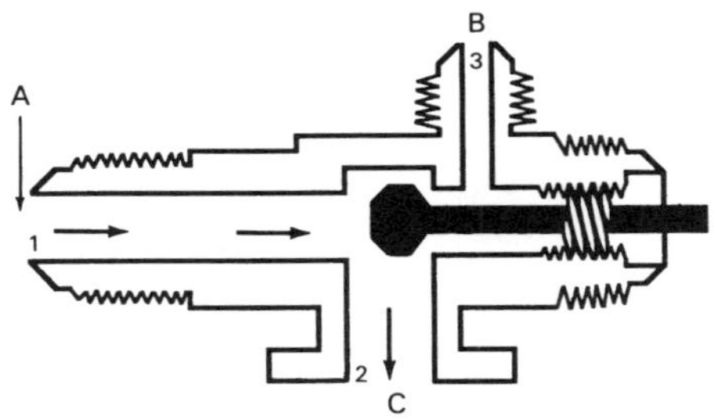

7. Are the two handwheels on the gauge manifold set normally open or normally closed? 7. ______________________________

8. A vacuum pump is used to evacuate _____ and _____ from the system. 8. ______________________________ ______________________________

9. The filter-drier is normally installed in the _____. 9. ______________________________
   a. suction line
   b. liquid line
   c. discharge line

10. *True or false?* The moisture indicator is located on the liquid receiver. 10. ______________________________

12

Refrigerants

Objectives

After studying this chapter in the text, the student will be able to:

- ❑ Recognize and identify different refrigerants.
- ❑ Use temperature-pressure charts.
- ❑ Determine superheat at any location in a refrigeration system.
- ❑ Recognize pumping ratio problems.
- ❑ Describe refrigerant applications within specific temperature ranges.
- ❑ Describe and use proper safety precautions with refrigerants.

Instructional Materials

Text: Pages 173–196
Lab Manual: Pages 95–98
Instructor's Manual:
Visual Masters, Pages 123–124
Chapter 12 Quiz, Page 125

Teaching Strategy

Text

1. Review objectives.
2. Discuss and assign chapter contents. Use these visual masters as appropriate:
   HC 12-1: Superheat and Subcooling in an R-22 System
   HC 12-2: Compressors
3. Direct students to complete Test Your Knowledge questions, page 195. After completion, discuss correct answers with students.

Lab Manual

1. Provide overview of Activities. Where necessary, stress safe laboratory practices and the importance of wearing protective clothing.
2. Assign Activities to individuals or pairs of students, as appropriate.
   Activity 12-1: Obtaining Refrigerant Information
   Activity 12-2: Alternative Refrigerant Information
   Activity 12-3: Finding Saturation, Superheat, and Subcooling
3. Observe student performance on Activities, providing guidance or correction as needed. After reviewing the student's work, initial and date the activity in the place provided.

Evaluation

1. Administer the quiz from this Instructor's Manual.
2. Grade the quiz, and return it with appropriate comments for the student.

Answers to Test Your Knowledge, Page 195

1. boiling point; safety features; Btu/lb.; operating pressures; density; compatibility with oil; noncorrosiveness; availability; cost
2. Refrigerant not circulating. Loss of refrigerant.
3. Freon-12 (introduced in 1931 by Dupont Chemical Company).
4. R-12 = white; R-22 = green; R-500 = yellow; R-502 = orchid
5. dichlorodifluoromethane
6. R-22
7. CHClF2
8. R-12 = –21.6°F (–29.7°C); R-22 = –41.4°F (–40.7°C)
9. halogens
10. predicted; controlled
11. R-500; R-502
12. 11/2 times
13. substitute; mix

14. 5°F (–15°C) evaporating temperature and 86°F (30°C) condensing temperature
15. R-12 = 14.64 psig; R-22 = 32.77 psig; R-500 = 19.70 psig; R-502 = 41.00 psig
16. –25°F
17. high; low
18. Divide the suction pressure (psia) into the condensing pressure (psia).
19. two-stage
20. high = air conditioner; medium = domestic refrigerator, store cooler; low = freezer

Answers to Lab Manual, Pages 95–98

Activity 12-1

Student response (answers will vary).

Activity 12-2

Student response (answers will vary).

Activity 12-3

1. A. 14.64 psig (101 kPa)
   B. 32.77 psig (226 kPa)
   C. 12 psig (83 kPa)
   D. 41 psig (283 kPa)
2. 10 degrees Fahrenheit (5.5 degrees Celsius)
3. 25 degrees Fahrenheit (14 degrees Celsius)
4. 0°F (–18°C)
5. 25°F (–4°C)
6. 195.9 psig (1352 kPa)
7. 93.3 psig (644 kPa)
8. 1.9 psig (13 kPa)
9. –25°F (–32°C)
10. 200°F (93°C)

Answers to Quiz

1. chlorofluorocarbon
2. R-22
3. false
4. b. heavier
5. R-500; R-502
6. false
7. superheated; subcooled
8. A. suction line
   B. first-stage compression
   C. interstage
   D. second-stage compression
   E. hot gas discharge to compressor
9. liquid nitrogen; liquid carbon dioxide
10. control

Superheat and Subcooling in an R-22 System

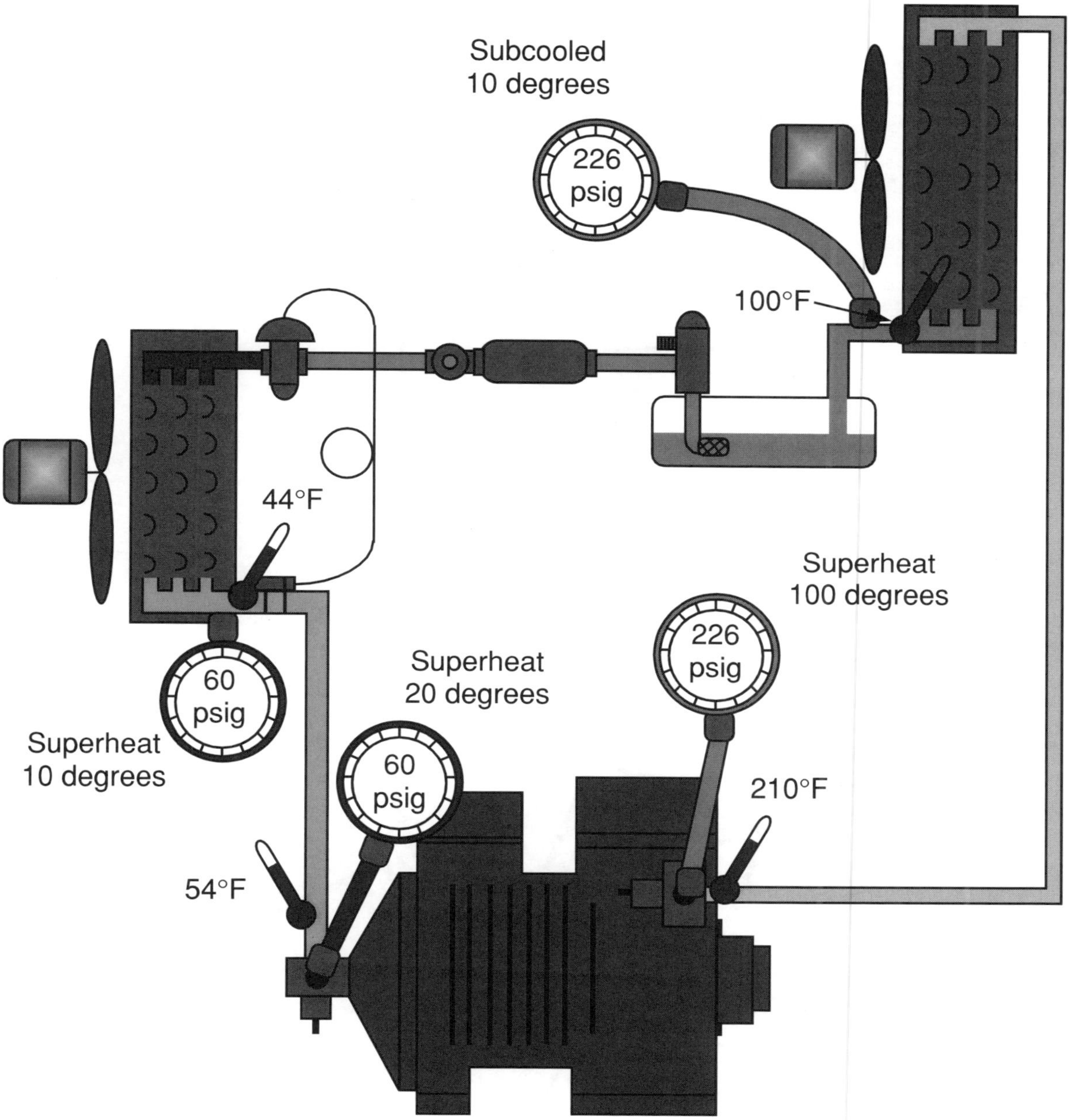

HC 12-1

Compressors

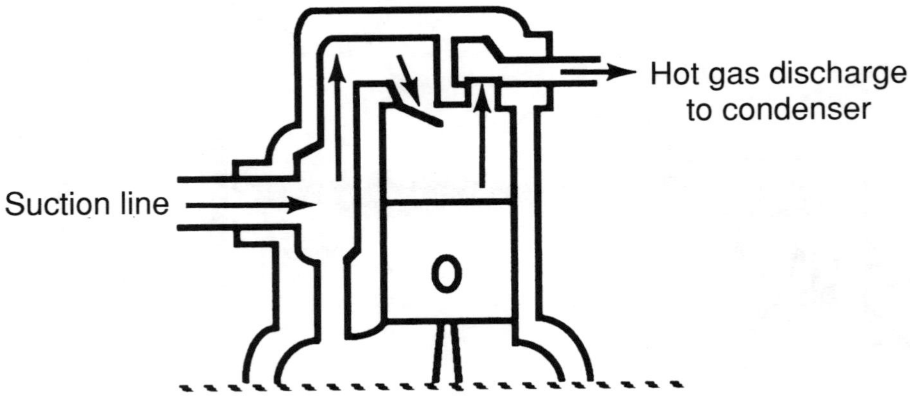

Single-stage Compressor

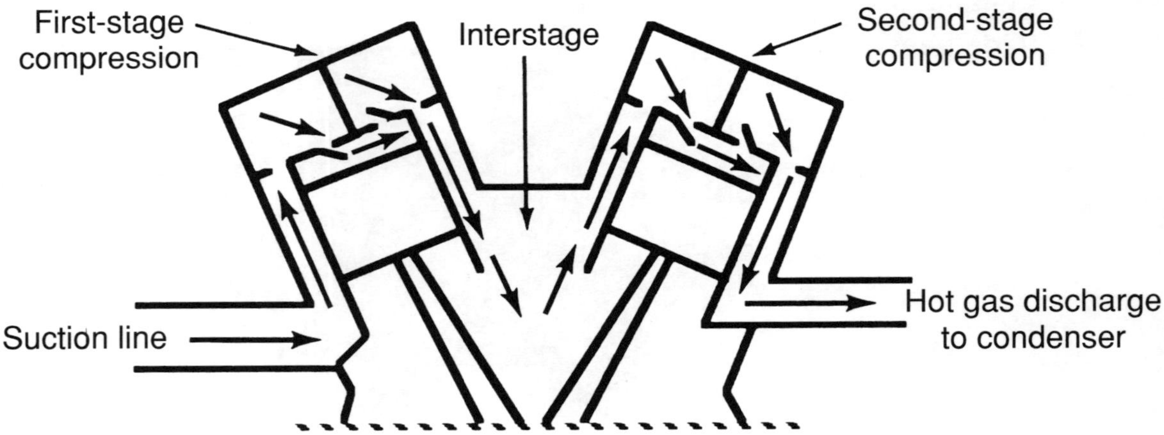

Two-stage Compressor

HC 12-2

Name ______________________________

Date ______________________________

Instructor ______________________________

Period______________________________

12 Refrigerants

Quiz

Instructions: Write your answers on the lines provided at the right-hand side of the page.

1. CFC is the abbreviation for what chemical name? 1. ______________
2. Monochlorodifluoromethane is the chemical name for the refrigerant _____. 2. ______________
3. *True or false?* The color of a cylinder is the primary means of identifying the refrigerant that it contains. 3. ______________
4. Fluorinated refrigerant vapors are _____ in weight when compared to an equal volume of air. 4. ______________
   a. lighter
   b. heavier
   c. about the same
5. Name two refrigerants that are azeotropic mixtures. 5. ______________ ______________
6. *True or false?* It is acceptable, in some cases, to substitute or mix refrigerants. 6. ______________
7. _____ is a word used to describe a gas whose temperature is above the saturation point, while _____ is a word used to describe a liquid whose temperature is below the saturation point. 7. ______________ ______________
8. Identify the areas of the two-stage compressor through which the refrigerant passes as it is compressed. 8. a. ______________ b. ______________ c. ______________ d. ______________ e. ______________

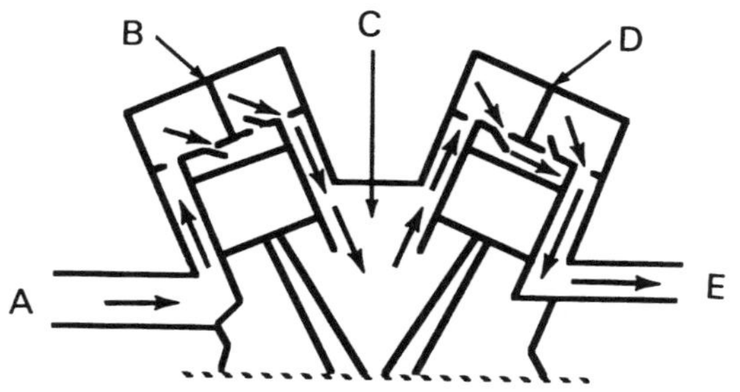

9. Name two commonly used expendable refrigerants. 9. ______________ ______________
10. Saturation charts contain the necessary information to predict and _____ the behavior of refrigerants. 10. ______________

13

Refrigerant Recovery and Recycling

Objectives

After studying this chapter in the text, the student will be able to:

- ❑ Explain the ozone depletion problem.
- ❑ Recognize CFCs, HCFCs, and HFCs.
- ❑ Comply with the 1990 Federal Clean Air Act.
- ❑ State equipment and technician certification requirements.
- ❑ Use recovery cylinders properly and safely.
- ❑ Recognize recovery and recycle equipment.
- ❑ Properly perform recovery and recycle procedures.

Instructional Materials

Text: Pages 197–214
Lab Manual: Pages 99–102
Instructor's Manual:
Visual Masters, Pages 129–130
Chapter 13 Quiz, Page 131

Teaching Strategy

Text

1. Review objectives.
2. Discuss and assign chapter contents. Use these visual masters as appropriate:
   HC 13-1: Refrigerant Recovery
   HC 13-2: Refrigerant Recycling
3. Direct students to complete Test Your Knowledge questions, pages 213–214. After completion, discuss correct answers with students.

Lab Manual

1. Provide overview of Activities. Where necessary, stress safe laboratory practices and the importance of wearing protective clothing.
2. Assign Activities to individuals or pairs of students, as appropriate.
   Activity 13-1: Definitions
   Activity 13-2: Using Refrigerant Recovery Equipment
   Activity 13-3: Using Refrigerant Recycling Equipment
3. Observe student performance on Activities, providing guidance or correction as needed. After reviewing the student's work, initial and date the activity in the place provided.

Evaluation

1. Administer the quiz from this Instructor's Manual.
2. Grade the quiz, and return it with appropriate comments for the student.

Answers to Test Your Knowledge, Pages 213–214

1. c. 10 to 30
2. d. all of the above
3. a. Chlorine.
4. b. ODP
5. c. HFC.
6. d. 1996
7. b. EPA.
8. d. all of the above
9. b. recover and recycle
10. false
11. a. liquid receiver
12. reuse
13. c. 10 in. Hg
14. true
15. b. gray bottom, yellow top
16. true
17. c. hydrostatic
18. b. 4 to 6 hours.

19. two
20. b. 15

Answers to Lab Manual, Pages 99–101

Activity 13-1

1. *Recover* means to remove refrigerant in any condition from a system and store it in an approved cylinder.
2. *Recycle* means to clean refrigerant for reuse by removing moisture and contaminants by repeatedly circulating the refrigerant through one or more filter-driers.
3. *Reclaim* means to restore refrigerant to a level equal to new product specifications as determined by chemical analysis.

Activity 13-2

Student response (answers will vary).

Activity 13-3

Student response (answers will vary).

Answers to Quiz

1. chlorine
2. b. Recover and recycle.
3. Environmental Protection Agency (EPA)
4. gray bottom, yellow top
5. R-22, R-123
6. cannot
7. A, I, II, and III
8. five
9. 80
10. Atmospheric air and moisture enter.

Refrigerant Recovery

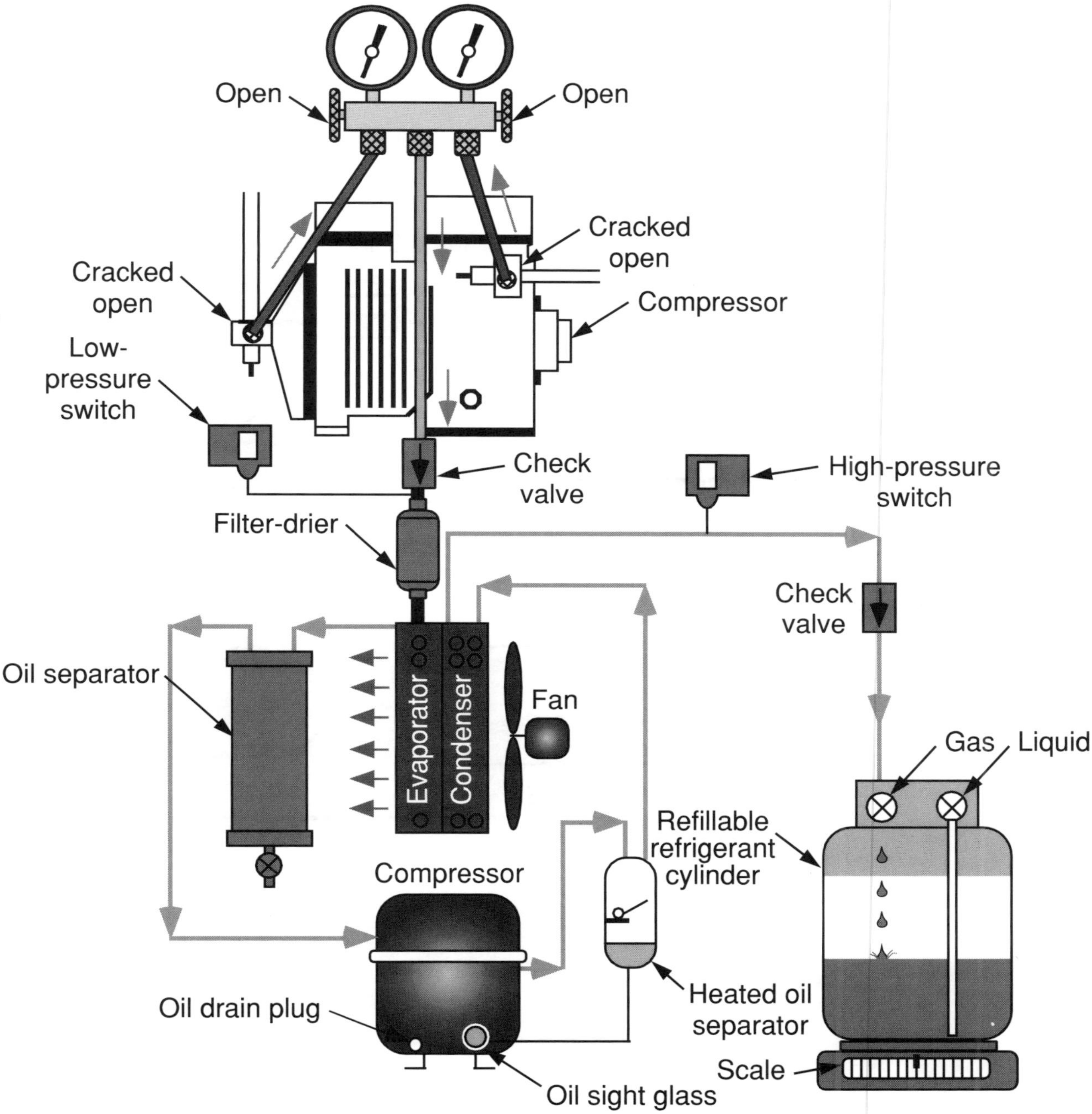

HC 13-1

Refrigerant Recycling

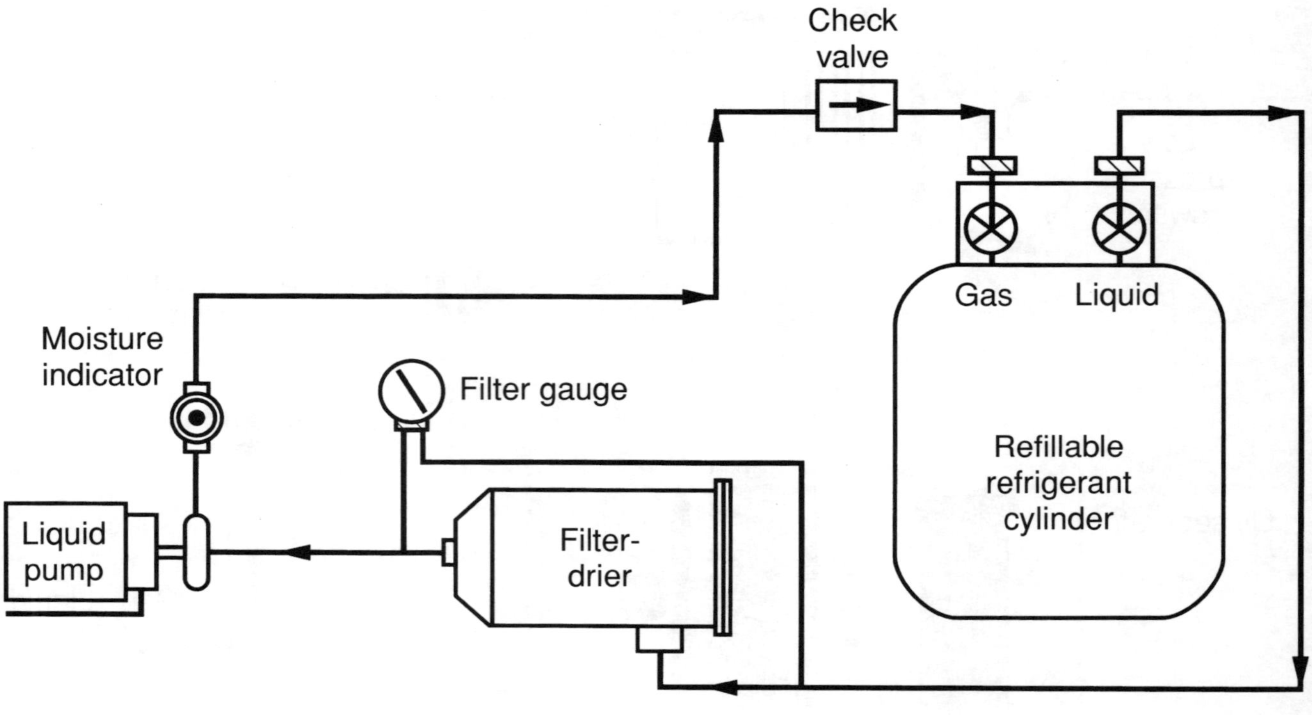

HC 13-2

Name ______________________________

Date ______________________________

Instructor __________________________

Period______________________________

13 Refrigerant Recovery and Recycling

Quiz

Instructions: Write your answers on the lines provided at the right-hand side of the page.

1. What component of CFC refrigerants destroys ozone? 1. ______________________

2. What does R & R mean? 2. ______________________
   a. Recycle and reclaim.
   b. Recover and recycle.
   c. Reclaim and recover.
   d. All of the above.

3. What government agency enforces and regulates the Clean Air Act? 3. ______________________

4. What is the color code for recovery cylinders? 4. ______________________

5. Name two examples of HCFCs. 5. ______________________
   ____

6. Recycled refrigerant (can, cannot) change ownership. 6. ______________________

7. What test sections must be passed to obtain Universal technician certification? 7. ______________________

8. Refillable cylinders must be retested and recertified every _____ years. 8. ______________________

9. Recovery cylinders are considered full at _____% of total capacity. 9. ______________________

10. What happens when the suction side is opened when in a vacuum? 10. ______________________

14

Zeotropic Blends

Objectives

After studying this chapter, the student will be able to:

- ❑ Distinguish between pure component refrigerants, azeotropic mixtures, and zeotropic blends.
- ❑ Explain fractionation.
- ❑ Determine temperature glide.
- ❑ Read and use a temperature-pressure card for blends.
- ❑ Define bubble point and dew point.
- ❑ Appropriately select and use lubricants for blends.
- ❑ Perform system retrofit to a blend.

Instructional Materials

Text: Pages 215–228
Lab Manual: Pages 103–106
Instructor's Manual:
Visual Masters, Page 135
Chapter 14 Quiz, Page 136

Teaching Strategy

Text

1. Review objectives.
2. Discuss and assign chapter contents. Use these visual masters as appropriate:
   HC 14-1: Controlling Evaporator Temperature
3. Direct students to complete Test Your Knowledge questions, page 228. After completion, discuss correct answers with students.

Lab Manual

1. Provide overview of Activities. Where necessary, stress safe laboratory practices and the importance of wearing protective clothing.
2. Assign Activities to individuals or pairs of students, as appropriate.
   Activity 14-1: Controlling Evaporator Temperature When Retrofitting from R-12 to R-401A
   Activity 14-2: Controlling Evaporator Temperature When Retrofitting from R-12 to R-409A
   Activity 14-3: Retrofitting from R-12 to R-401A
3. Observe student performance on Activities, providing guidance or correction as needed. After reviewing the student's work, initial and date the activity in the place provided.

Evaluation

1. Administer the quiz from this Instructor's Manual.
2. Grade the quiz, and return it with appropriate comments for the student.

Answers to Test Your Knowledge, Page 228

1. a. R-134a.
2. d. R-507.
3. c. R-123.
4. b. HCFC.
5. a. pure
6. a. R-502.
7. d. R-22.
8. true
9. b. Zeotropic blends.
10. c. bubble point
11. d. all of the above
12. d. all of the above
13. false
14. a. Mineral oil.
15. b. 5
16. d. all of the above
17. d. 100
18. a. acid

19. true
20. d. All of the above.

Answers to Lab Manual, Pages 103–105

Activity 14-1

1. 10 degrees Fahrenheit
2. 5°F
3. 0°F; 13 psig
4. 10°F; 13 psig
5. 20°F (10 degrees above dew point)

Activity 14-2

1. Suction pressure: 16.8 psig
2. Bubble point: 0°F
3. Dew point: 15°F
4. At thermal bulb: 25°F

Activity 14-3

Student response (answers will vary).

Answers to Quiz

1. true
2. mineral
3. pure
4. zeotropic
5. azeotropic
6. d. AZ50
7. fractionation
8. vapor
9. liquid
10. three-part

Controlling Evaporator Temperature

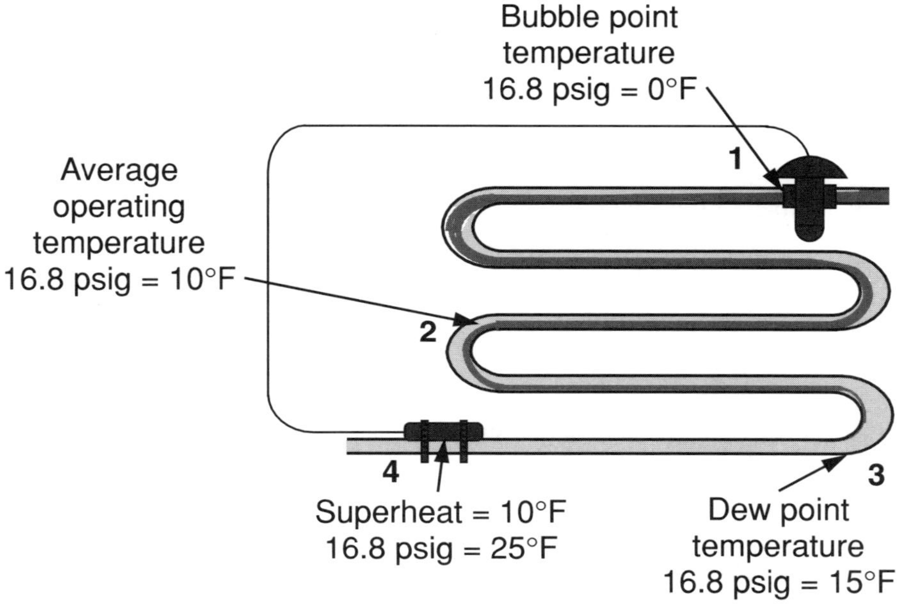

HC 14-1

Name ______________________________

Date ______________________________

Instructor ______________________________

Period______________________________

14 Zeotropic Blends

Quiz

Instructions: Write your answers on the lines provided at the right-hand side of the page.

1. *True or false*? The same gauge manifold can be used with all refrigerants except ammonia. 1. ______________

2. HFC refrigerants are not compatible with which oil? 2. ______________

3. R-134a is an example of a (pure, zeotropic) refrigerant. 3. ______________

4. Temperature glide only occurs with what type of refrigerants? 4. ______________

5. R-502 is an example of a(n) (zeotropic, azeotropic) refrigerant. 5. ______________

6. Which one of the following refrigerants is not a blend? 6. ______________
   a. R-401A
   b. R-409A
   c. MP39
   d. AZ50

7. What term describes the separation of refrigerant components? 7. ______________

8. Bubble point describes the first bubble of (liquid, vapor). 8. ______________

9. Dew point describes the first bubble of (liquid, vapor). 9. ______________

10. A ternary blend is a (two-part, three-part) blend of refrigerants. 10. ______________

15

Troubleshooting Refrigerant Problems

Objectives

After studying this chapter in the text, the student will be able to:

- ❑ Identify and correct contamination problems.
- ❑ Detect refrigerant leaks using approved methods.
- ❑ Use refrigerant cylinders properly.
- ❑ Predict and correct low-side pressures.
- ❑ Recognize and eliminate causes of high head pressure.
- ❑ Use electronic scales and graduated cylinders for charging.

Instructional Materials

Text: Pages 229–252
Lab Manual: Pages 107–118
Instructor's Manual:
Visual Masters, Pages 139–144
Chapter 15 Quiz, Page 145-146

Teaching Strategy

Text

1. Review objectives.
2. Discuss and assign chapter contents. Use these visual masters as appropriate:
   HC 15-1: Moisture Removal
   HC 15-2: Leak Testing Under Pressure
   HC 15-3: Leak Detection (Halide Torch)
   HC 15-4: Leak Detection (Electronic Tester)
   HC 15-5: Charging Methods
   HC 15-6: Charging Cylinder
3. Direct students to complete Test Your Knowledge questions, page 252. After completion, discuss correct answers with students.

Lab Manual

1. Provide overview of Activities. Where necessary, stress safe laboratory practices and the importance of wearing protective clothing.
2. Assign Activities to individuals or pairs of students, as appropriate.
   Activity 15-1: Leak Detection Using Soap Bubbles
   Activity 15-2: Leak Detection Using a Halide Torch
   Activity 15-3: Electronic Leak Detection
   Activity 15-4: Evacuating a Refrigeration System
   Activity 15-5: Vapor-charging a Refrigeration System
   Activity 15-6: Low-side Temperatures and Pressures
   Activity 15-7: High-side Temperatures and Pressures
3. Observe student performance on Activities, providing guidance or correction as needed. After reviewing the student's work, initial and date the activity in the place provided.

Evaluation

1. Administer the quiz from this Instructor's Manual.
2. Grade the quiz, and return it with appropriate comments for the student.

Answers to Test Your Knowledge, Page 252

1. When oxygen combines with oil in the system, it becomes explosive.
2. vacuum
3. bubble method; halide torch; electronic leak detector
4. 150; 300

5. dry nitrogen; carbon dioxide
6. year
7. To allow for hydrostatic expansion.
8. Disposable
9. R-12; R-22; R-502
10. air
11. air; product
12. 10
13. 10
14. no
15. 20 degrees Fahrenheit
16. evaporator = 26°F; refrigerant = 16°F
17. heat; liquid
18. ambient
19. Add 30 to 35 degrees Fahrenheit to the ambient air temperature and then convert to pressure. The head pressure should fall within the two readings.
20. 15; 20
21. air in the system; overcharge of refrigerant; dirty condenser; no air or water movement; high low-side pressure
22. between 245.8 psig and 262.7 psig
23. between 99.8 psig and 108.2 psig
24. vapor charging
25. liquid

Answers to Lab Manual, Pages 112–117

Activities 15-4 and 15-5

Student response (answers will vary).

Activity 15-6

1. 12°F (–11°C)
2. 54°F (12°C)
3. 0.5 psig (3.5 kPa)
4. 16°F (–9°C); 18.4 psig (127 kPa)
5. 0.6 psig (4.1 kPa)
6. 4°F (–15°C)
7. 68.5 psig (473 kPa)
8. 45 psig (310 kPa)
9. fall
10. 68.5 psig (473 kPa)

Activity 15-7

1. 147 psig to 158 psig (1014 kPa to 1090 kPa)
2. 70 psig to 77 psig (483 kPa to 531 kPa); no
3. 278 psig to 297 psig (1918 kPa to 2049 kPa)
4. 111 psig to 121 psig (766 kPa to 835 kPa); no
5. 216 psig to 232 psig (1490 kPa to 1600 kPa)
6. 81 psig to 89 psig (559 kPa to 614 kPa); no
7. 317 psig to 337 psig (2187 kPa to 2325 kPa); yes
8. 196 psig to 211 psig (1352 kPa to 1456 kPa)
9. up
10. down

Answers to Quiz

1. evacuate
2. oxides
3. soap bubbles; halide torch; electronic
4. 150; 300
5. a. hydrostatic expansion
6. fall
7. true
8. a. 55°F (13°C)
   b. 45°F (7°C)
   c. 65°F (18°C)
9. pressure increases
10. a. 125°F (52°C)

Moisture Removal

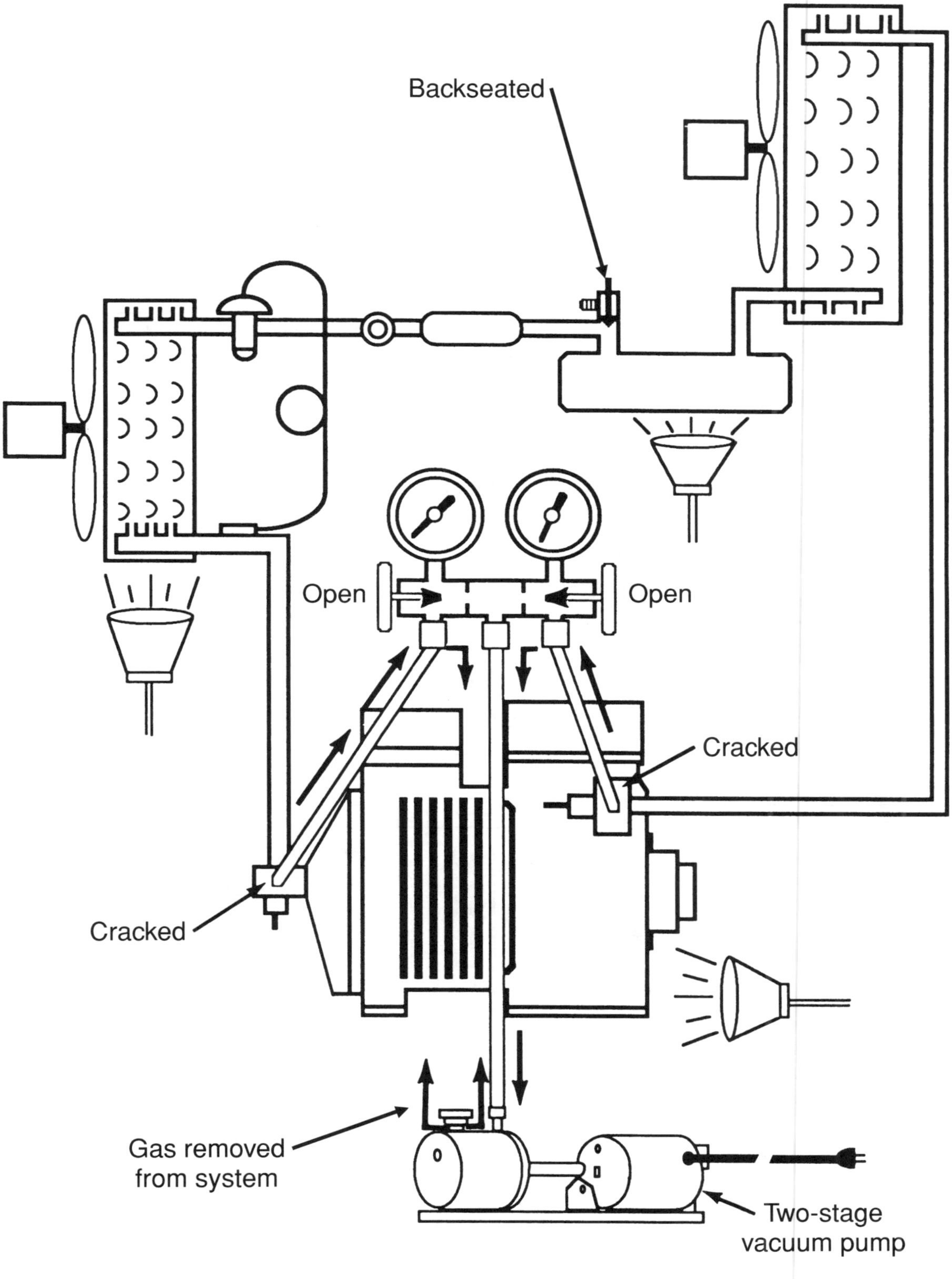

HC 15-1

Leak Testing Under Pressure

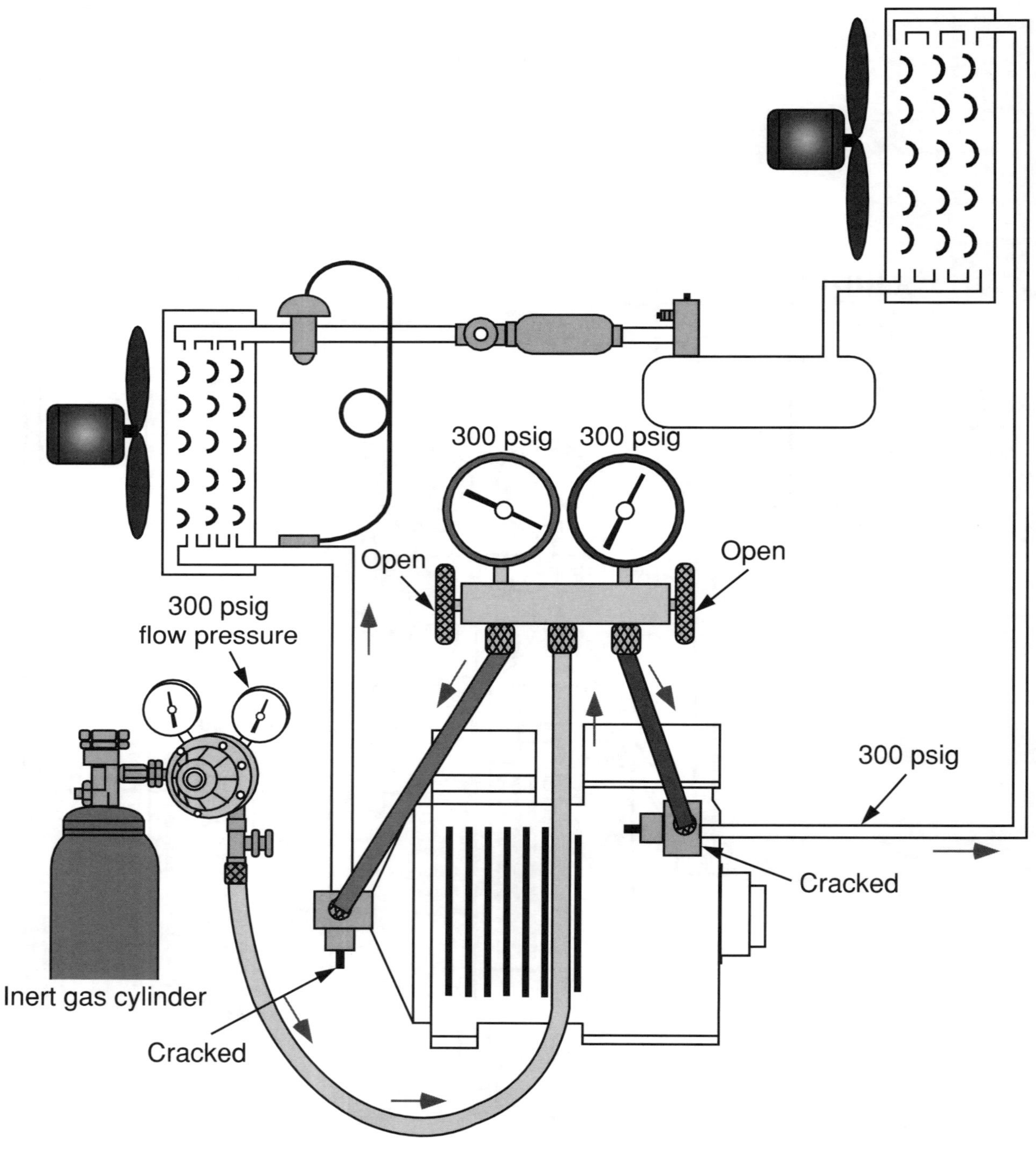

HC 15-2

Leak Detection (Halide Torch)

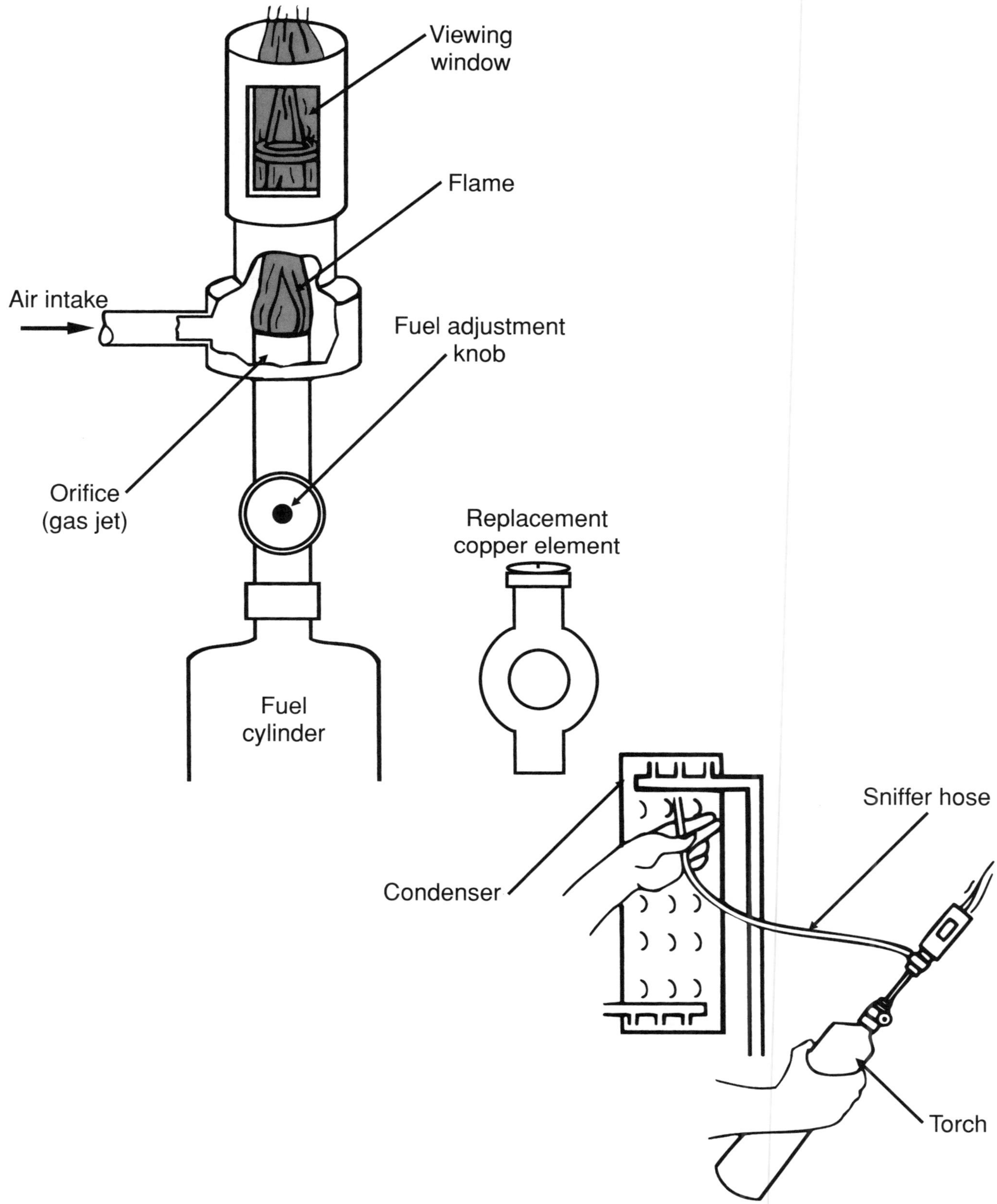

HC 15-3

Leak Detection (Electronic Tester)

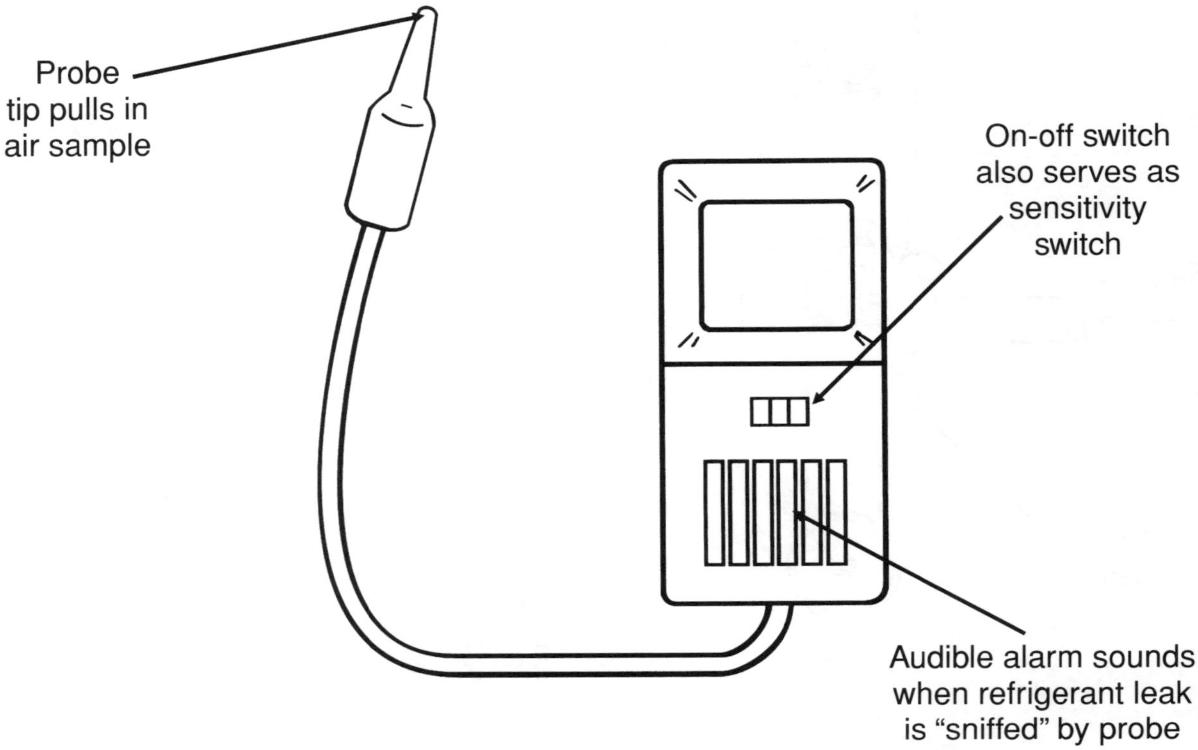

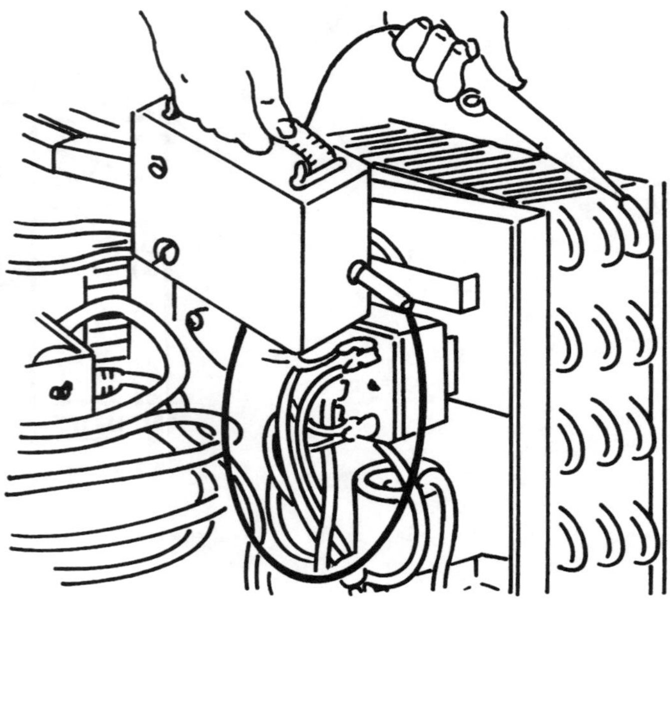

HC 15-4

Charging Methods

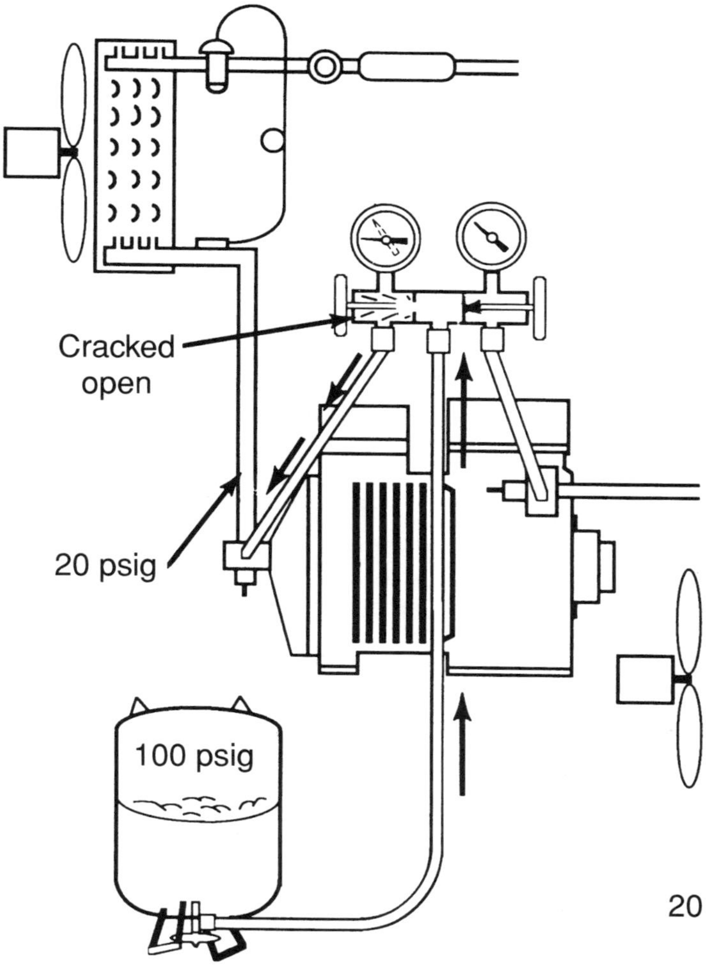

Liquid

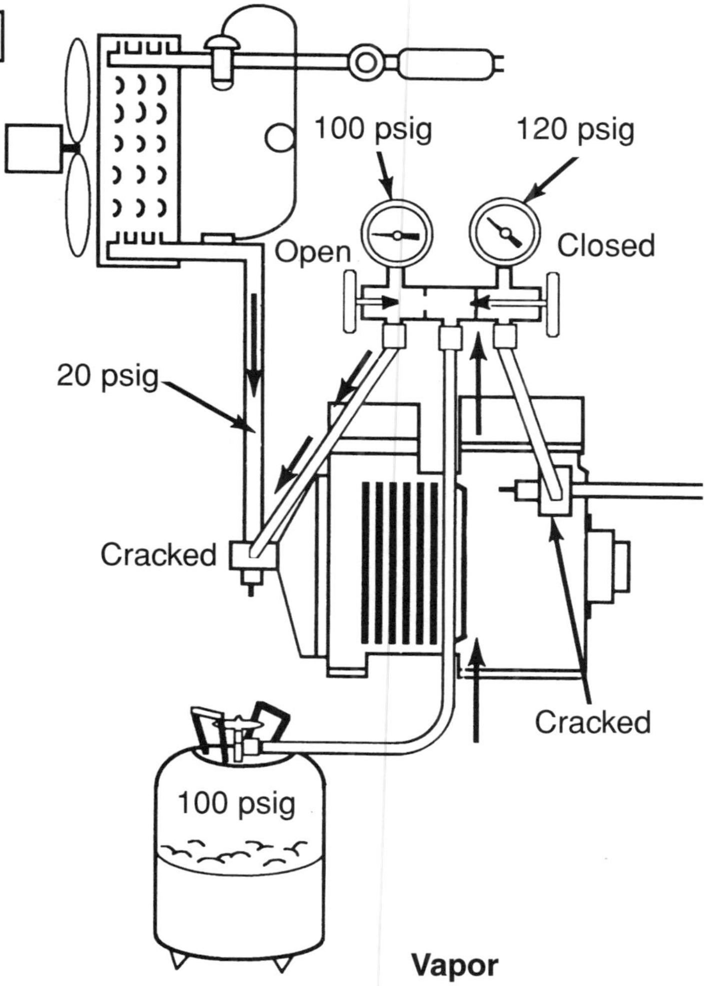

Vapor

HC 15-5

Charging Cylinder

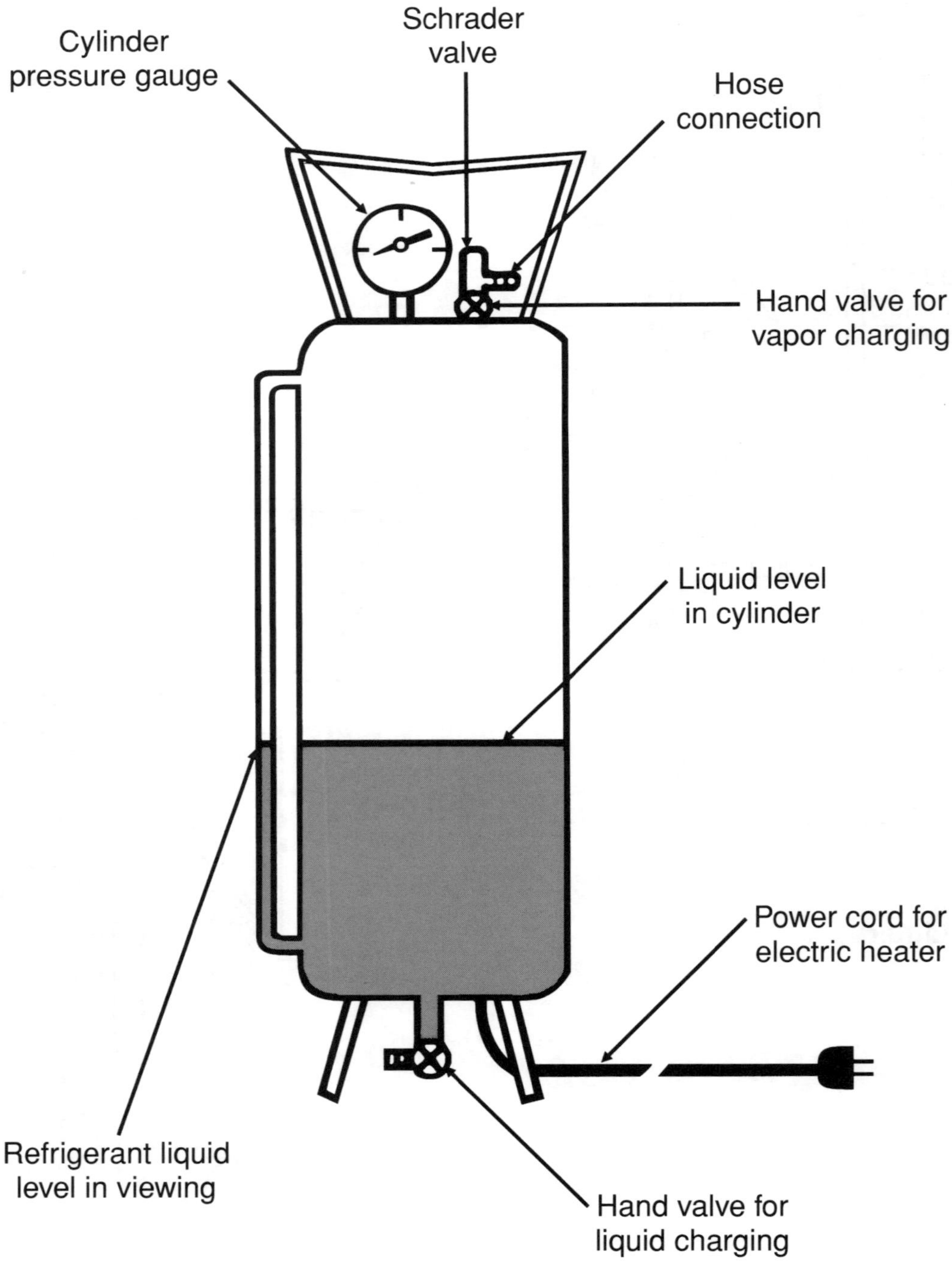

HC 15-6

Name ______________________________

Date ______________________________

Instructor ______________________________

Period______________________________

15 Troubleshooting Refrigerant Problems

Quiz

Instructions: Write your answers on the lines provided at the right-hand side of the page.

1. A vacuum pump is used to _____ air and moisture from the refrigeration system. 1. ______________________

2. Sweeping the system with an inert gas prevents black _____ from forming on the inside of the tubing when brazing. 2. ______________________

3. Name three common methods used for leak detection. 3. ______________________
____
____

4. Leak testing with soap bubbles should be done with pressures between _____ psig and _____ psig. 4. ______________________
____

5. Fifteen percent of a refrigerant cylinder's volume should be left empty for _____. 5. ______________________

   a. hydrostatic expansion
   b. hydraulic extrusion
   c. pneumatic inversion

6. Insufficient airflow through the evaporator resulting from dirty filters, frost, or a defective blower motor will cause suction pressures to _____. 6. ______________________

7. *True or false?* An evaporator temperature difference can exist only when the system is running. 7. ______________________

continued

8. Using the drawing of the evaporator, identify the temperature at each point indicated.

8. a. ______________________

b. ______________________

c. ______________________

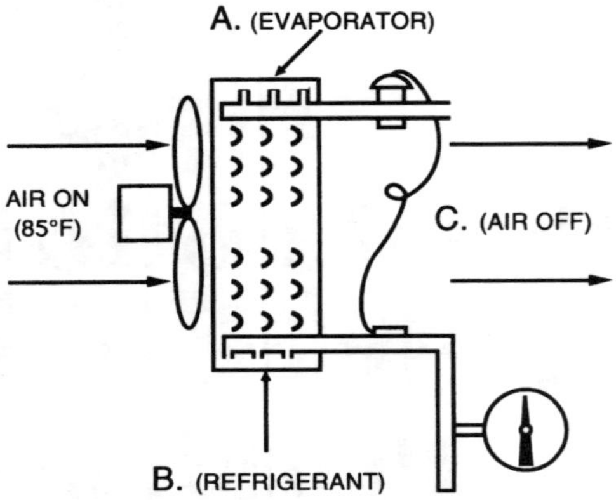

9. What happens to the high-side pressure in an air conditioning system when the outdoor temperature rises?

9. ______________________

10. Refrigerant cylinders should never be heated above _____.

   a. 125°F (52°C)

   b. 150°F (66°C)

   c. 100°F (38°C)

10. ______________________

16

Working with Refrigerant Controls

Objectives

After studying this chapter in the text, the student will be able to:

- ❑ Install and adjust automatic expansion valves.
- ❑ Install and service capillary tubes.
- ❑ Install and adjust thermostatic expansion valves.
- ❑ Determine when and how to use internally and externally equalized valves.
- ❑ Identify and describe the operation of multiplexed evaporators.

Instructional Materials

Text: Pages: 253–268
Lab Manual: Pages 119–128
Instructor's Manual:
Visual Masters, Pages 149-153
Chapter 16 Quiz, Page 154

Teaching Strategy

Text

1. Review objectives.
2. Discuss and assign chapter contents. Use these visual masters as appropriate:
   HC 16-1: Forces Acting on the AEV
   HC 16-2: Temperature Difference
   HC 16-3: Adding Service Valves to a Capillary Tube System
   HC 16-4: Pressures Acting on the TEV
   HC 16-5: TEV Types
3. Direct students to complete Test Your Knowledge questions, pages 267-268. After completion, discuss correct answers with students.

Lab Manual

1. Provide overview of Activities. Where necessary, stress safe laboratory practices and the importance of wearing protective clothing.
2. Assign Activities to individuals or pairs of students, as appropriate.
   Activity 16-1: Adjusting the Automatic Expansion Valve
   Activity 16-2: Checking a Capillary Tube System
   Activity 16-3: Replacing a Filter-drier on a Capillary Tube System
   Activity 16-4: Using a Graduated Cylinder to Charge a Capillary Tube System
   Activity 16-5: Adjusting the Thermostatic Expansion Valve
   Activity 16-6: Mounting the TEV Thermal Bulb
3. Observe student performance on Activities, providing guidance or correction as needed. After reviewing the student's work, initial and date the activity in the place provided.

Evaluation

1. Administer the quiz from this Instructor's Manual.
2. Grade the quiz, and return it with appropriate comments for the student.

Answers to Test Your Knowledge, Pages 267-268

1. To control the amount of liquid entering the evaporator.
2. AEV; capillary tube; metering orifice; TEV
3. evaporator
4. To maintain a constant pressure in the evaporator when the system is running.
5. close
6. spring pressure; evaporator pressure
7. restriction
8. true

9. no moving parts; low cost; permits bleedover during off cycle
10. the thermostat
11. easily plugged; amount of refrigerant is critical
12. capillary tube
13. It maintains a constant superheat at the evaporator outlet when the system is running.
14. close
15. sensing bulb (diaphragm); evaporator; spring; high-side
16. sensing bulb
17. bottom
18. downstream
19. increase
20. 10

Answers to Lab Manual, Pages 119–128

Activity 16-1

1.–10. Student response (answers will vary).
11. It goes up.
12. Student response (answers will vary).
13. none (or no effect)

Activity 16-2

1.–4. Student response (answers will vary).
5. Locate the leak, repair it, and recharge the system.
6. Repair or replace the defective compressor.
7. Student response (answers will vary).
8. a. When discharge pressure goes up, so does the suction pressure.
9. Student response (answers will vary).
10. It goes down.

Activity 16-5

1.–8. Student response (answers will vary).
9. It remains unchanged.
10. It decreases.

Activity 16-6

Student response (answers will vary).

Answers to Quiz

1. To control the amount of liquid refrigerant entering the evaporator.
2. pressure
3. b. restriction
4. high; low
5. a. hermetic systems
6. superheat
7. starve
8. 10 degrees Fahrenheit (6 degrees Celsius)
9. false
10. multiplexing

Forces Acting on the AEV

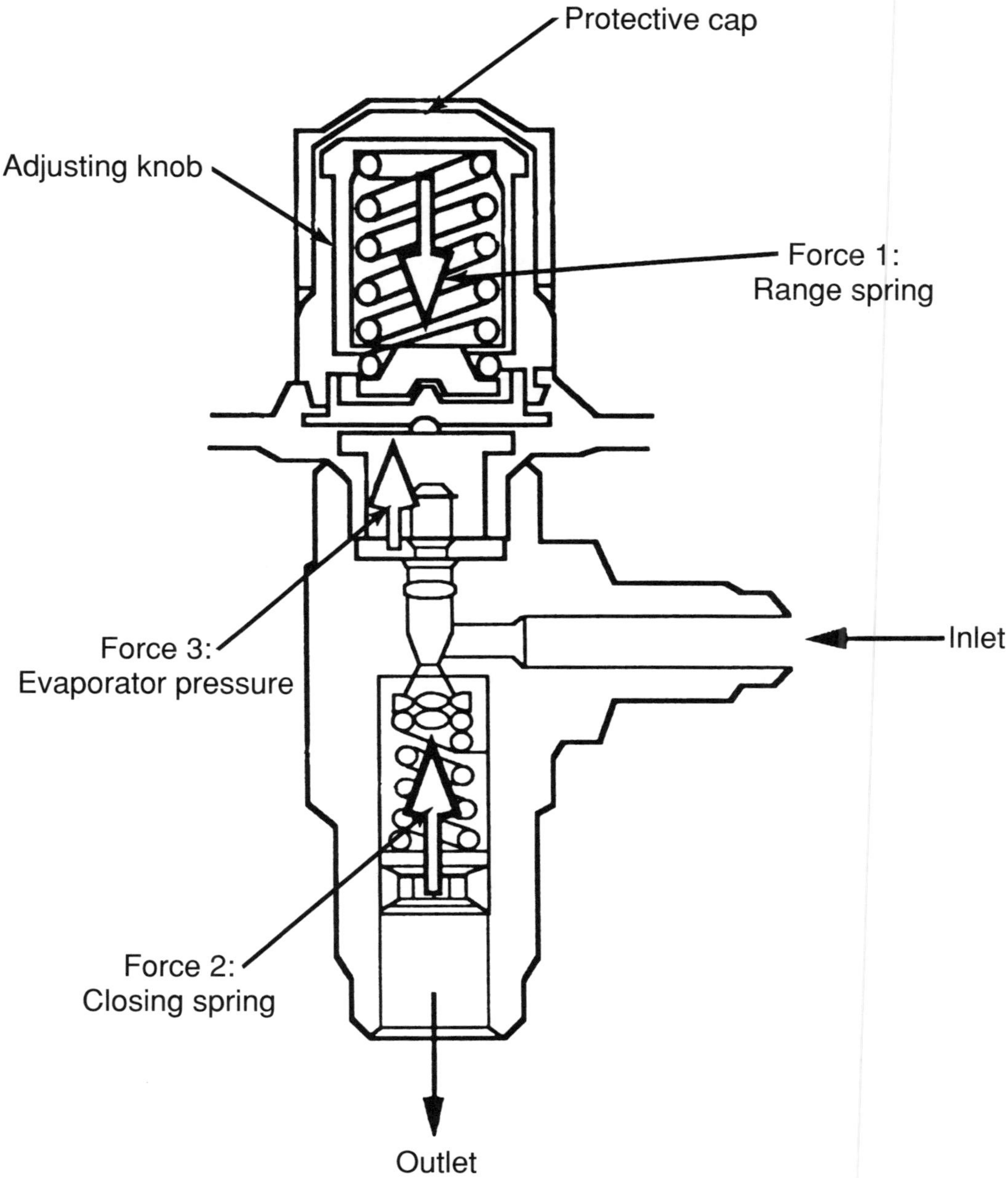

HC 16-1

Temperature Difference

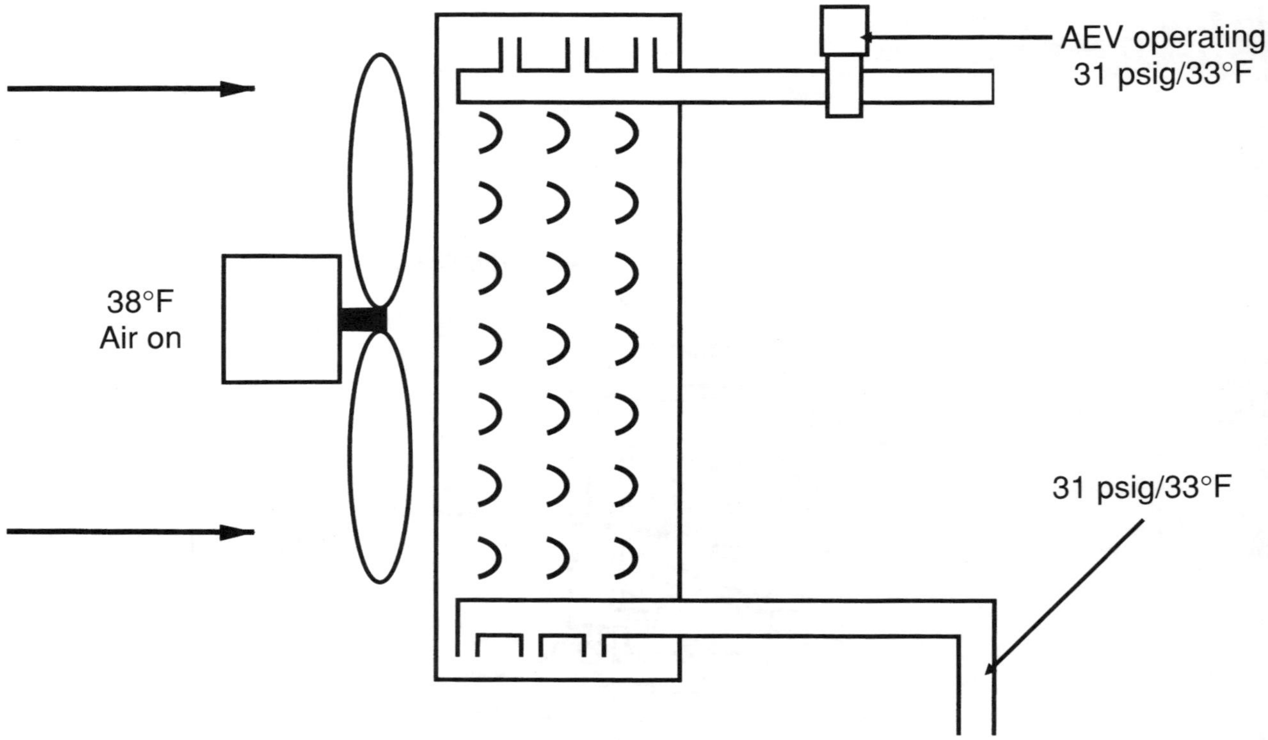

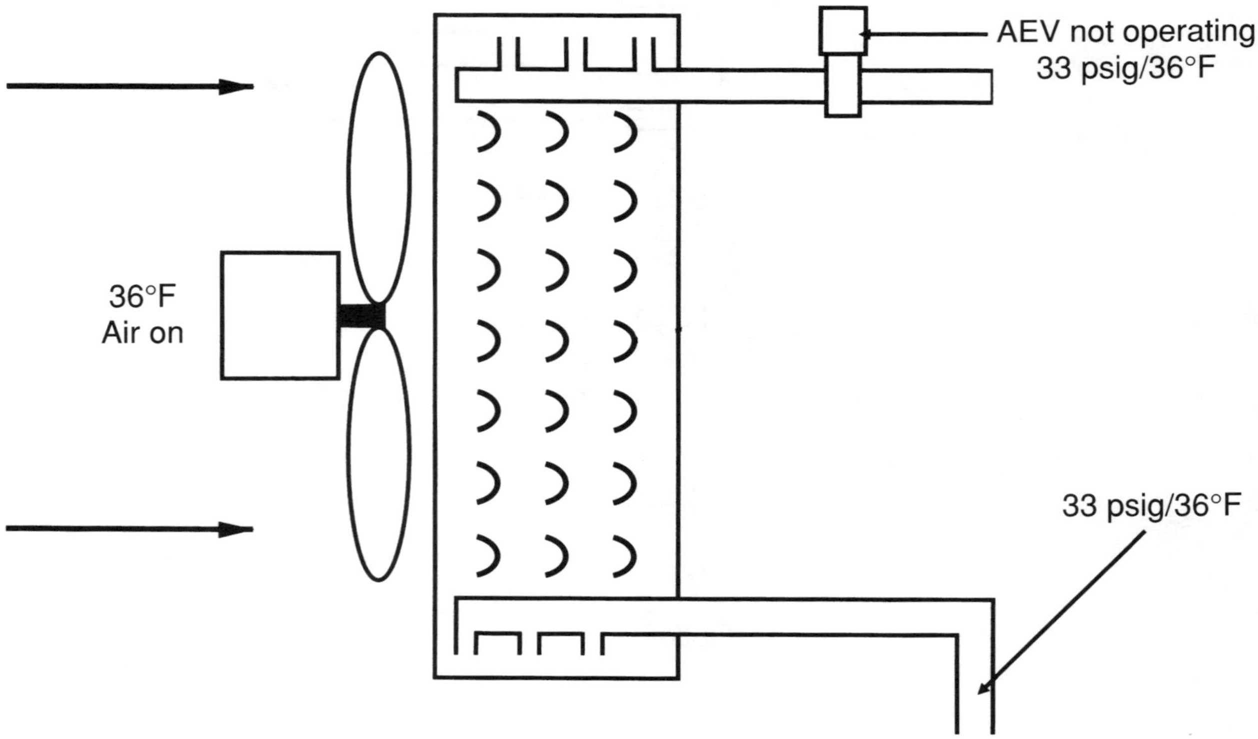

HC 16-2

Adding Service Valves to a Capillary Tube System

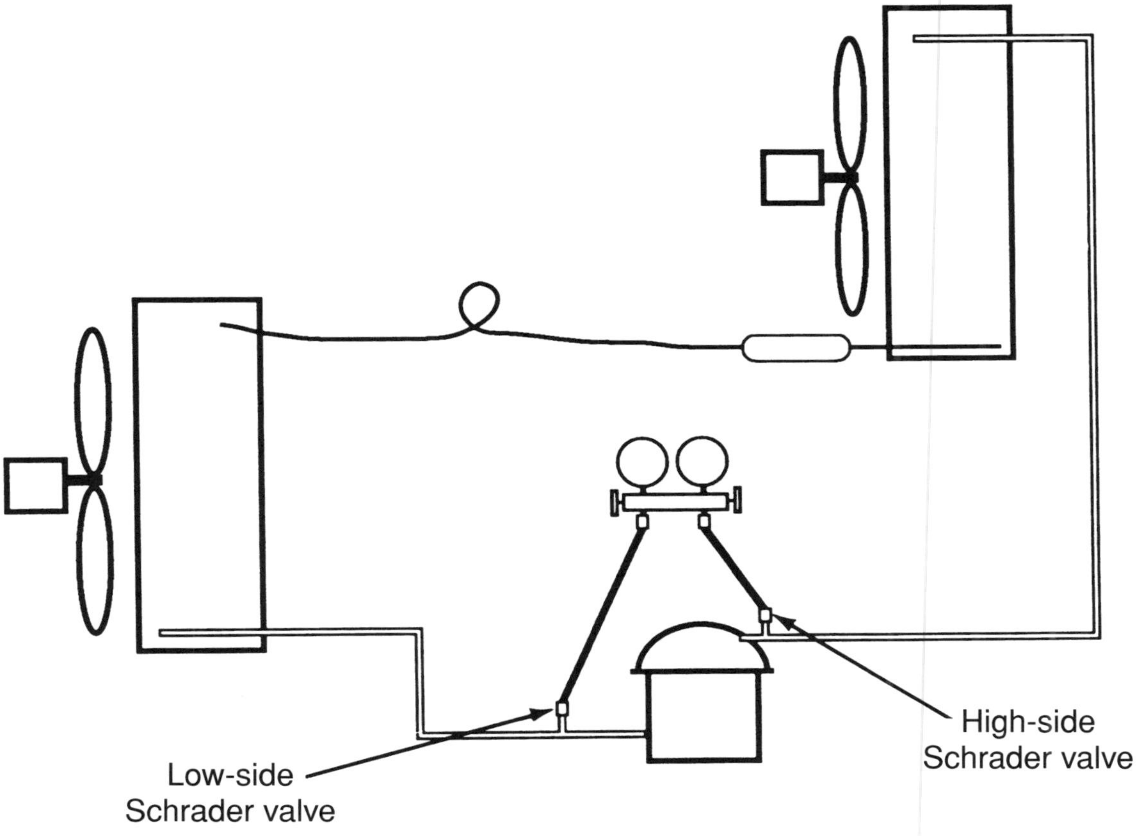

HC 16-3

Pressures Acting on the TEV

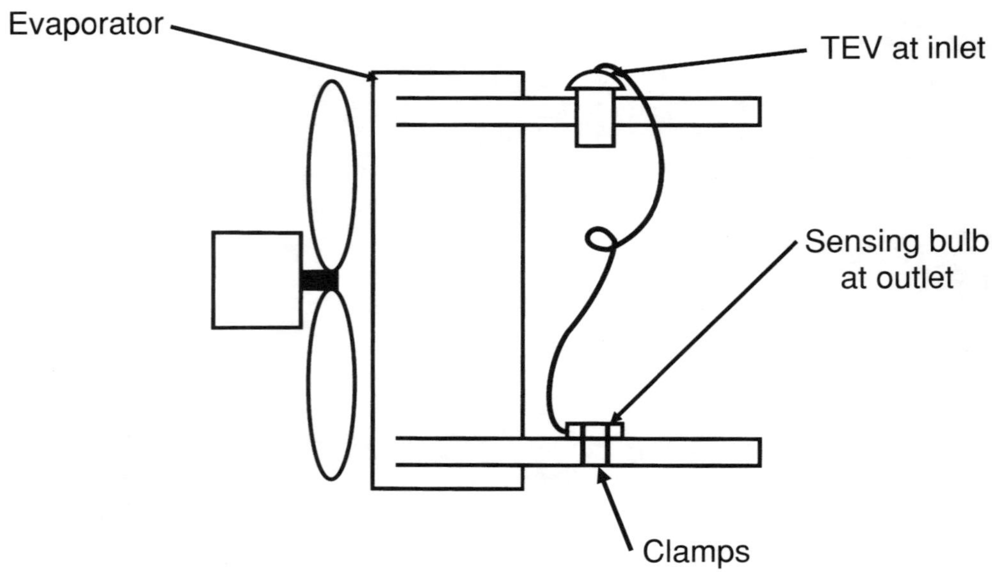

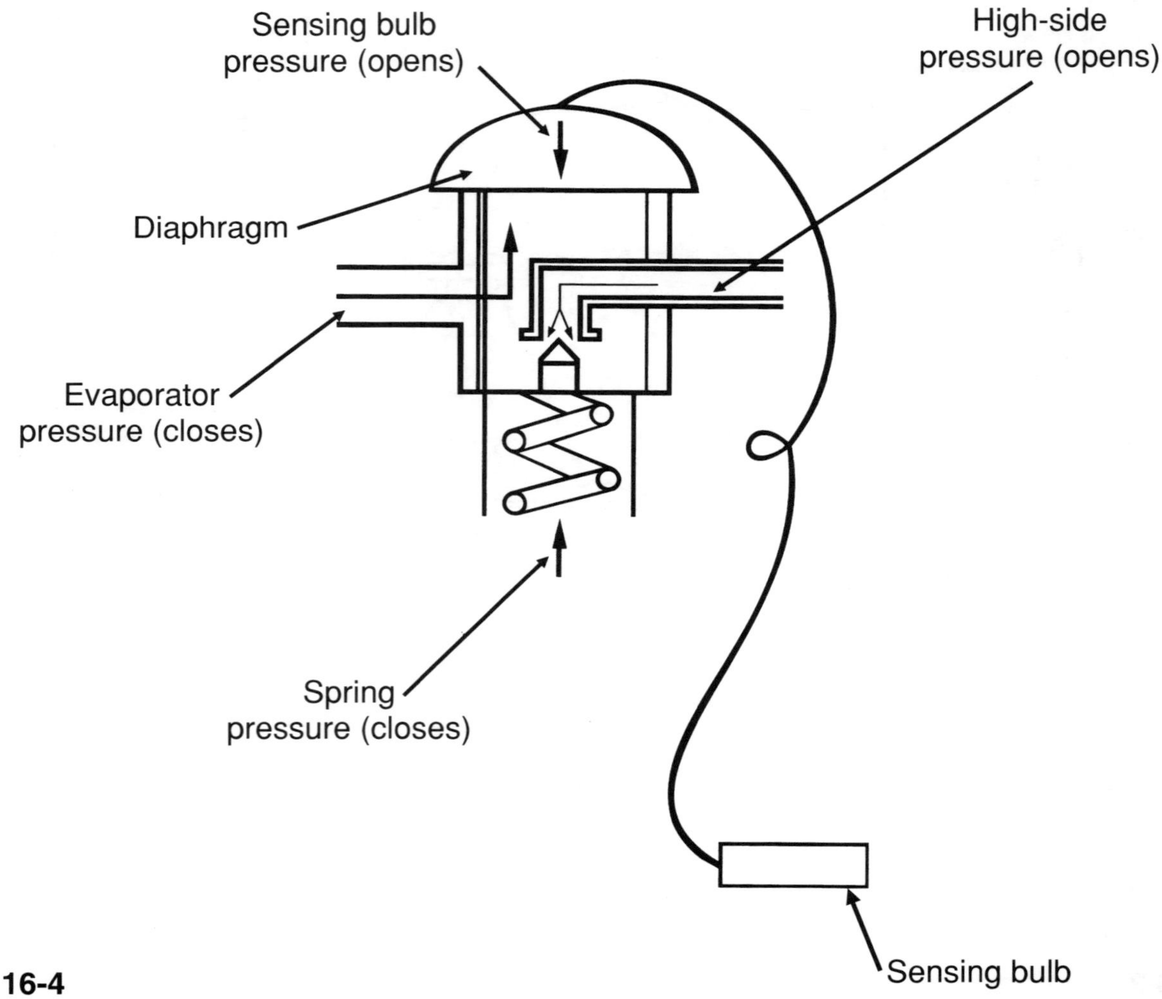

HC 16-4

TEV Types

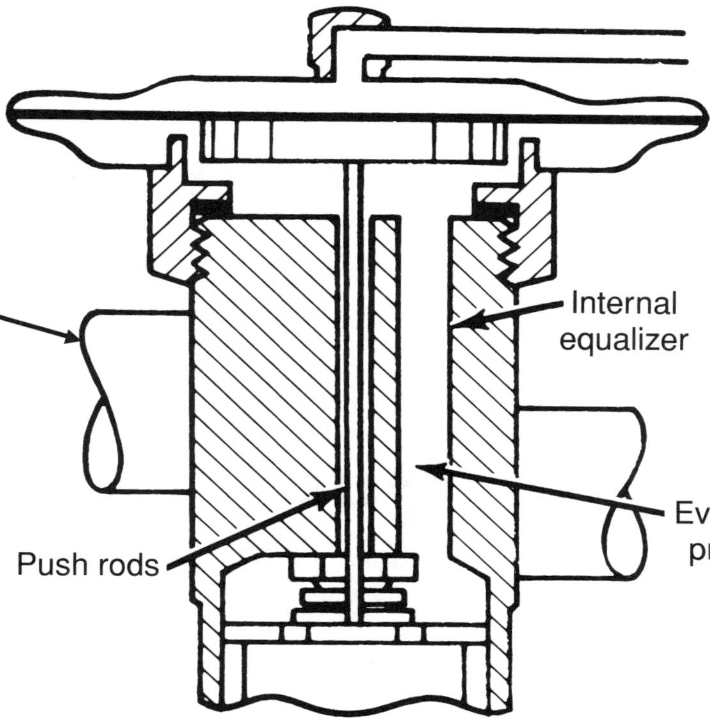

Internally Equalized

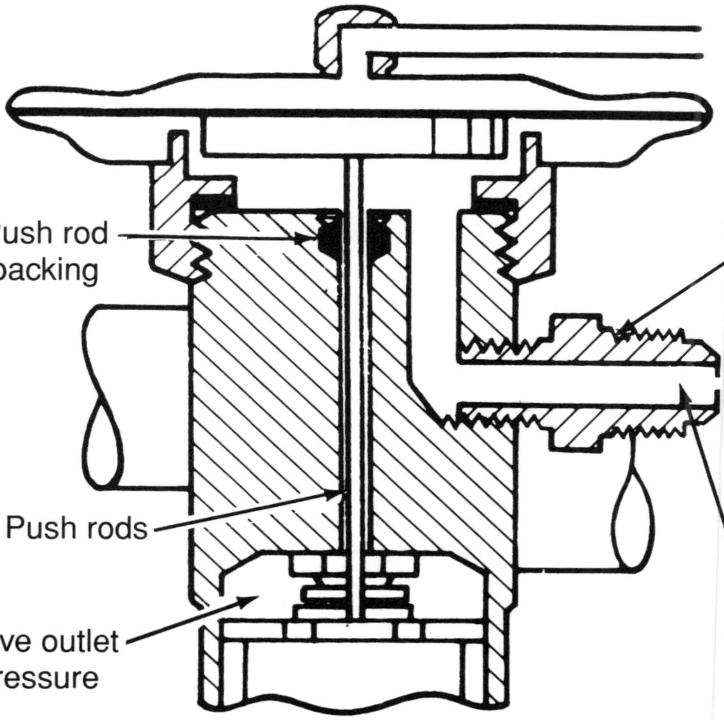

Externally Equalized

HC 16-5

Name ______________________________

Date ______________________________

Instructor ______________________________

Period______________________________

16 Working with Refrigerant Controls

Quiz

Instructions: Write your answers on the lines provided at the right-hand side of the page.

1. What is the purpose of the refrigerant control? 1. ______________

2. The AEV maintains a constant _____ in the evaporator when the system is running. 2. ______________

3. The capillary tube operates on the principle of _____. 3. ______________
   a. superheat
   b. restriction
   c. subcooling

4. An overcharge causes _____ head pressure, while an undercharge causes _____ head pressure. 4. ______________ ______________

5. Capillary tubes are normally found on _____. 5. ______________
   a. hermetic systems
   b. semihermetic systems
   c. open-type systems

6. The TEV is designed to maintain a constant _____ at the evaporator outlet. 6. ______________

7. Does a high TEV superheat setting tend to starve or flood the evaporator? 7. ______________

8. What is the normal superheat setting for a TEV? 8. ______________

9. *True or false?* An internally equalized TEV can be used to replace an externally equalized TEV. 9. ______________

10. Connecting two or more evaporators to a single compressor is called _____. 10. ______________

17

Special-purpose Valves

Objectives

After studying this chapter in the text, the student will be able to:

- ❑ Identify hand valves and describe their installation and use.
- ❑ Identify check valves and describe their installation.
- ❑ Install solenoid valves.
- ❑ Install and adjust EPR and CPR valves.
- ❑ Install and adjust head pressure control valves.
- ❑ Install and adjust hot gas bypass valves.

Instructional Materials

Text: Pages 269–286
Lab Manual: Pages 129–136
Instructor's Manual:
Visual Masters, Pages 157–162
Chapter 17 Quiz, Page 163

Teaching Strategy

Text

1. Review objectives.
2. Discuss and assign chapter contents. Use these visual masters as appropriate:
   HC 17-1: Hand Valves
   HC 17-2: EPR Valve
   HC 17-3: Two-temperature Systems
   HC 17-4: Multiplexed Systems
   HC 17-5: Desuperheating TEV
   HC 17-6: Adjusting Head Pressure
3. Direct students to complete Test Your Knowledge questions, page 286. After completion, discuss correct answers with students.

Lab Manual

1. Provide overview of Activities. Where necessary, stress safe laboratory practices and the importance of wearing protective clothing.
2. Assign Activities to individuals or pairs of students, as appropriate.
   Activity 17-1: Adjusting an EPR Valve
   Activity 17-2: Observing the Operation of a Pumpdown Solenoid Valve
   Activity 17-3: Adjusting Crankcase Pressure Regulator
   Activity 17-4: Adjusting Discharge Bypass Valve
   Activity 17-5: Setting EPR Valves
3. Observe student performance on Activities, providing guidance or correction as needed. After reviewing the student's work, initial and date the activity in the place provided.

Evaluation

1. Administer the quiz from this Instructor's Manual.
2. Grade the quiz, and return it with appropriate comments for the student.

Answers to Test Your Knowledge, Page 286

1. diaphragm; globe; ball
2. reverse (or back)
3. evaporator
4. two-temperature
5. suction
6. true
7. in supermarkets
8. electricity (or electromagnetism)
9. open
10. pumpdown cycle
11. compressor
12. discharge bypass valve; compressor cylinder unloaders

13. To protect the compressor from overheating due to high suction superheat.
14. R-12 = 100 psig; R-22 = 200 psig; R-502 = 180 psig
15. increase

Answers to Lab Manual, Pages 129–136

Activity 17-1

1.–6. Student response (answers will vary).
7. warmer
8. colder
9. close
10. It remains in the same state.
11. b. 45 psig

Activity 17-2

Student response (answers will vary).

Activity 17-3

1.–8. Student response (answers will vary).
9. b. 18 psig to 20 psig
10. b. Limits suction pressure entering the compressor.

Activity 17-4

1.–4. Student response (answers will vary).
5. 60 psig to 61 psig
6. Lower pressures would permit supply air to become too cold.

Activity 17-5

1. Cabinet 1: 30.9
   Cabinet 2: 41.7
2. Cabinet 1: 27.8
   Cabinet 2: 29.5
3. Cabinet 1: 12.1
   Cabinet 2: 22.6

Answers to Quiz

1. false
2. b. suction line
3. true
4. entering
5. solenoid
6. pumpdown
7. exiting
8. false
9. hot gas bypass or discharge bypass
10. 200 psig (1379 kPa)

Hand Valves

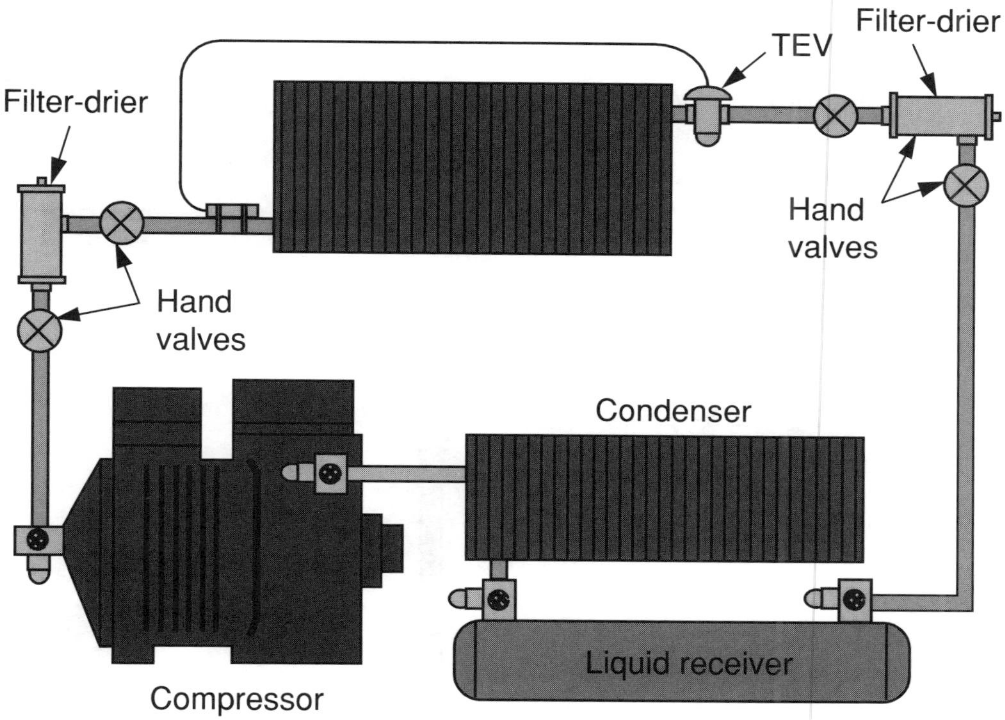

HC 17-1

EPR Valve

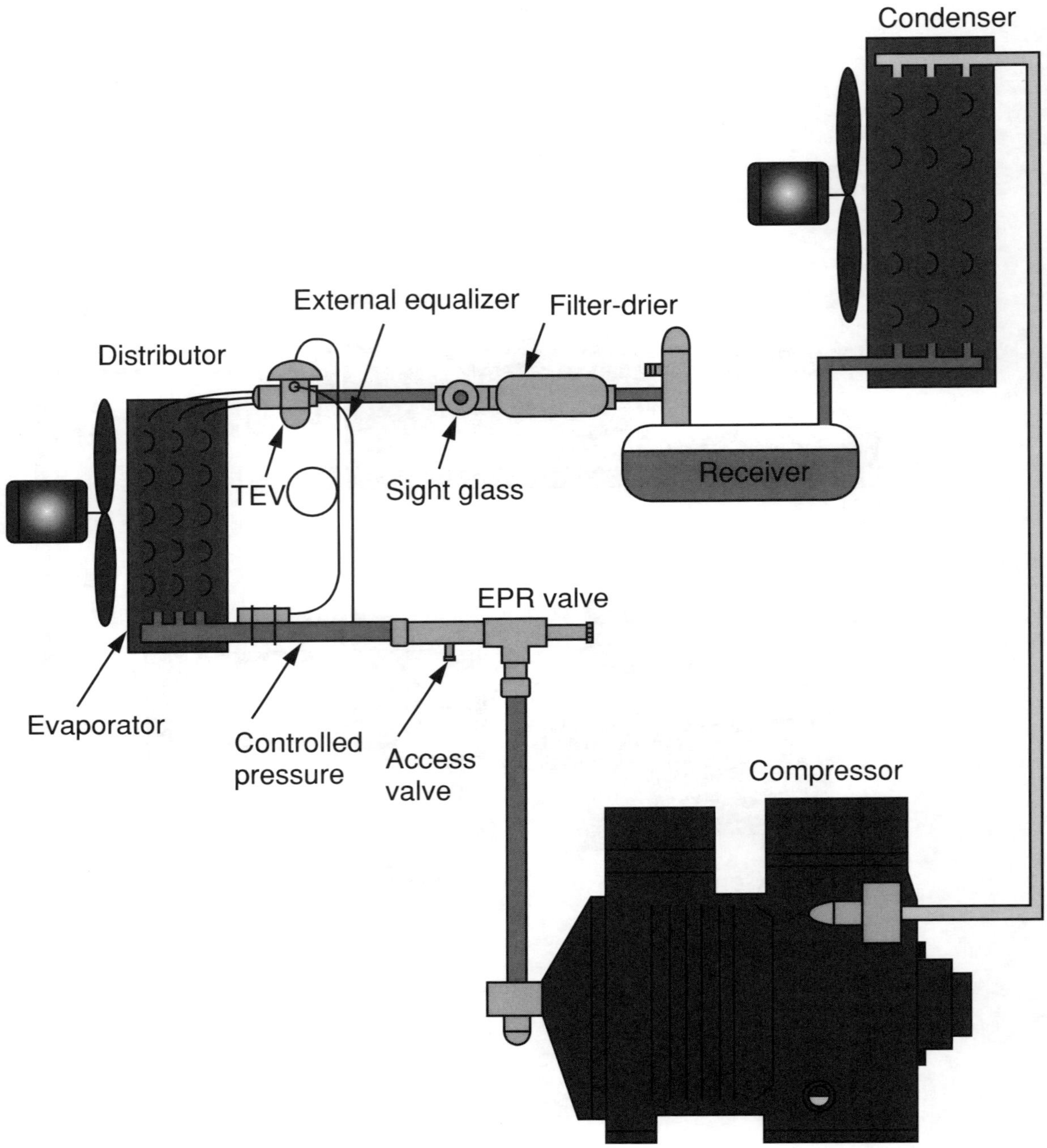

HC 17-2

Two-temperature Systems

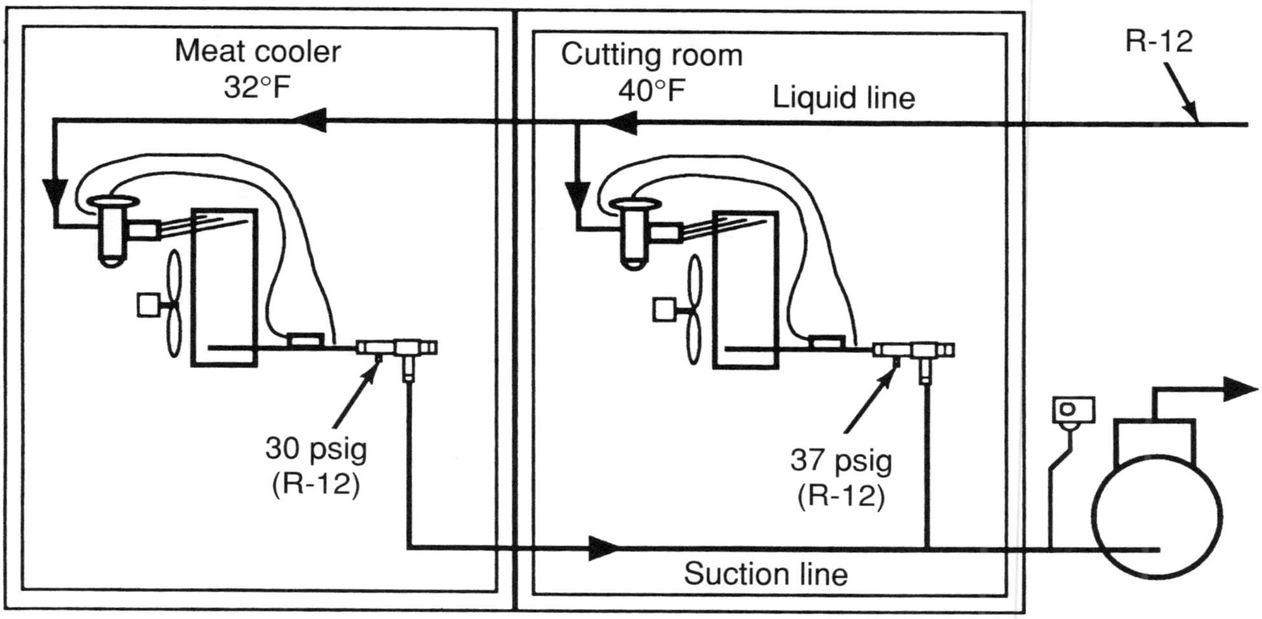

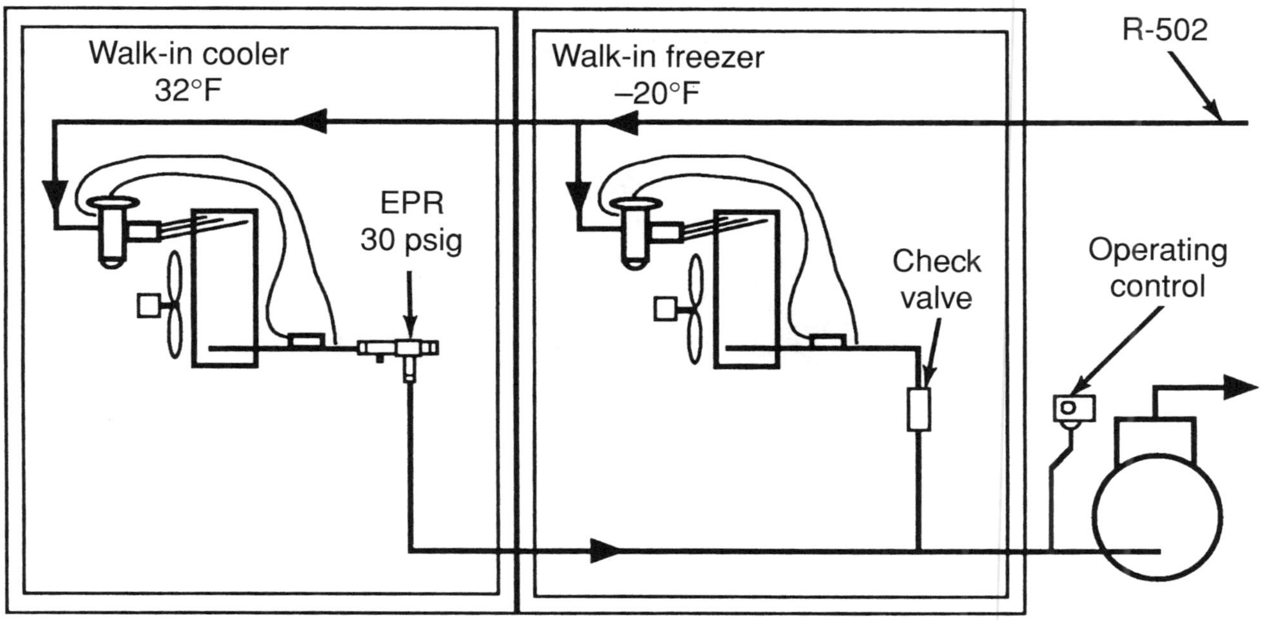

HC 17-3

Multiplexed Systems

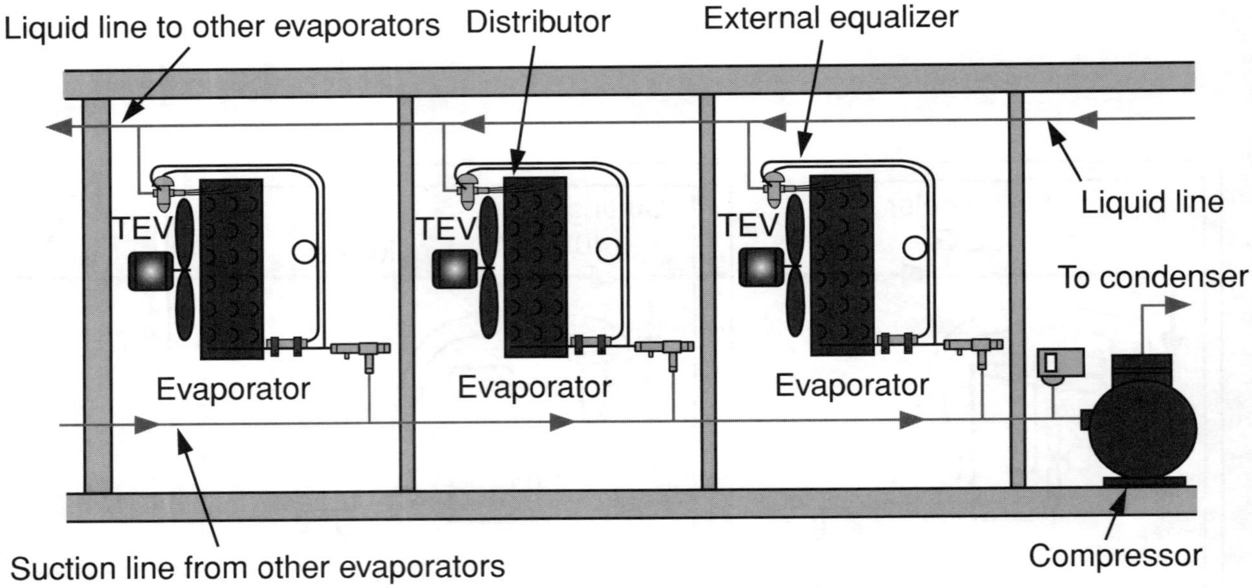

Evaporators

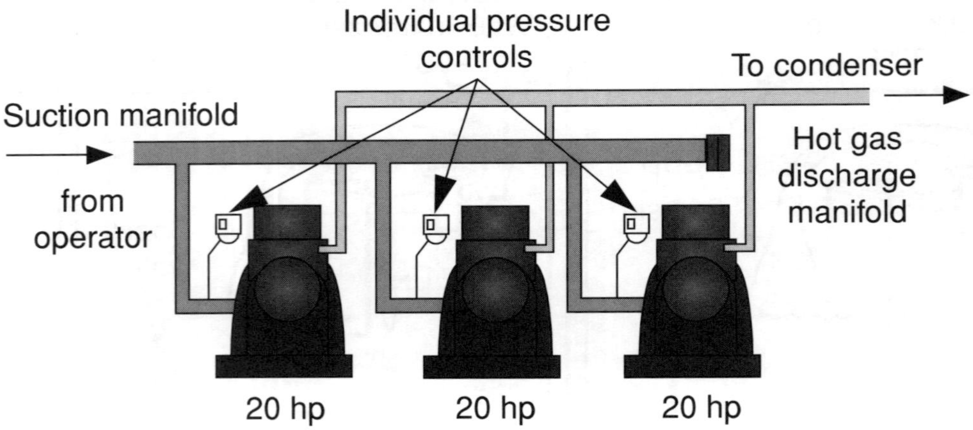

HC 17-4

Desuperheating TEV

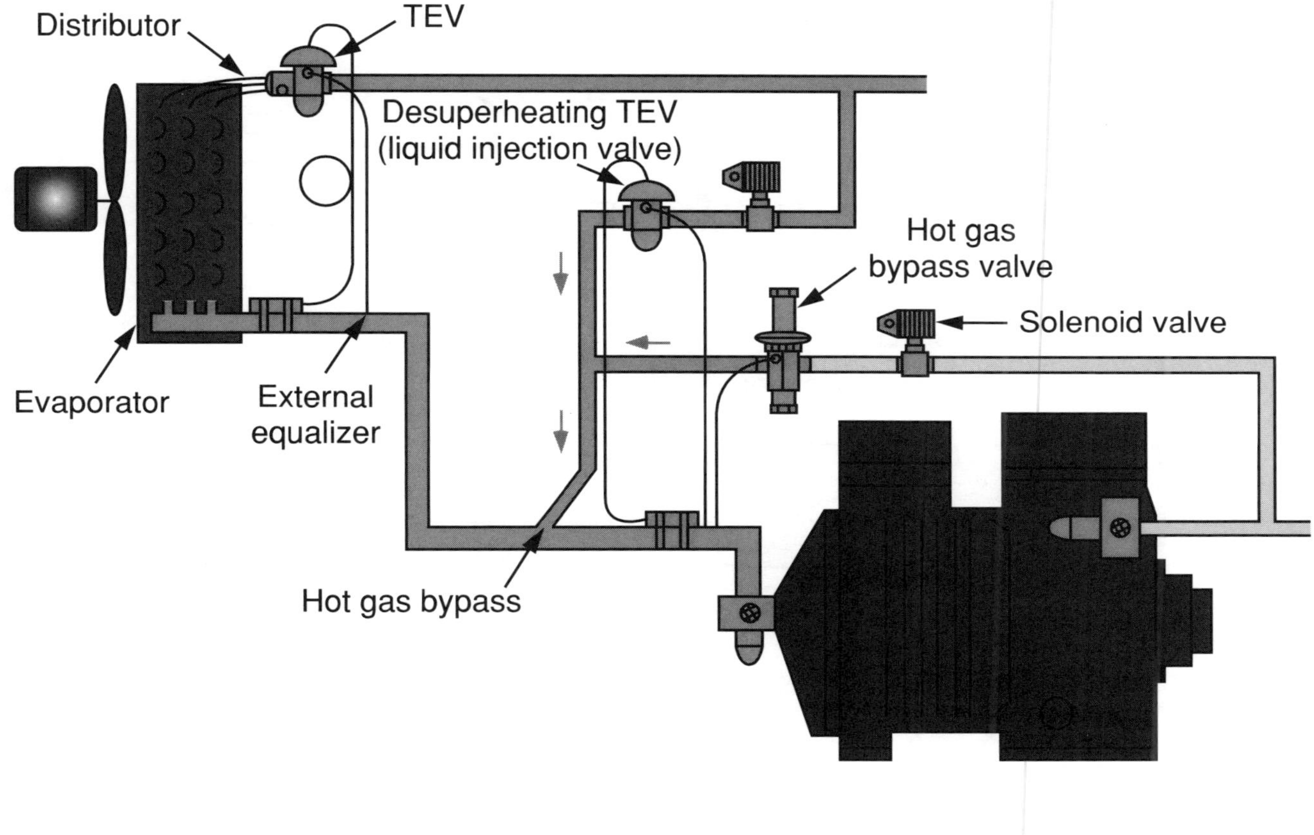

HC 17-5

Adjusting Head Pressure

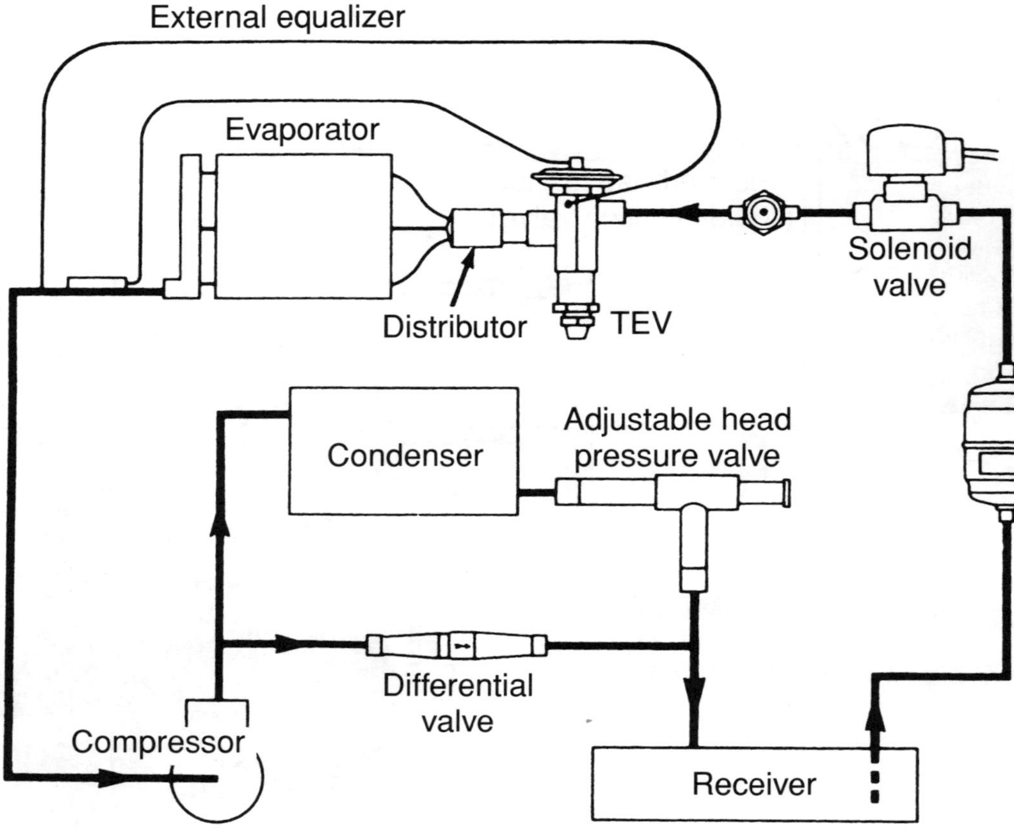

HC 17-6

Name ________________________________

Date ________________________________

Instructor ____________________________

Period________________________________

17 Special-purpose Valves

Quiz

Instructions: Write your answers on the lines provided at the right-hand side of the page.

1. *True or false?* Diaphragm, globe, and ball valves are seldom used as hand shut-off valves. 1. ____________

2. EPR valves are installed in the _____. 2. ____________
   a. liquid line
   b. suction line
   c. discharge line

3. *True or false?* An EPR prevents evaporator pressure from falling below a specific setpoint. 3. ____________

4. Does the EPR valve control the pressure entering or exiting the valve? 4. ____________

5. An electrically operated valve is called a _____ valve. 5. ____________

6. A _____ cycle requires the use of a solenoid valve in the liquid line. 6. ____________

7. Does the CPR valve control the pressure entering or exiting the valve? 7. ____________

8. *True or false?* Head pressure control valves protect the system against high head pressure. 8. ____________

9. Capacity control during periods of low load can be achieved by using a _____ valve. 9. ____________

10. What is the minimum head pressure for R-22? 10. ____________

18

Troubleshooting Flow Controls

Objectives

After studying this chapter in the text, the student will be able to:

- ❑ Diagnose and correct problems in AEV or TEV systems.
- ❑ Identify and correct capillary tube problems.
- ❑ Recognize and properly use various types of hand valves.
- ❑ Diagnose EPR and CPR valve problems and make necessary adjustments.
- ❑ Identify problems and make repairs needed for correct operation of solenoid valves.
- ❑ Diagnose and correct problems in head pressure control and discharge bypass valves.

Instructional Materials

Text: Pages 287–298
Lab Manual: Pages 137–142
Instructor's Manual:
Visual Masters, Pages 165–167
Chapter 18 Quiz, Page 168

Teaching Strategy

Text

1. Review objectives.
2. Discuss and assign chapter contents. Use these visual masters as appropriate:
   HC 18-1: Troubleshooting the AEV
   HC 18-2: Troubleshooting the Capillary Tube System
   HC 18-3: Troubleshooting the TEV
   HC 18-4: Crankcase Heaters
3. Direct students to complete Test Your Knowledge questions, pages 297-298. After completion, discuss correct answers with students.

Lab Manual

1. Provide overview of Activities. Where necessary, stress safe laboratory practices and the importance of wearing protective clothing.
2. Assign Activities to individuals or pairs of students, as appropriate.
   Activity 18-1: Troubleshooting the AEV System
   Activity 18-2: Troubleshooting the Capillary Tube System
   Activity 18-3: Troubleshooting the TEV System
   Activity 18-4: Troubleshooting Solenoid Valves
   Activity 18-5: Troubleshooting Head Pressure Controls
3. Observe student performance on Activities, providing guidance or correction as needed. After reviewing the student's work, initial and date the activity in the place provided.

Evaluation

1. Administer the quiz from this Instructor's Manual.
2. Grade the quiz, and return it with appropriate comments for the student.

Answers to Test Your Knowledge, Pages 297-298

1. refrigerant not moving properly; undercharged
2. low head; low suction; bubbles in sight glass
3. low; high
4. high head
5. true
6. restriction; undercharge; inefficient compressor
7. File a notch, and bend the tubing back and forth at the notch until it breaks off.
8. Check TEV superheat.
9. evaporator airflow
10. filter-drier
11. heat

12. screwdriver
13. snap
14. plunger
15. failure to open; failure to close

Answers to Lab Manual, Pages 137–142

Activity 18-1

1. A foreign substance is fouling the valve seat. Pour hot water over the AEV to melt the contaminant. Observe any change in suction pressure.
2. Defective compressor. Use cardboard to block the condenser, and observe any change in discharge pressure. Minimal change indicates a defective compressor.
3. Dirty condenser. Check the condenser and airflow.
4. No problem. How long have the soft drink containers been stored in this cooler? How warm was the product when it arrived?

Activity 18-2

1. High discharge pressure. Check for a dirty condenser or a defective condenser fan.
2. Defective compressor or loss of refrigerant. Install saddle valves and a gauge manifold. Actual pressure readings would reveal the true problem.
3. Low suction pressure. Low discharge pressure. Poor cooling. Partially frosted evaporator. Would give false indication of loss of refrigerant.

Activity 18-3

1. Undercharged system. Charge refrigerant into the system, and observe the results.
2. TEV is restricted. Heat the valve with hot water, and watch for an increase in suction pressure. Check the filter screen at the TEV inlet.
3. Defective TEV. The thermal bulb lost its charge. Examine the capillary tube and thermal bulb for damage and proper mounting. Pour hot water on the thermal bulb, and watch for a change in suction pressure.
4. Low evaporator load. Check for dirty air filters.

Activity 18-4

1. The pumpdown solenoid valve is not energized. Use a small screwdriver to test the solenoid coil for magnetism.
2. The pumpdown solenoid valve is not opening properly. Remove and disassemble the solenoid valve. Inspect for damage or fouled seat.
3. The pumpdown solenoid valve seat is leaking. Shut off electrical power to the solenoid valve. Install a gauge on the SSV, and perform a manual pumpdown to a deep vacuum. Frontseat the SSV, and shut off the compressor. If the compressor holds the vacuum, the solenoid valve seat is leaking. If the vacuum is lost, the compressor discharge reeds are leaking.
4. The indoor thermostat is turned off or not set properly. Check the indoor thermostat for proper settings.

Activity 18-5

1. Defective head pressure control. Use your hands to check for hot gas bypass. Stop the condenser fan, or use cardboard to stop airflow across the condenser. If pressures become normal, head pressure control is defective.
2. Undercharged (low on refrigerant). Add some refrigerant, and recheck sight glass and superheat.
3. Compressor failure was due to liquid slugging and/or liquid migration during the off cycle. Recommend installation of a head pressure control and a crankcase heater to protect the compressor during cold ambient.

Answers to Quiz

1. Troubleshooting
2. low; high
3. crankcase heater
4. decrease
5. erratic
6. liquid
7. b. Poor cooling, high head, high suction.
8. increase
9. false
10. airflow

Troubleshooting the AEV

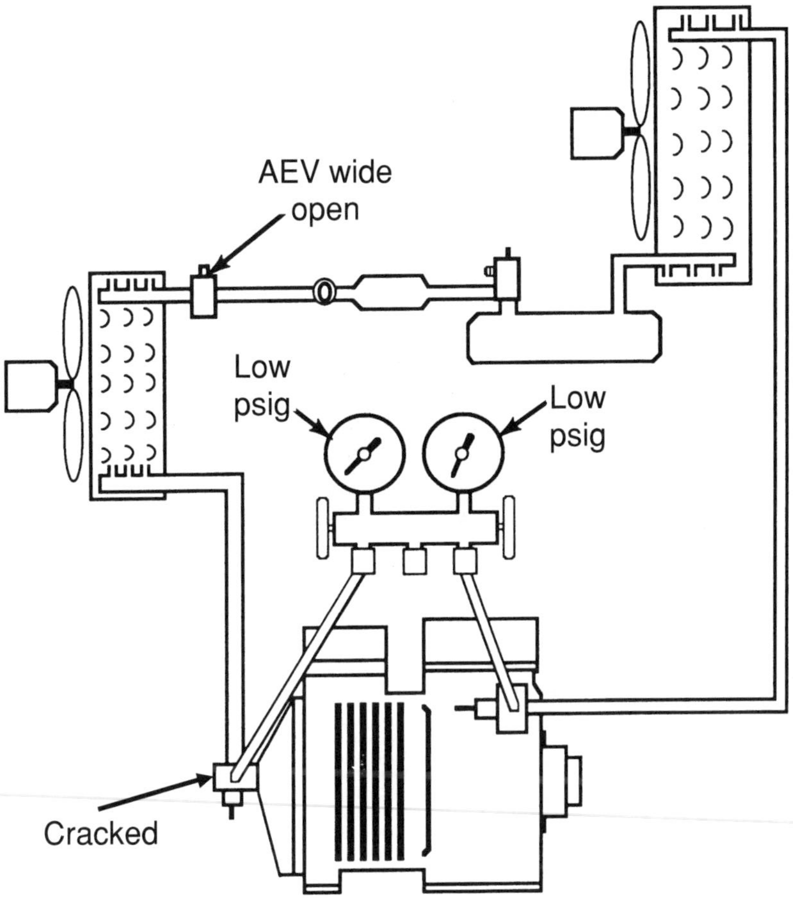

Undercharged System

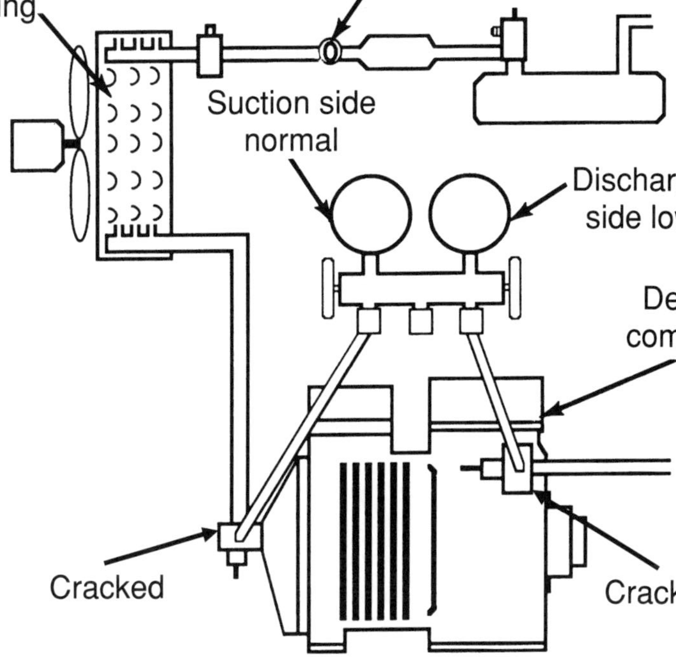

Defective Compressor

HC 18-1

Troubleshooting the Capillary Tube System

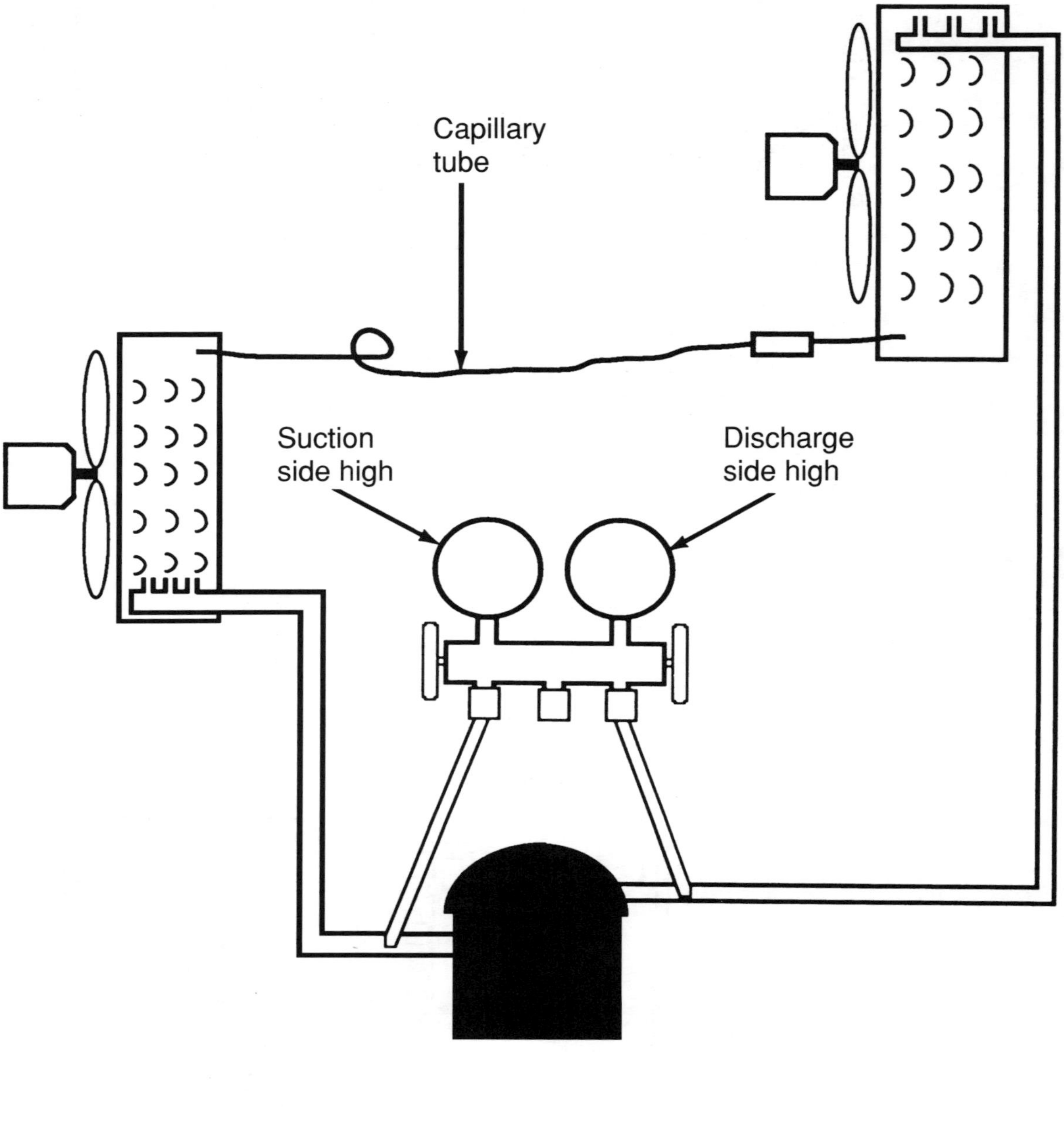

HC 18-2

Troubleshooting the TEV

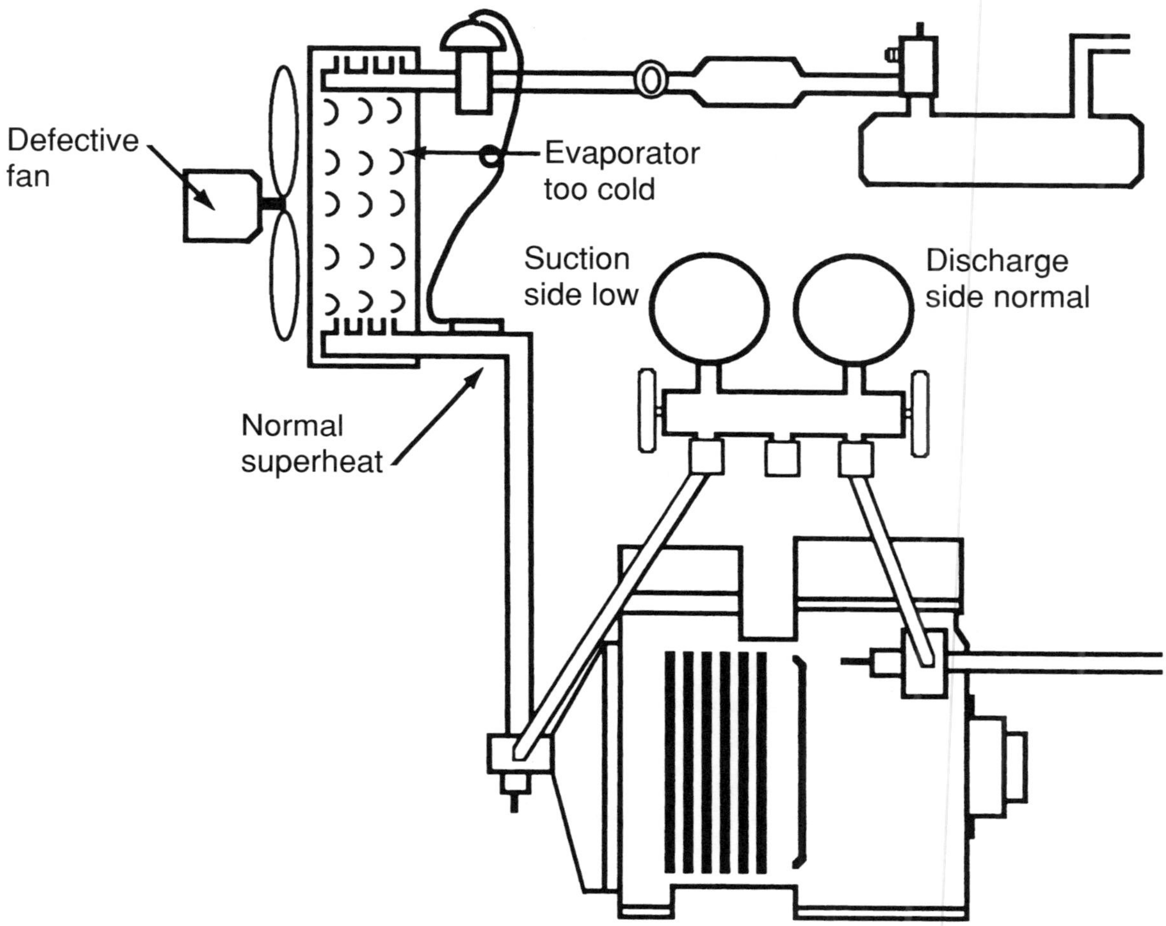

HC 18-3

Name ______________________________

Date ______________________________

Instructor ______________________________

Period______________________________

18

Troubleshooting Flow Controls

Quiz

Instructions: Write your answers on the lines provided at the right-hand side of the page.

1. _____ is the term used to describe the process of gathering information needed to make an intelligent servicing decision. 1. ______________________

2. An inefficient compressor is usually indicated by _____ head pressure and _____ suction pressure. 2. ______________________ ______________________

3. Any compressor located in a cold ambient should be equipped with a _____. 3. ______________________

4. Does poor airflow across the evaporator cause the suction pressure to increase or decrease? 4. ______________________

5. A valve that does not maintain a particular setting for very long is said to be _____. 5. ______________________

6. No refrigerant control can feed properly if sufficient _____ is not available. 6. ______________________

7. How does a dirty condenser effect a capillary tube system? 7. ______________________
   a. Poor cooling, low head, low suction.
   b. Poor cooling, high head, high suction.
   c. Poor cooling, low head, high suction.

8. If the system is overcharged, will head pressure increase or decrease? 8. ______________________

9. *True or false?* A high superheat reading indicates a defective TEV. 9. ______________________

10. An evaporator freeze-up is normally caused by inadequate _____. 10. ______________________

19

Compressors

Objectives

After studying this chapter in the text, the student will be able to:

- ❑ Identify five types of compressors.
- ❑ List the advantages of the compliant scroll compressor.
- ❑ Identify the three basic compressor designs.
- ❑ Discuss the operation of each type of compressor.
- ❑ Properly align pulleys and flywheels.
- ❑ Select, size, and replace V-belts.
- ❑ Identify the components of a reciprocating compressor.
- ❑ Describe the advantages of the disc-type valve design.

Instructional Materials

Text: Pages 299–314
Lab Manual: Pages 143–154
Instructor's Manual:
Visual Masters, Pages 173-176
Chapter 19 Quiz, Page 177

Teaching Strategy

Text

1. Review objectives.
2. Discuss and assign chapter contents. Use these visual masters as appropriate:
   HC 19-l: Reciprocating Compressor
   HC 19-2: Rotary, Screw-type, and Centrifugal Compressors
   HC 19-3: Scroll Compressor Operation
   HC 19-4: Open-type Compressor
3. Direct students to complete Test Your Knowledge questions, page 313. After completion, discuss correct answers with students.

Lab Manual

1. Provide overview of Activities. Where necessary, stress safe laboratory practices and the importance of wearing protective clothing.
2. Assign Activities to individuals or pairs of students, as appropriate.
   Activity 19-1: Identifying Types of Compressors
   Activity 19-2: Dismantling and Reassembling a Reciprocating Compressor
   Activity 19-3: Dismantling and Reassembling an Open-type Compressor
   Activity 19-4: Installing and Aligning V-belts
   Activity 19-5: Opening and Examining a Hermetic Compressor
   Activity 19-6: Replacing a Hermetic Compressor
   Activity 19-7: Examining Discus Compressor Valve Plate
   Activity 19-8: Examining a Six-cylinder Compressor
3. Observe student performance on Activities, providing guidance or correction as needed. After reviewing the student's work, initial and date the activity in the place provided.

Evaluation

1. Administer the quiz from this Instructor's Manual.
2. Grade the quiz, and return it with appropriate comments for the student.

Answers to Test Your Knowledge, Page 313

1. reciprocating; rotary; screw; centrifugal; scroll
2. rotating vane; stationary vane
3. augers
4. fixed scroll; orbiting scroll
5. scroll compressor
6. open-type; hermetic; semihermetic

7. V-belts
8. hermetic
9. false
10. open-type
11. discharge chamber in head
12. valve plate
13. Suction; discharge
14. 25
15. 10

Answers to Lab Manual, Pages 143–148

Activity 19-1

A. reciprocating
B. screw
C. rotary
D. centrifugal
E. scroll

Activity 19-4

Student response (answers will vary).

Answers to Quiz

1. reciprocating; rotary; screw; centrifugal; scroll
2. A. volute
   B. impeller vanes
   C. discharge (to condenser)
   D. suction intake
3. rotary
4. b. Totally sealed.
5. For ease of brazing to copper tubing.
6. gasket
7. valve plate
8. true
9. reciprocating
10. scroll

Reciprocating Compressor

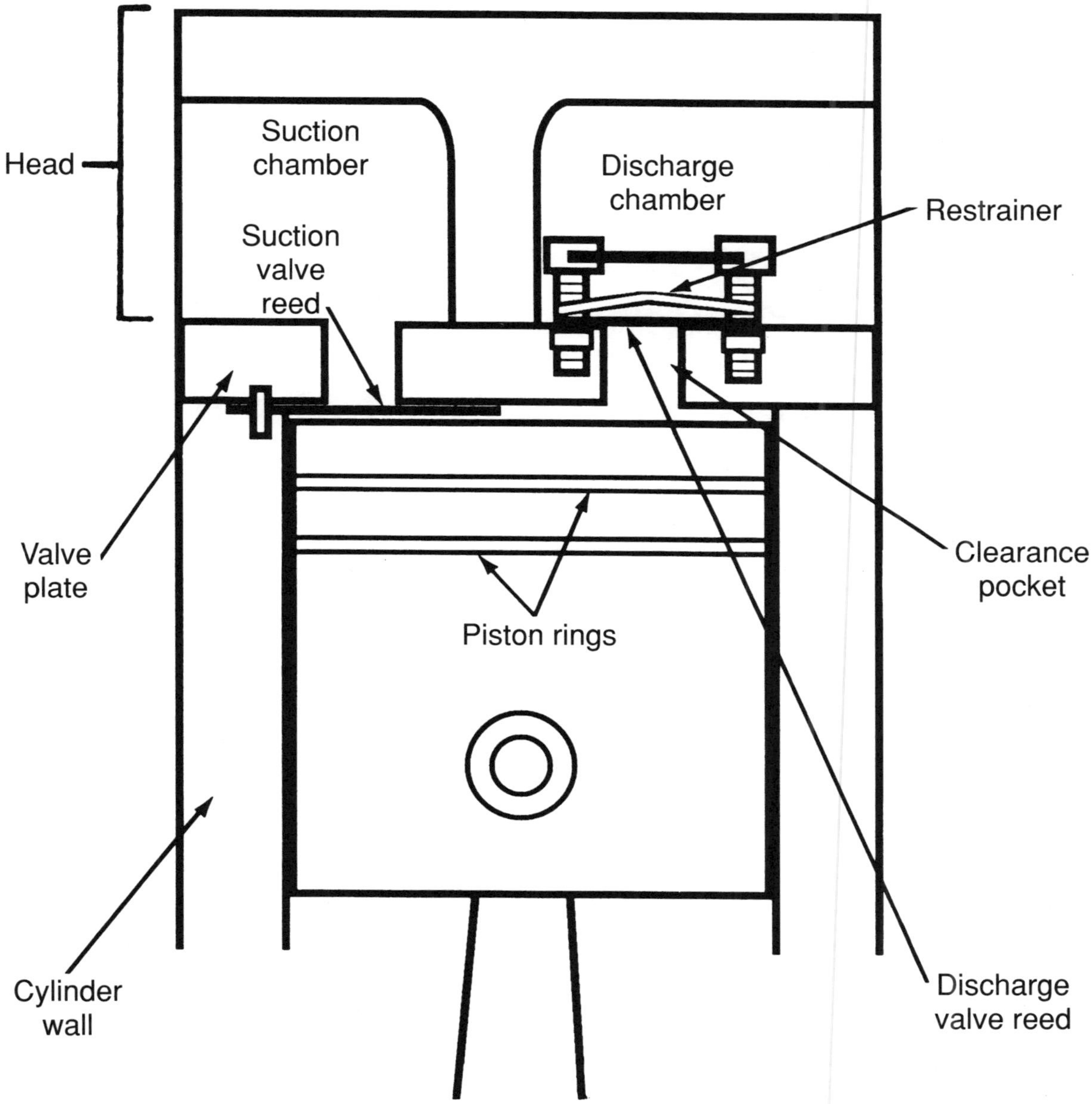

HC 19-1

Rotary, Screw-type, and Centrifugal Compressors

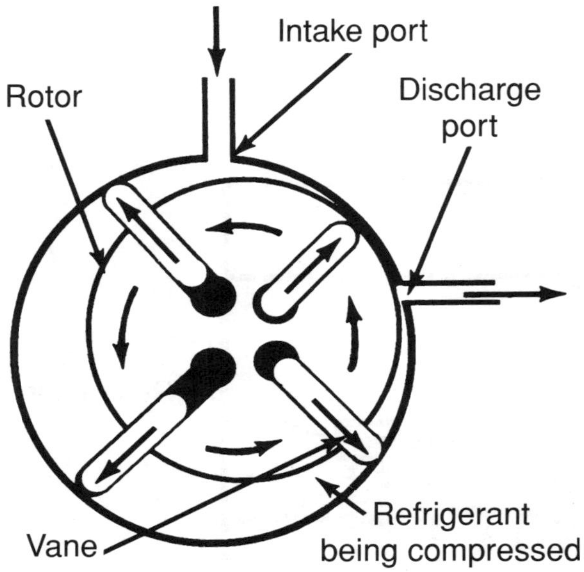

Rotary

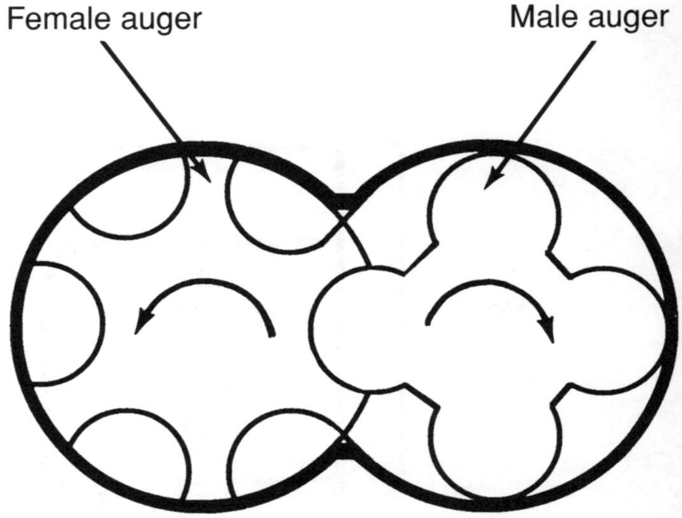

Screw-type

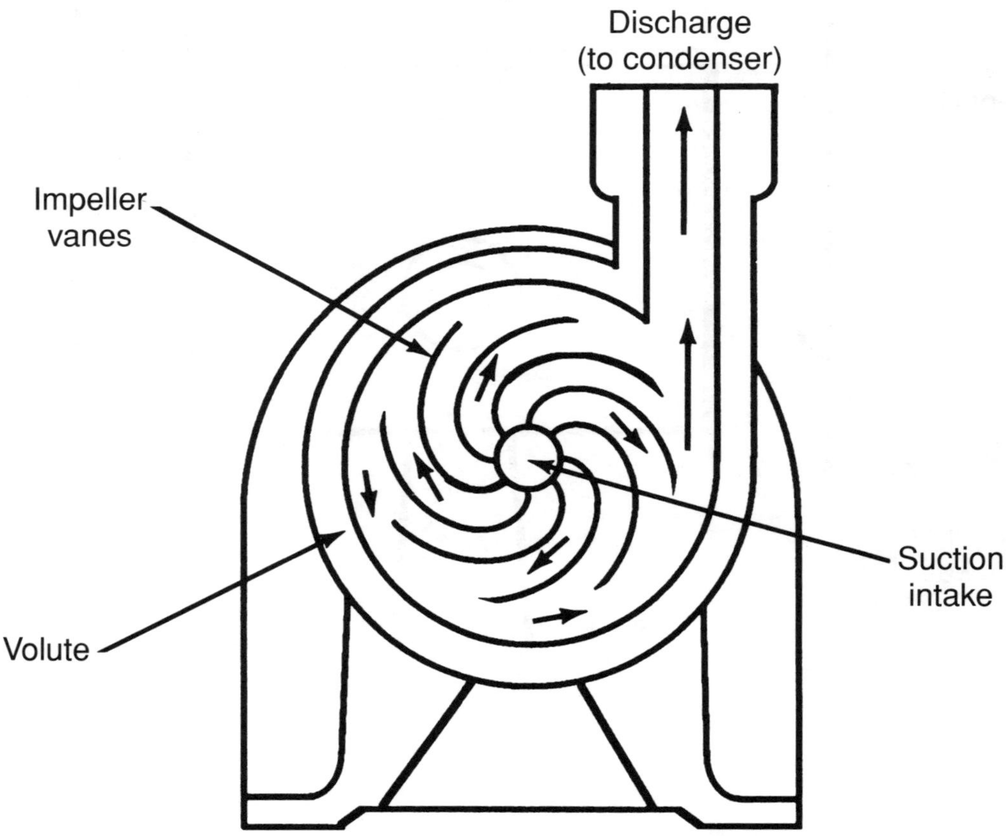

Centrifugal

HC 19-2

Scroll Compressor Operation

Scroll Gas Flow

1

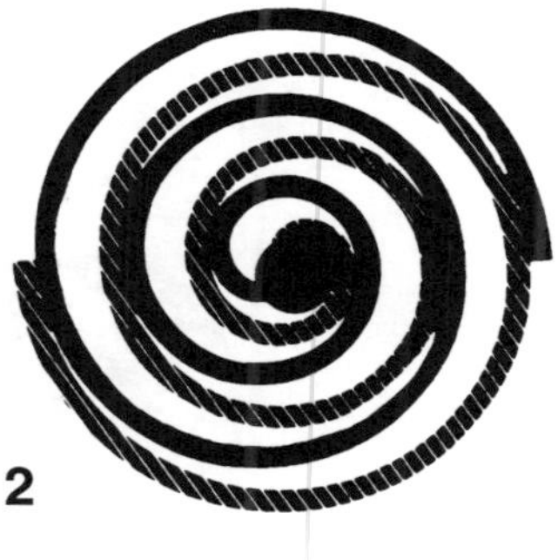

2

Compression in the scroll is created by the interaction of an orbiting spiral and a stationary spiral. Gas enters an outer opening as one of the spirals orbits.

The passage is sealed off as gas is drawn into the spiral.

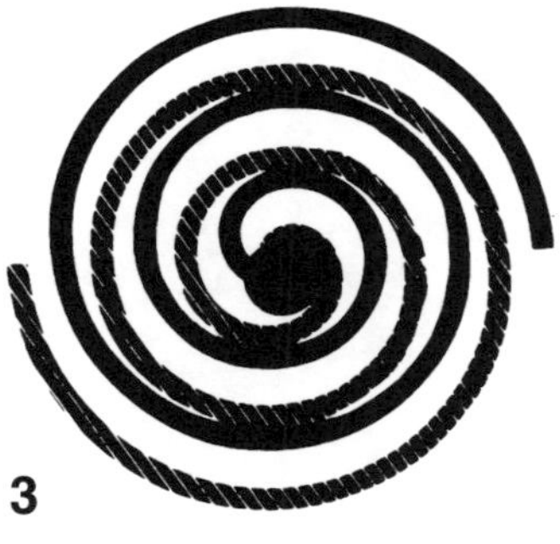

3

4

5

As the spiral continues to orbit, the gas is compressed into an increasingly smaller pocket.

By the time the gas arrives at the center port, discharge pressure has been reached.

Actually, during operation, all six gas passages are in various stages of compression at all times, resulting in nearly continuous suction and discharge.

HC 19-3

Open-type Compressor

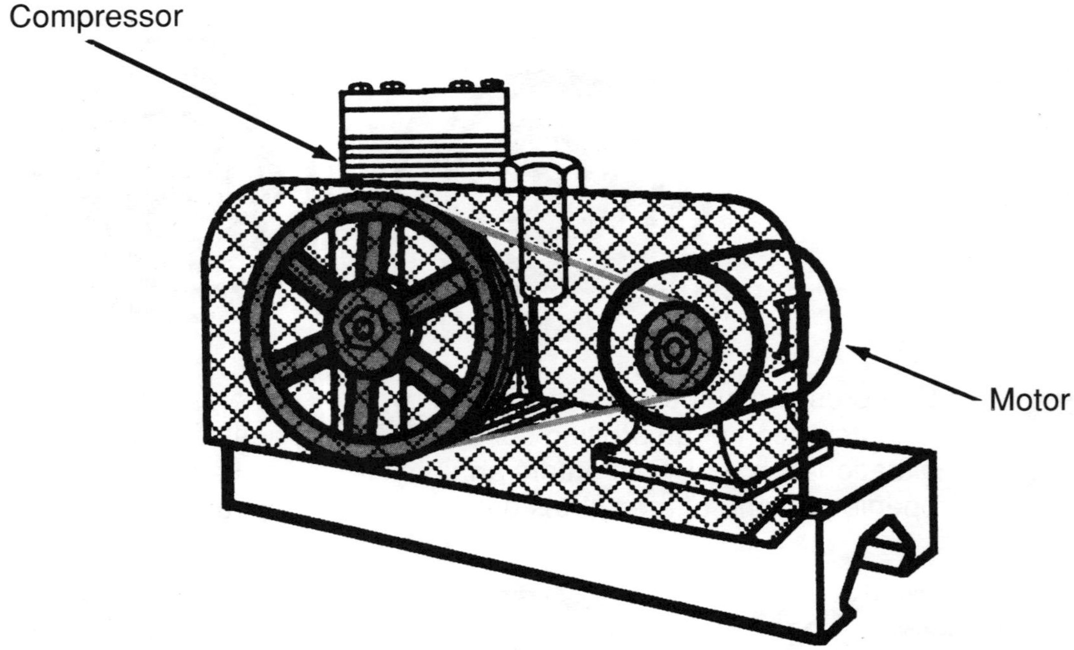

HC 19-4

Name ______________________________

Date ______________________________

Instructor ______________________________

Period______________________________

19

Compressors

Quiz

Instructions: Write your answers on the lines provided at the right-hand side of the page.

1. Name the five basic types of compressors.

1. ______________________________
____
____
____
____

2. Identify the parts of this centrifugal compressor.

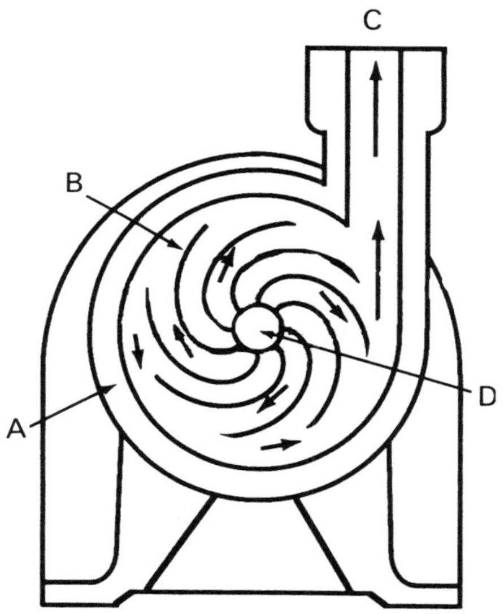

2. a. ______________________________
b. ______________________________
c. ______________________________
d. ______________________________

3. What type of compressor uses either a stationary vane or a rotating vane?

3. ______________________________

Continued

4. When used to describe a compressor, the word *hermetic* means that the compressor is: 4. ____________
   a. Repairable.
   b. Totally sealed.
   c. Semiaccessible.

5. Why are hermetic compressor stubs copper-coated? 5. ____________

6. To obtain leakproof seals, semihermetic compressors require a _____ between the mating parts. 6. ____________

7. The suction and discharge valve reeds are located on the _____ of a reciprocating compressor. 7. ____________

8. *True or false?* The disc-type valve design increases compressor capacity by up to 25%. 8. ____________

9. On _____ compressors, pistons travel up and down inside cylinders. 9. ____________

10. Which one of the five types of compressors will not be damaged by liquid slugging? 10. ____________

20

Compressor Lubrication and Accessories

Objectives

After studying this chapter in the text, the student will be able to:

- ❑ Select the proper oil for each system.
- ❑ Identify the proper oil level in a compressor.
- ❑ Describe the operation of oil separators.
- ❑ Add or remove oil from a compressor.
- ❑ Discuss the factors that can affect compressor capacity.
- ❑ Describe the means to control liquid migration in a system.
- ❑ Identify and install vibration eliminators.
- ❑ Interpret and use information from compressor data plates.

Instructional Materials

Text: Pages 315–332
Lab Manual: Pages 155–166
Instructor's Manual:
Visual Masters, Pages 181–184
Chapter 20 Quiz, Page 185

Teaching Strategy

Text

1. Review objectives.
2. Discuss and assign chapter contents. Use these visual masters as appropriate:
   HC 20-1: Compressor Lubrication
   HC 20-2: Oil Pressure Safety Control
   HC 20-3: Two-stage Compressor
   HC 20-4: Unloading System
3. Direct students to complete Test Your Knowledge questions, page 332. After completion, discuss correct answers with students.

Lab Manual

1. Provide overview of Activities. Where necessary, stress safe laboratory practices and the importance of wearing protective clothing.
2. Assign Activities to individuals or pairs of students, as appropriate.
   Activity 20-1: Checking Oil Pressure
   Activity 20-2: Examining an Oil Separator
   Activity 20-3: Removing Oil from a Semihermetic Compressor; Checking for Acid
   Activity 20-4: Replacing Oil in a Semihermetic Compressor
   Activity 20-5: Examining a Compressor Unloader
   Activity 20-6: Changing a Hermetic Compressor Crankcase Heater
   Activity 20-7: Changing a Semihermetic Compressor Crankcase Heater
   Activity 20-8: Installing a Suction Vibration Eliminator
3. Observe student performance on Activities, providing guidance or correction as needed. After reviewing the student's work, initial and date the activity in the place provided.

Evaluation

1. Administer the quiz from this Instructor's Manual.
2. Grade the quiz, and return it with appropriate comments for the student.

Answers To Test Your Knowledge, Page 332

1. mineral
2. moisture (or contaminants)
3. splash; oil pump (or force feed)
4. oil pump; crankcase
5. refrigerant; crankcase

6. true
7. To improve volumetric efficiency of the compressor.
8. suction; head
9. increase
10. ultralow
11. Unloaders
12. air; refrigerant; water
13. liquid refrigerant
14. false
15. on the data plate

Answers to Lab Manual, Pages 155–163

Activity 20-1

Student response (answers will vary).

Activity 20-5

4. for capacity control
5. solenoid coil

Activity 20-6

Student response (answers will vary).

Activity 20-7

Student response (answers will vary).

Answers to Quiz

1. To prevent absorption of moisture from atmospheric air. To prevent contamination.
2. c. Ester-based.
3. splash system; forced feed (pumped)
4. oil pressure safety control
5. false
6. in the oil
7. capacity
8. 250°F (121°C)
9. air; water; refrigerant
10. false

Compressor Lubrication

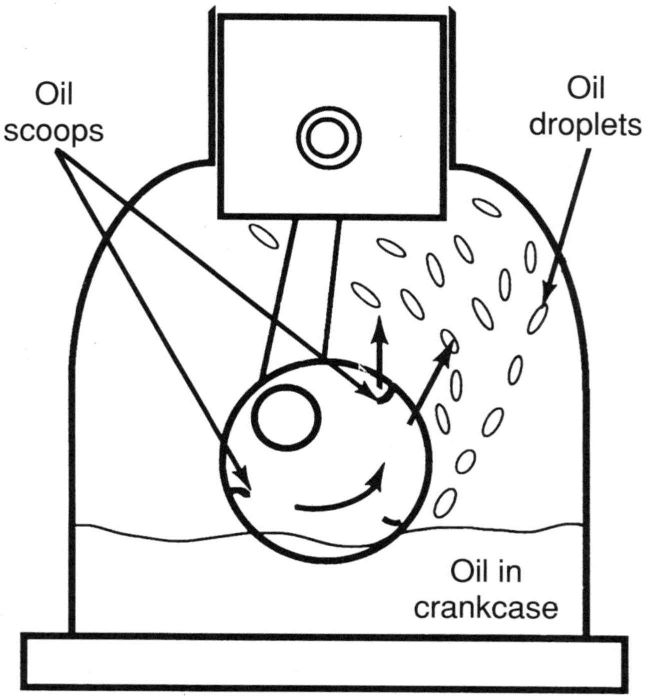

Splash System

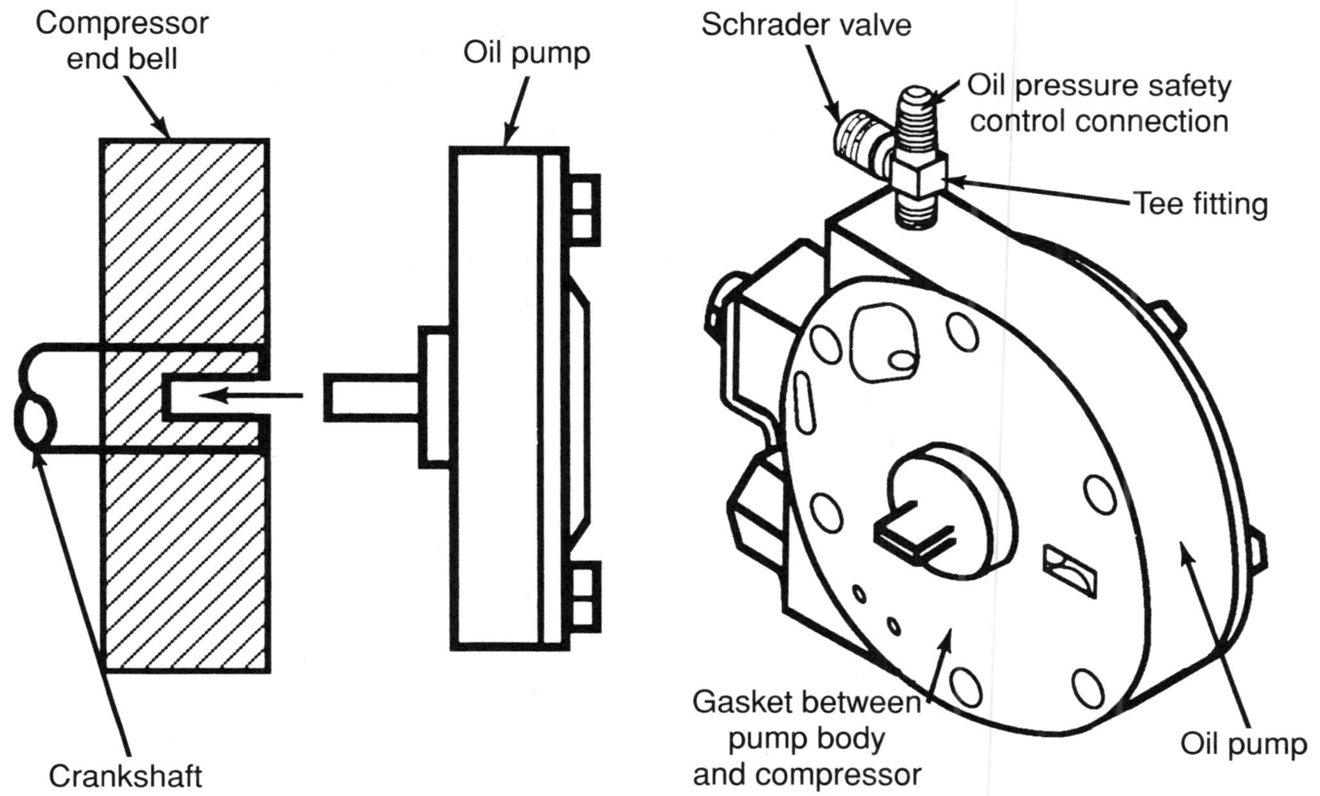

Forced-feed (Oil Pump) System

HC 20-1

Oil Pressure Safety Control

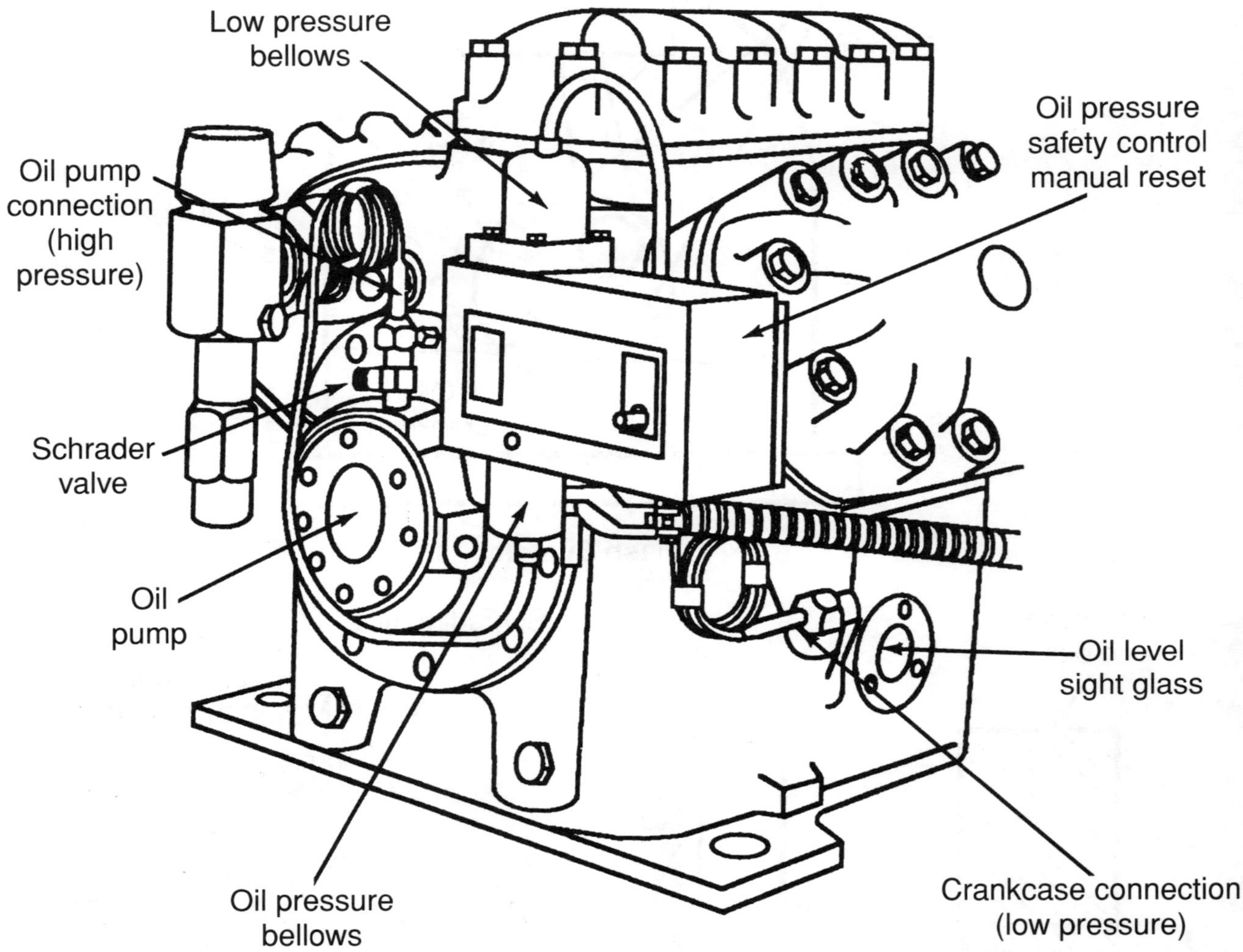

HC 20-2

Two-stage Compressor

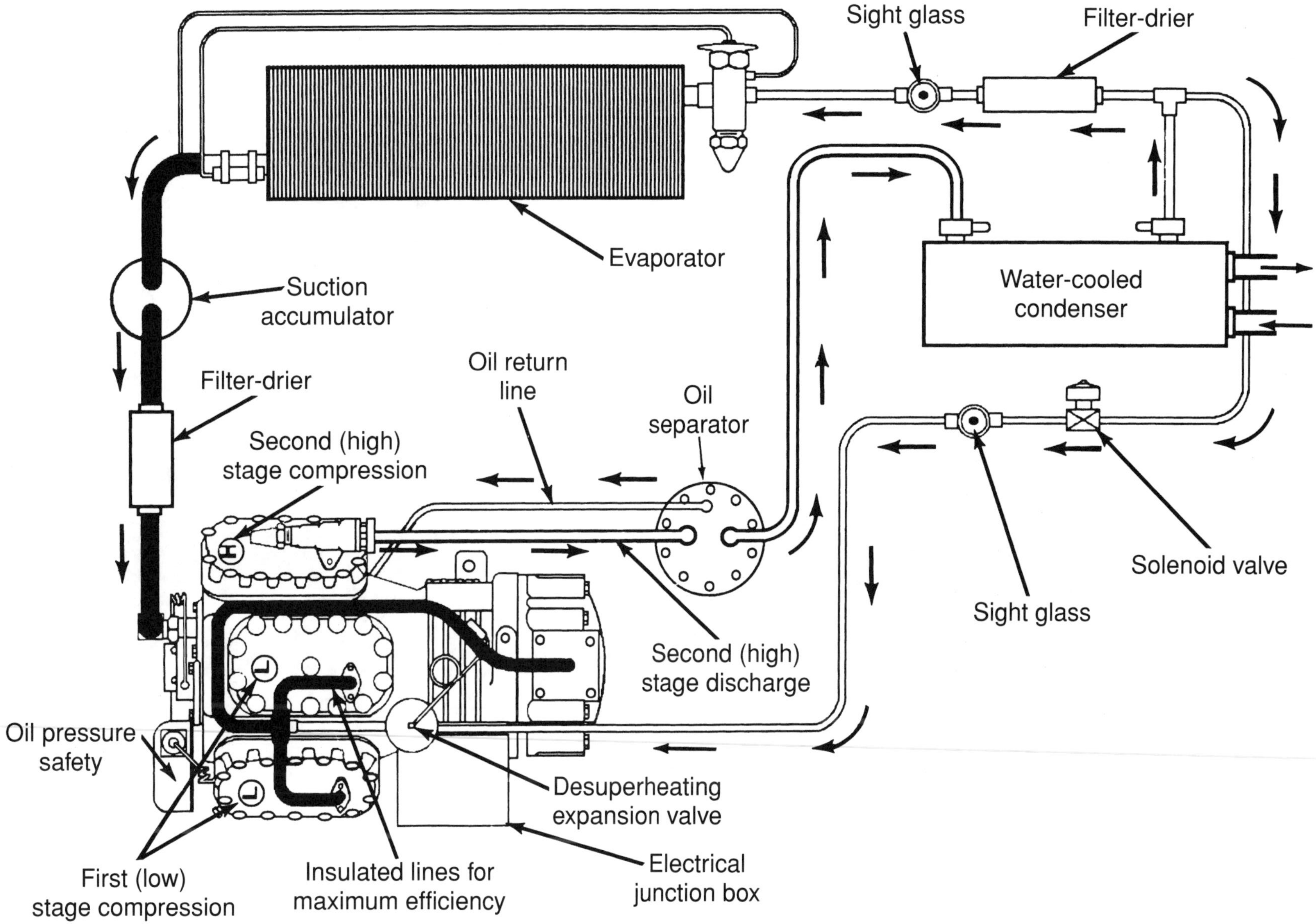

HC 20-3

Unloading System

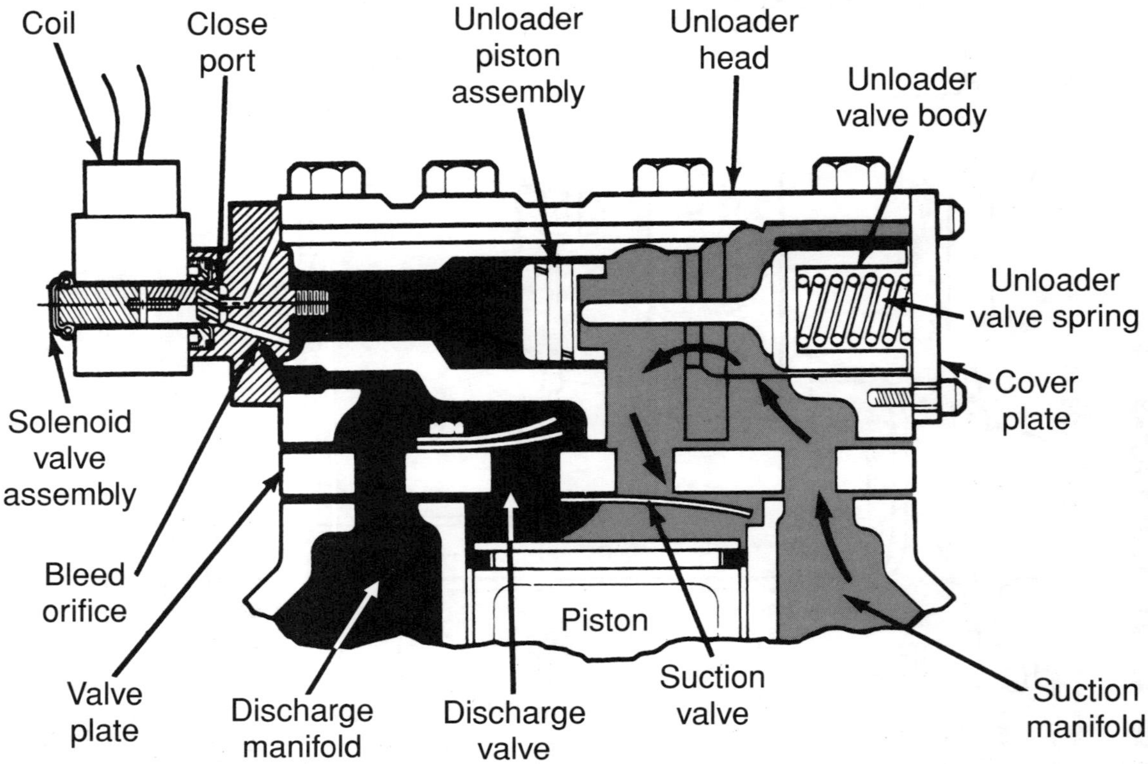

Loaded Operation

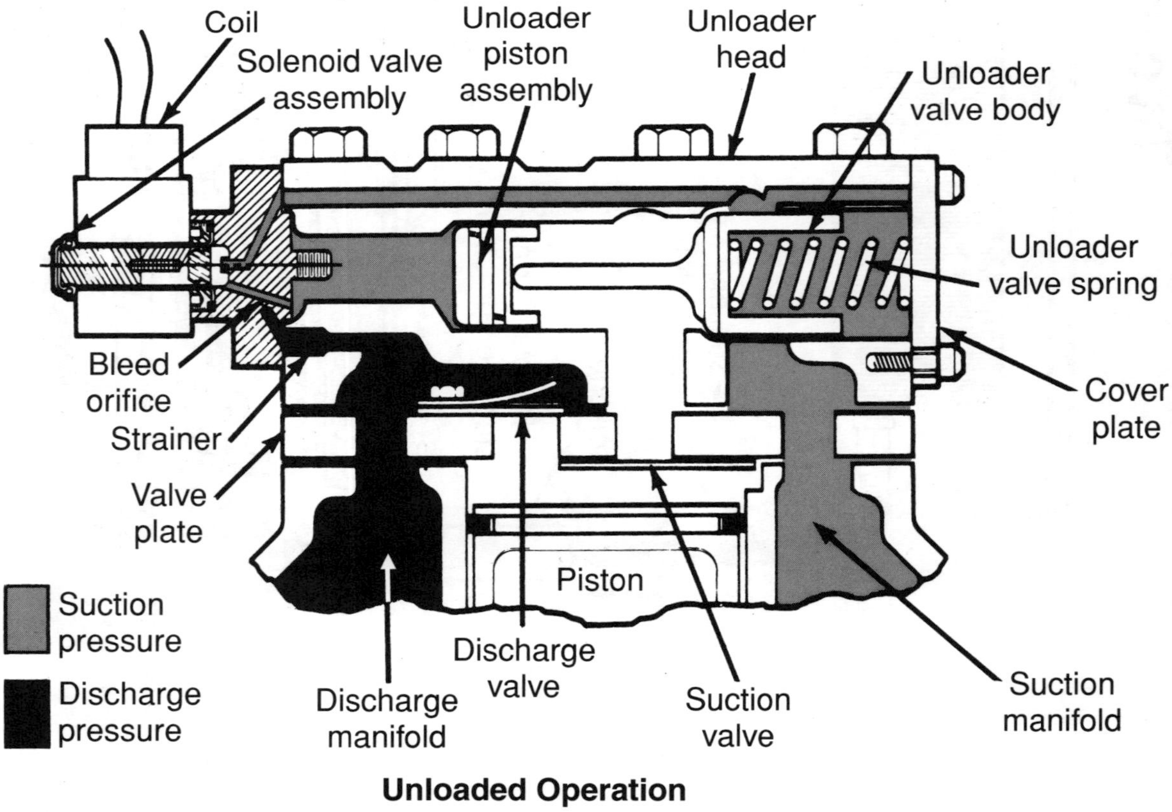

Unloaded Operation

HC 20-4

Name ______________________________

Date ______________________________

Instructor ______________________________

Period______________________________

20

Compressor Lubrication and Accessories

Quiz

Instructions: Write your answers on the lines provided at the right-hand side of the page.

1. Why should refrigeration oil be stored in sealed containers? 1. ____________________

2. What compressor oil is best for systems using refrigerant R-134a? 2. ____________________
   a. Napthenic mineral oil.
   b. Polyalkaline glycol.
   c. Ester-based.

3. Name two methods used for compressor lubrication. 3. ____________________

4. What control is used to protect the compressor against lack of lubrication? 4. ____________________

5. *True or false?* Oil separators are 100% efficient. 5. ____________________

6. When acid is present in a refrigeration system, where can it be detected? 6. ____________________

7. Unloaders are used to reduce system _____. 7. ____________________

8. In a refrigeration system, the maximum allowable discharge line temperature is _____, measured six inches from the discharge service valve. 8. ____________________

9. Name three liquids or gases used for compressor cooling. 9. ____________________

10. *True or false?* Vibration eliminators can be installed anywhere in the suction or discharge lines. 10. ____________________

21

Water Chillers

Objectives

After studying this chapter in the text, the student will be able to:

- ❑ Explain the operating principles of vapor-compression-type chillers.
- ❑ Describe the operation of a centrifugal chiller.
- ❑ Recognize open-loop and closed-loop systems.
- ❑ Explain temperature difference in a closed loop.
- ❑ Describe the operation of a low-pressure chiller.
- ❑ Explain the operating principles of absorption-type chillers.
- ❑ Explain the operating principles of water-ammonia absorption-type refrigerators.

Instructional Materials

Text: Pages 333–350
Lab Manual: Pages 167–170
Instructor's Manual:
Visual Masters, Pages 189–192
Chapter 21 Quiz, Page 193

Teaching Strategy

Text

1. Review objectives.
2. Discuss and assign chapter contents. Use these visual masters as appropriate:
   HC 21-1: Evaporator
   HC 21-2: Open Loop
   HC 21-3: Centrifugal Low-pressure Chiller
   HC 21-4: Absorption-type Refrigerator
3. Direct students to complete Test Your Knowledge questions, pages 349-350. After completion, discuss correct answers with students.

Lab Manual

1. Provide overview of Activities. Where necessary, stress safe laboratory practices and the importance of wearing protective clothing.
2. Assign Activities to individuals or pairs of students, as appropriate.
   Activity 21-1: Evaporator
   Activity 21-2: Low-pressure Chillers
   Activity 21-3: Vapor-compression Chillers
3. Observe student performance on Activities, providing guidance or correction as needed. After reviewing the student's work, initial and date the activity in the place provided.

Evaluation

1. Administer the quiz from this Instructor's Manual.
2. Grade the quiz, and return it with appropriate comments for the student.

Answers to Test Your Knowledge, Pages 349-350

1. a. Chill a liquid.
2. compression; absorption
3. absorption
4. open
5. cooling tower
6. multiple water tubes
7. d. All of the above.
8. closed
9. delta tee
10. centrifugal
11. true
12. air and moisture
13. b. Iron.
14. by centrifugal liquid pumps
15. four-pipe
16. oxygen-deprivation sensor

17. d. All of the above.
18. refrigerant sensor
19. absorption
20. true

Answers to Lab Manual, Pages 167–170

Activity 21-1

1. Because the water tubes are submerged in liquid refrigerant.
2. b. tubing bundle
3. The liquid refrigerant level in the evaporator just covers the tubing bundle, as shown in the illustration.

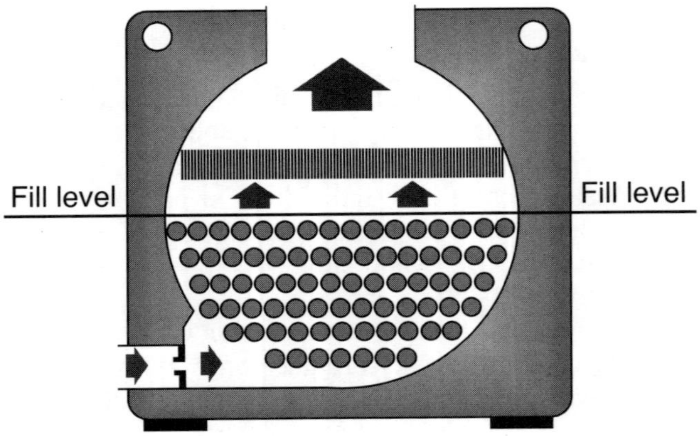

4. water; refrigerant; chills; vaporizes

Activity 21-2

A. condenser
B. evaporator
C. control panel
D. suction intake
E. volutes
F. purge tank

Activity 21-3

1. To remove water vapor and noncondensibles.
2. 10 degrees Fahrenheit (6 degrees Celsius)
3. open loop
4. closed loop
5. To control the amount of vapor entering the impeller.
6. R-11; R-123
7. 10 psig (69 kPa)
8. To prevent freezing the water inside the tubes.

Answers to Quiz

1. absorption
2. open
3. closed
4. evaporator
5. All of the above.
6. centrifugal
7. true
8. absorption
9. all of the above.
10. purge unit

Evaporator

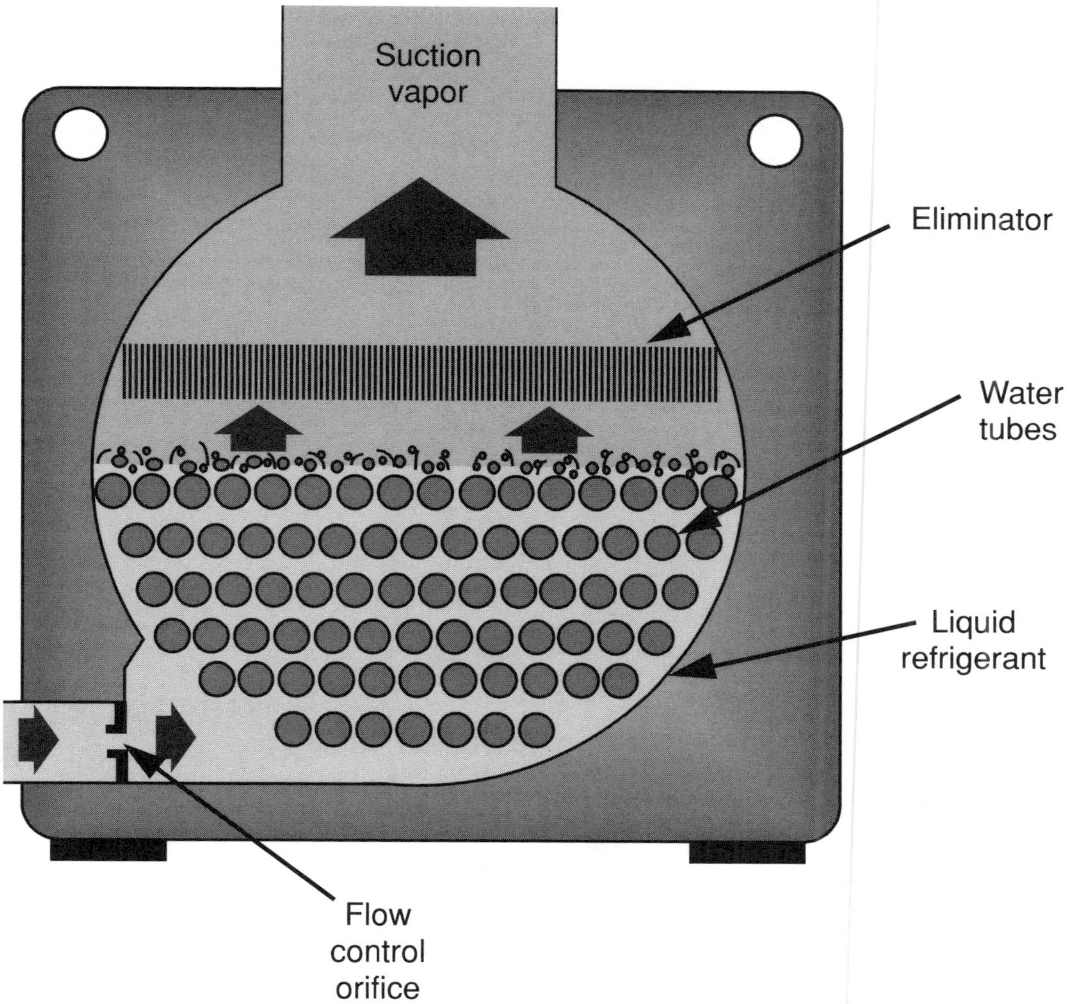

HC 21-1

Open Loop

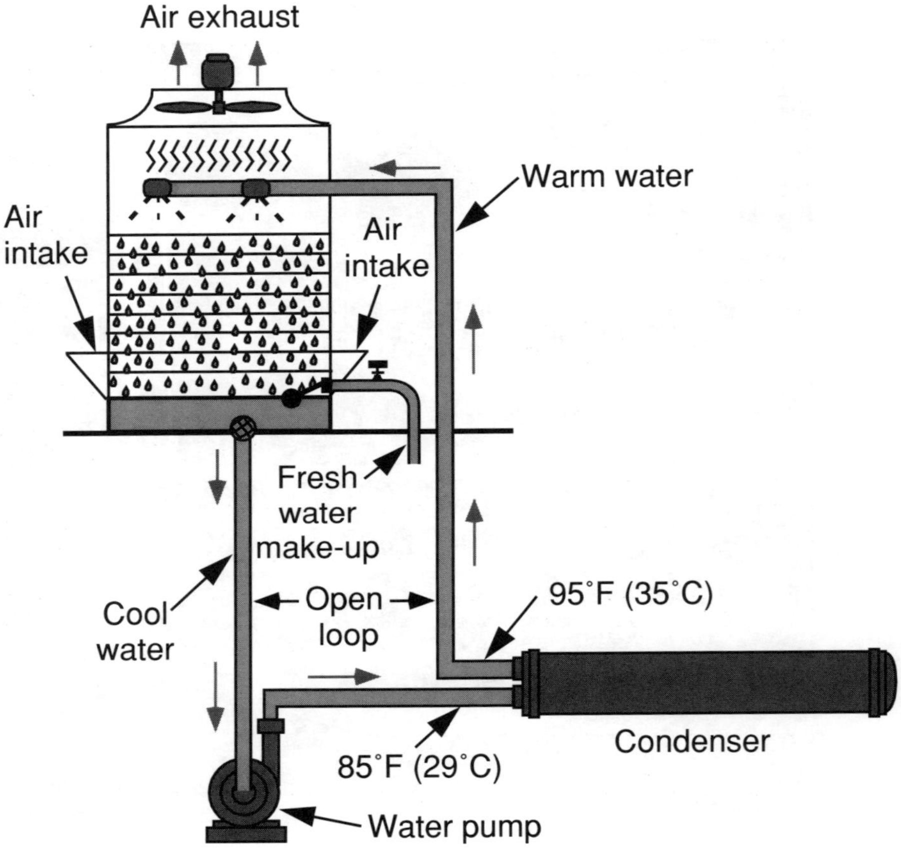

HC 21-2

Closed Loop

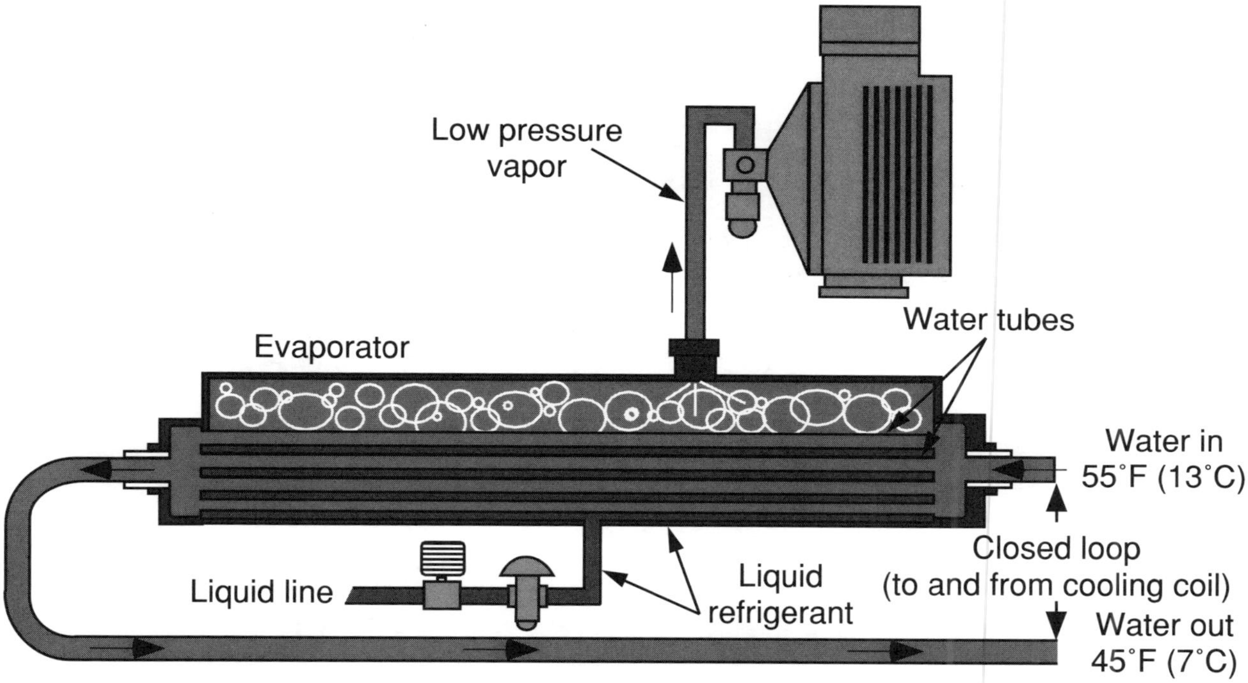

HC 21-3

Absorption-type Refrigerator

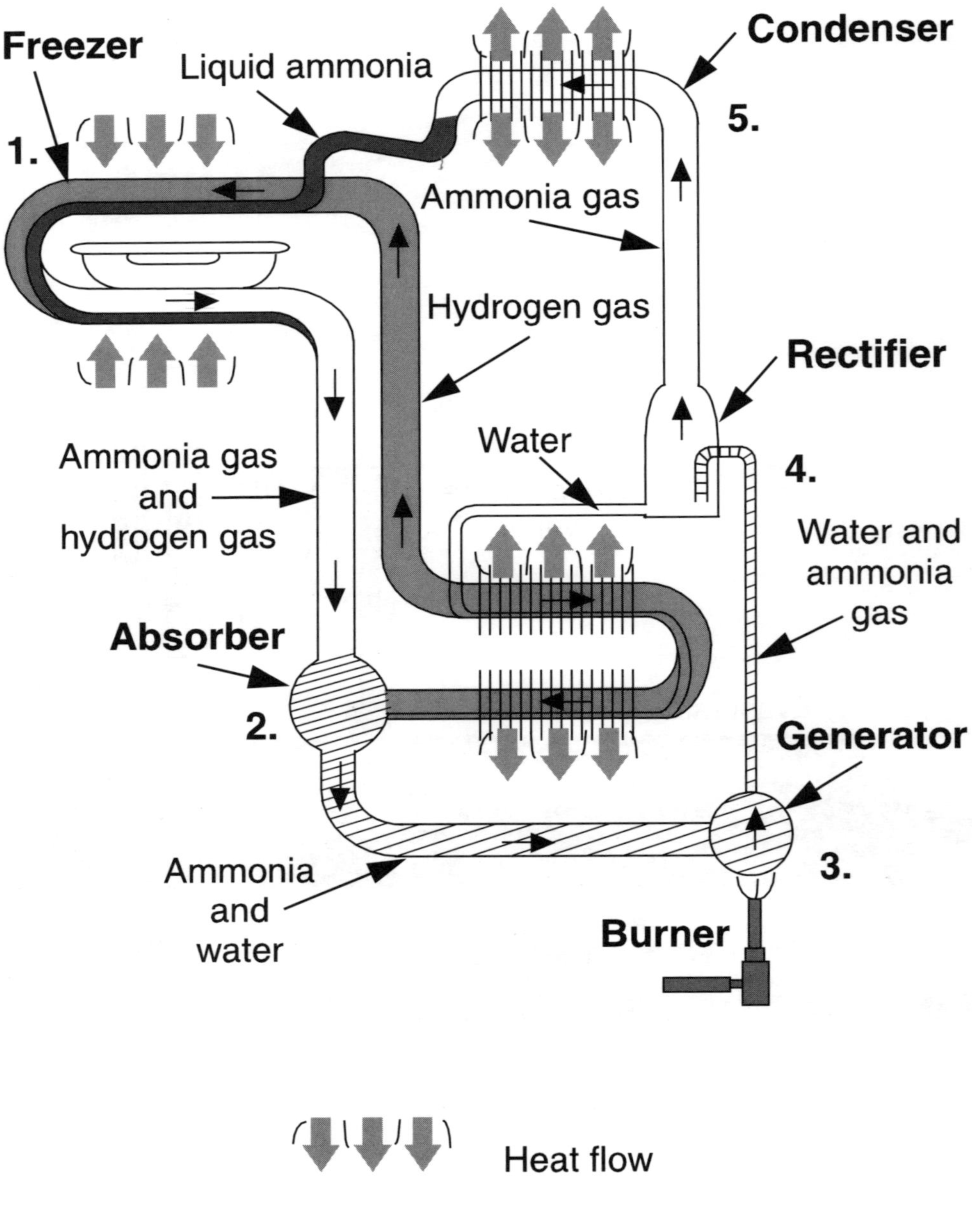

HC 21-4

Name ____________________

Date ____________________

Instructor ____________________

Period ____________________

21

Water Chillers

Quiz

Instructions: Write your answers on the lines provided at the right-hand side of the page.

1. What type of chiller does not use a compressor? 1. ____________

2. A(n) (open, closed) loop of insulated steel pipe carries water to and from the chiller condenser. 2. ____________

3. What loop is used for air conditioning? 3. ____________

4. What does a "water box" refer to? 4. ____________

5. What type of liquids are used in the closed loop? 5. ____________
   a. Water.
   b. Brine.
   c. Glycol.
   d. All of the above.

6. What type of chiller uses a volute? 6. ____________

7. *True or false?* Low-pressure chillers operate in a vacuum. 7. ____________

8. What type of chiller uses water-lithium bromide for cooling? 8. ____________

9. Decomposition of A1 refrigerants causes: 9. ____________
   a. hydrochloric acid.
   b. hydrofluoric acid.
   c. phosgene gas.
   d. all of the above.

10. What device removes air and moisture from a low-pressure chiller? 10. ____________

22

What is Electricity?

Objectives

After studying this chapter in the text, the student will be able to:

- ❑ Define electricity.
- ❑ Demonstrate proper use of electrical test meters.
- ❑ Describe the application of Ohm's Law.
- ❑ Identify open circuits and short circuits.
- ❑ Determine proper wire size for various uses.
- ❑ Select and use proper insulation on conductors.
- ❑ Describe how to connect loads and switches in electrical circuits.
- ❑ Describe the functions of various types of plugs, receptacles, and switches.
- ❑ Demonstrate how to safely connect portable electric power tools.

Instructional Materials

Text: Pages 351–378
Lab Manual: Pages 171–180
Instructor's Manual:
Visual Masters, Pages 197–201
Chapter 22 Quiz, Page 202

Teaching Strategy

Text

1. Review objectives.
2. Discuss and assign chapter contents. Use these visual masters as appropriate:
   HC 22-1: Ohm's Law
   HC 22-2: Ohmmeter Readings
   HC 22-3: Switches in Circuits
   HC 22-4: Connections to Ground
   HC 22-5: Switch Types
3. Direct students to complete Test Your Knowledge questions, page 378. After completion, discuss correct answers with students.

Lab Manual

1. Provide overview of Activities. Where necessary, stress safe laboratory practices and the importance of wearing protective clothing.
2. Assign Activities to individuals or pairs of students, as appropriate.
   Activity 22-1: Using a Voltage Tester
   Activity 22-2: Using an Ammeter
   Activity 22-3: Using an Ohmmeter
   Activity 22-4: Selecting Conductors
   Activity 22-5: Wiring a Circuit
   Activity 22-6: Wiring a Duplex (Wall) Receptacle
   Activity 22-7: Checking Wiring Connections
3. Observe student performance on Activities, providing guidance or correction as needed. After reviewing the student's work, initial and date the activity in the place provided.

Evaluation

1. Administer the quiz from this Instructor's Manual.
2. Grade the quiz, and return it with appropriate comments for the student.

Answers To Test Your Knowledge, Page 378

1. electricity
2. voltmeter
3. ammeter
4. true
5. ohmmeter
6. open; short
7. infinity
8. unlimited
9. watts
10. silver; copper; aluminum
11. size; length; location; and type of wire; the quantity of insulation

12. #12
13. white or gray
14. an energy conversion device
15. two
16. in series
17. true
18. fuses; circuit breakers
19. no
20. false

Answers to Lab Manual, Pages 171–177

Activity 22-1

1.– 4. Student response (answers will vary).
5. The neutral wire is broken somewhere between the lightbulb and the source.
6. EMF; voltage; potential difference; electrical pressure

Activity 22-2

1.– 4. Student response (answers will vary).
5. A. 33
   B. 15
   C. 2
   D. 1
   E. 15
   F. 15
   G. 18

Activity 22-3

2. A. zero
   B. infinity
   C. other
   D. other
   E. other

Activity 22-4

1. A. 1, THWN
   B. 10, THWN
   C. 8, THWN
   D. 6, THWN
2. A. 4
   B. 4
   C. 8
   D. 12
   E. 12
   F. 8

Activity 22-5

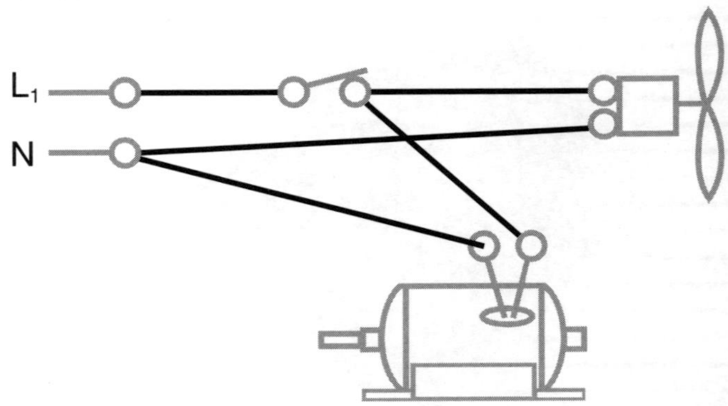

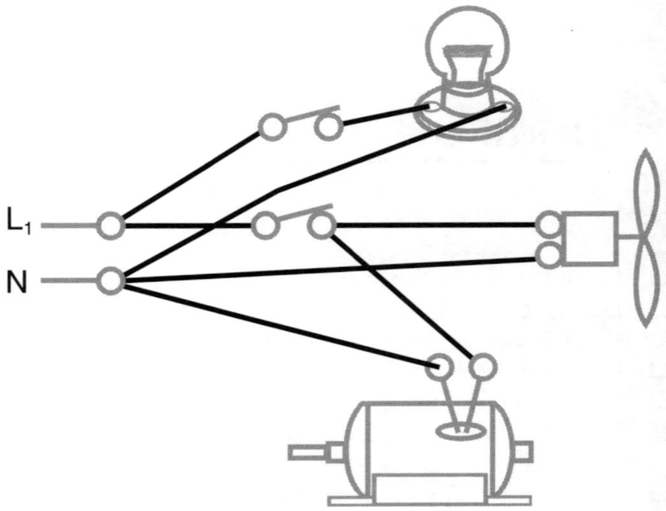

Answers to Quiz

1. electrons
2. voltage; electromotive force; potential; electrical pressure
3. ammeter
4. resistance
5. d. all of the above.
6. #12
7. parallel; series
8. voltage
9. green or bare (uninsulated) copper
10. fuses; circuit breakers

Ohm's Law

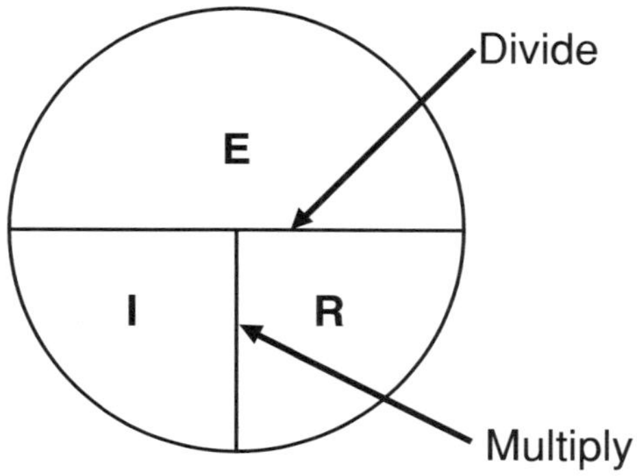

E = Voltage
I = Amperage (current)
R = Resistance (Ω)

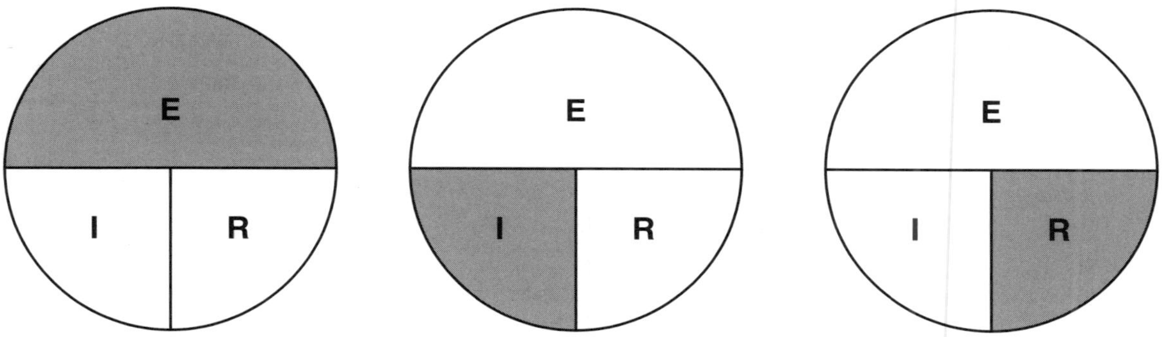

E = I x R I = E ÷ R R = E ÷ I

HC 22-1

Ohmmeter Readings

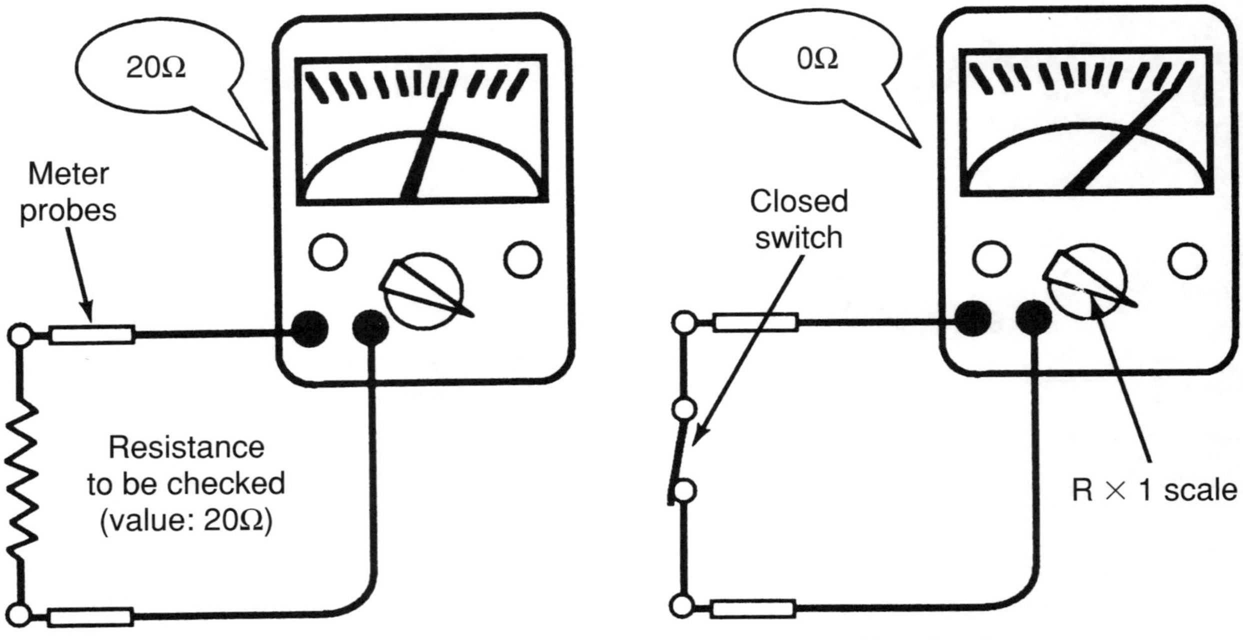

Resistance

Continuity

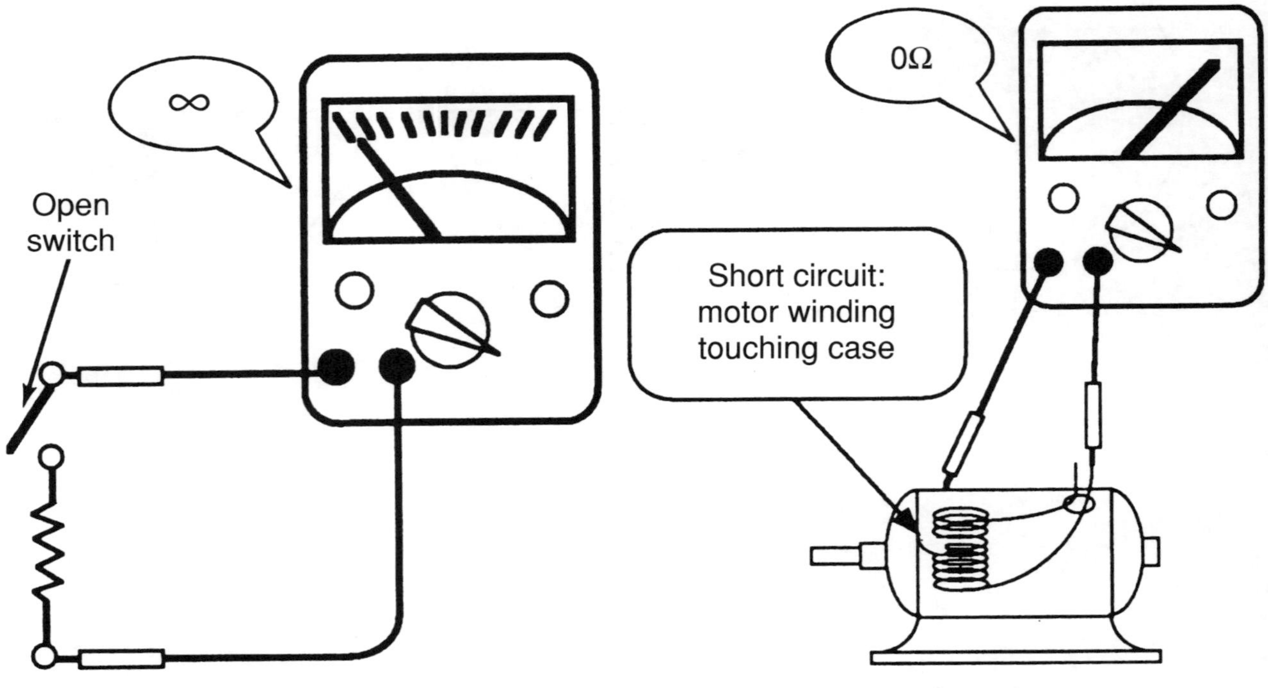

Open Circuit

Short Circuit

HC 22-2

Switches in Circuits

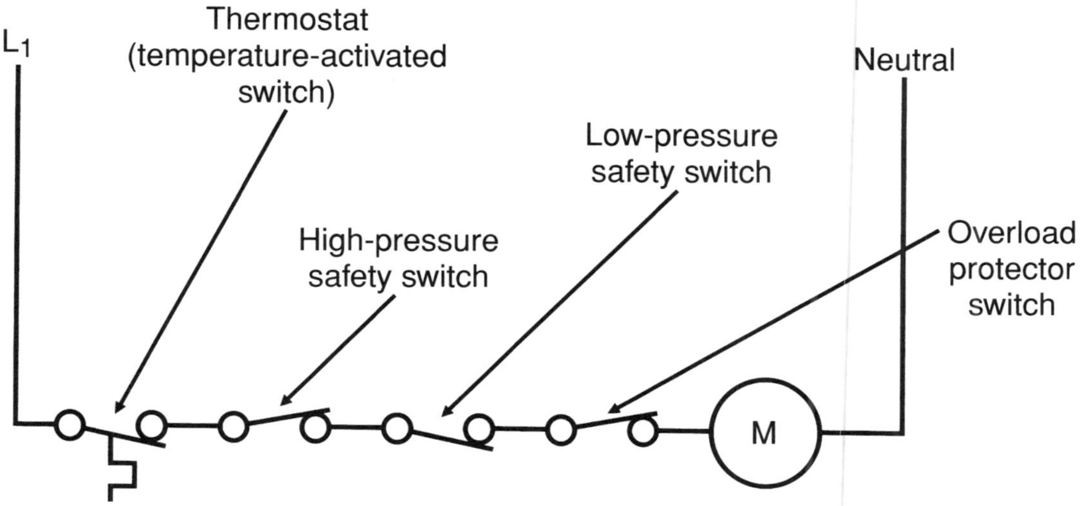

Series Circuit

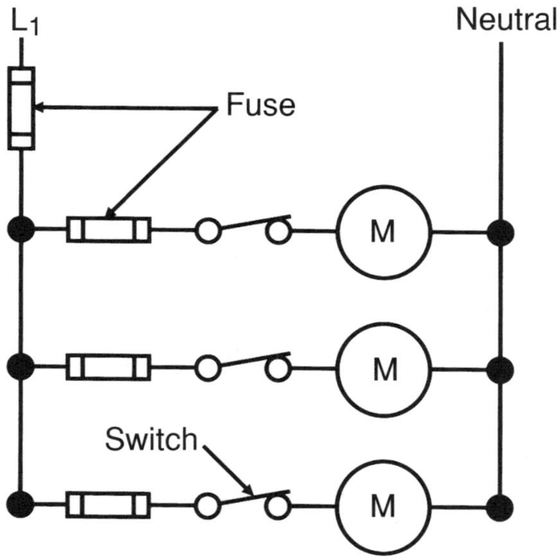

Parallel Circuit

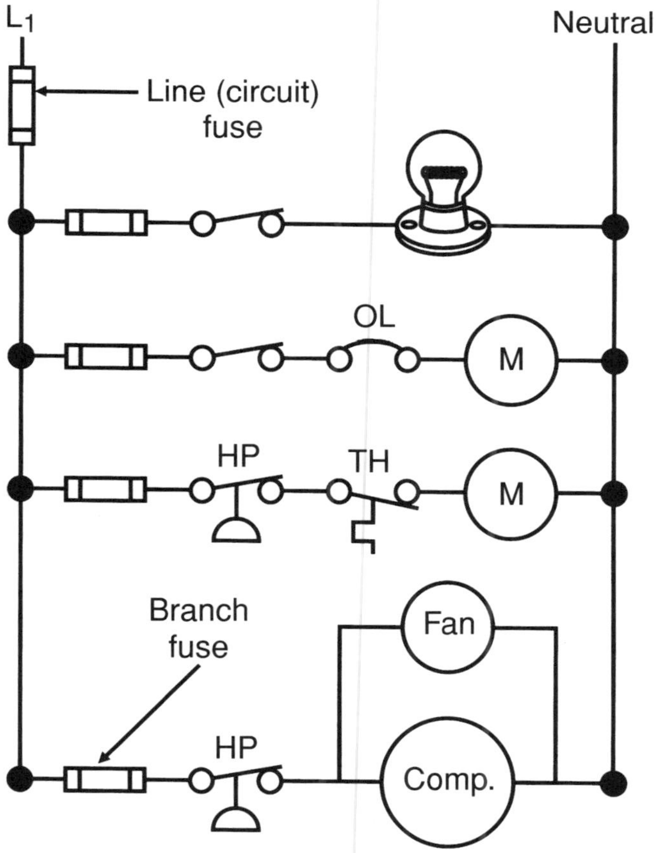

Series-Parallel Circuit

HC 22-3

Connections to Ground

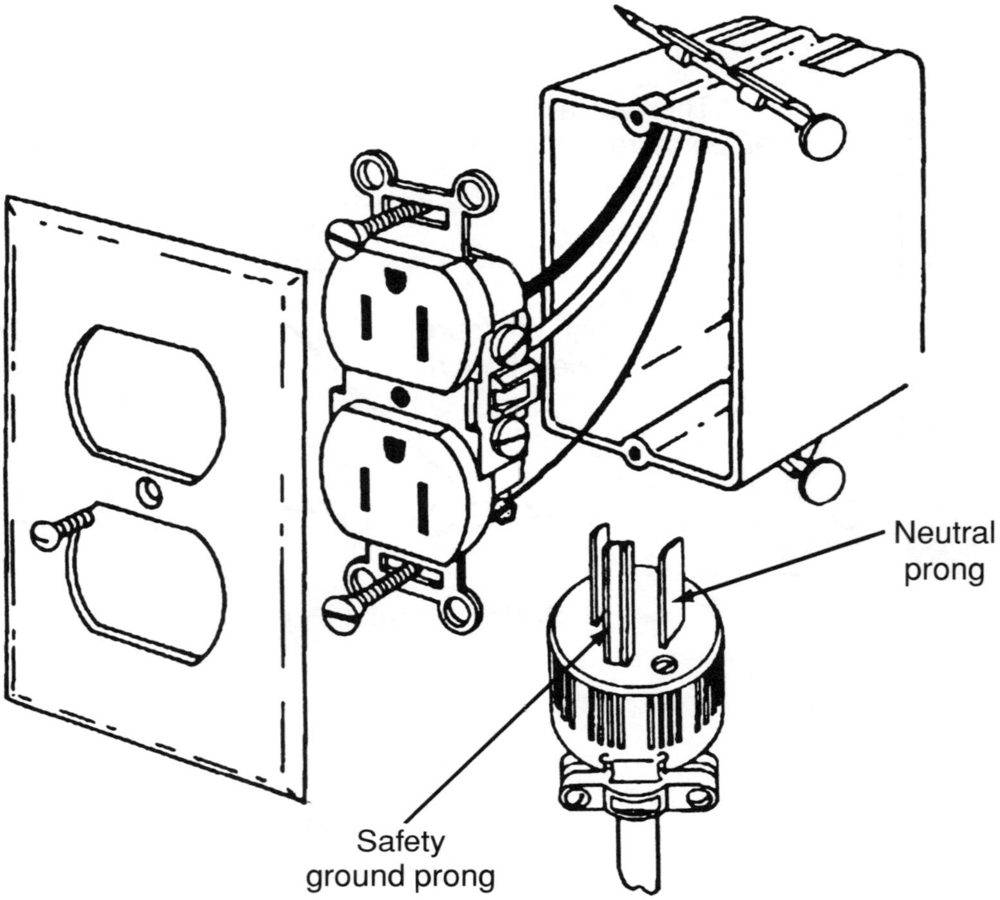

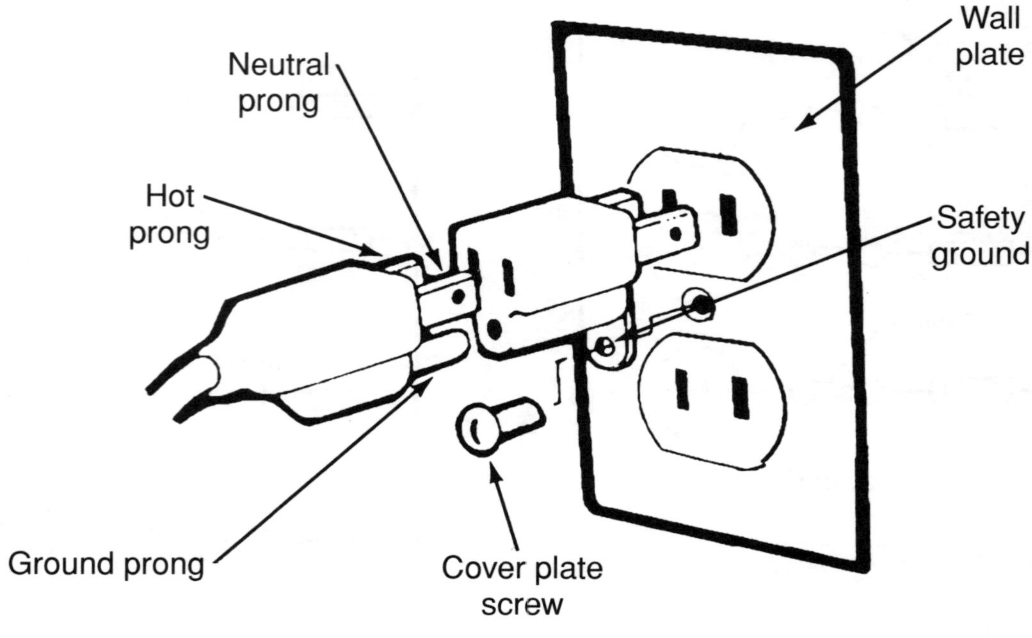

HC 22-4

Switch Types

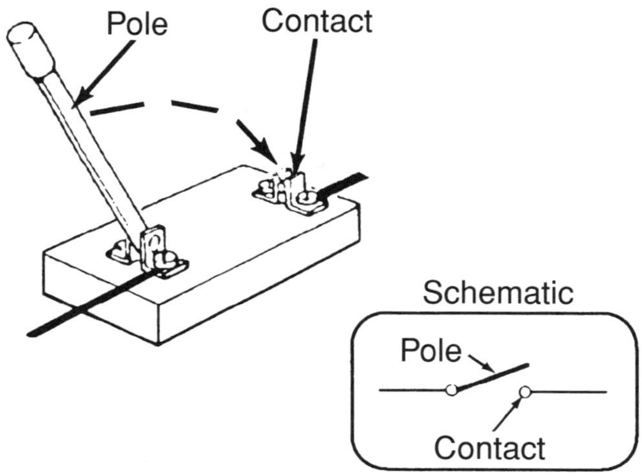

SPST

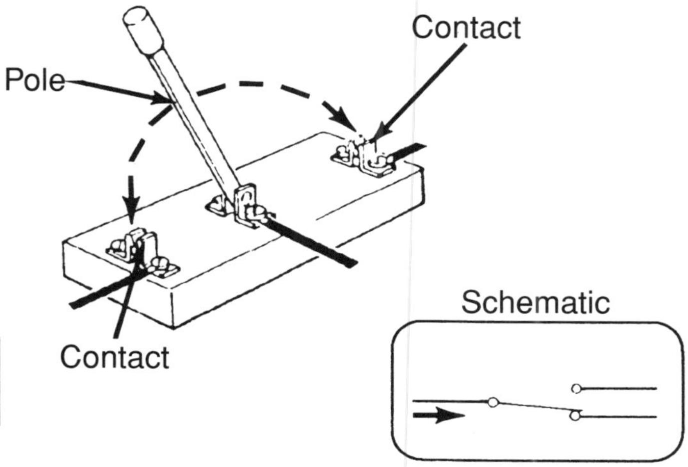

SPDT

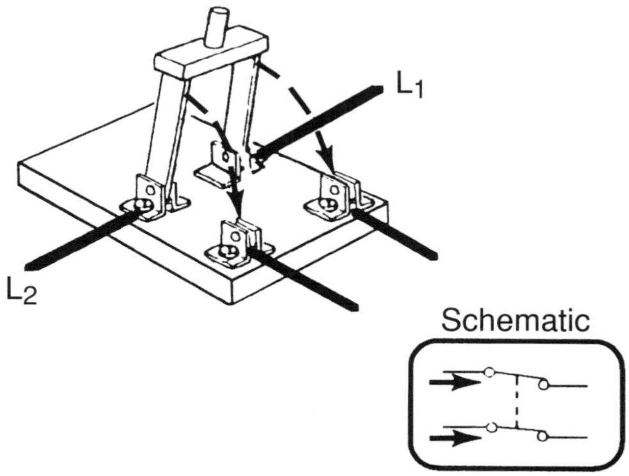

DPST

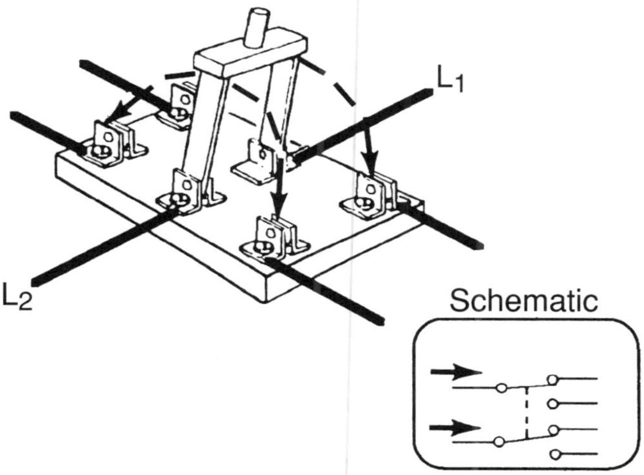

DPDT

Name ______________________________

Date ______________________________

Instructor ______________________________

Period______________________________

22 What is Electricity?

Quiz

Instructions: Write your answers on the lines provided at the right-hand side of the page.

1. Electricity involves the movement of _____ from one atom to another. 1. ____________________
2. List four terms used to describe the force that causes electrons to move. 2. ____________________ ____________________ ____________________ ____________________
3. What type of meter would you use to measure the movement of electrons? 3. ____________________
4. Various types of _____ are used to convert electrical energy to another form of energy. 4. ____________________
5. An ohmmeter is used to check: 5. ____________________
   a. resistance
   b. continuity
   c. short circuits
   d. all of the above
6. What size copper wire is used more than any other? 6. ____________________
7. Most loads are designed to be connected in _____; switches are designed to be connected in _____ with a load. 7. ____________________ ____________________
8. The neutral wire is a current-carrying conductor but has no _____. 8. ____________________
9. The color code for the safety ground wire is _____. 9. ____________________
10. Name two devices used to protect wires against excess current flow. 10. ____________________ ____________________

23

Power Transmission and Circuits

Objectives

After studying this chapter in the text, the student will be able to:

- ❑ Select and connect transformers.
- ❑ Identify common voltages and their uses.
- ❑ Size and use circuit protectors properly.
- ❑ Make good terminal connections.

Instructional Materials

Text: Pages 379–392
Lab Manual: Pages 181–186
Instructor's Manual:
Visual Masters, Pages 205–208
Chapter 23 Quiz, Page 209

Teaching Strategy

Text

1. Review objectives.
2. Discuss and assign chapter contents. Use these visual masters as appropriate:
   HC 23-1: Three-phase Wiring System
   HC 23-2: Residential Service
   HC 23-3: Transformers
   HC 23-4: GFCI
3. Direct students to complete Test Your Knowledge questions, page 391. After completion, discuss correct answers with students.

Lab Manual

1. Provide overview of Activities. Where necessary, stress safe laboratory practices and the importance of wearing protective clothing.
2. Assign Activities to individuals or pairs of students, as appropriate.
   Activity 23-1: Connecting a Transformer in a Circuit
   Activity 23-2: Fuses and Circuit Breakers
   Activity 23-3: Sizing Fuses in Circuits
   Activity 23-4: Working with Terminal Connections
3. Observe student performance on Activities, providing guidance or correction as needed. After reviewing the student's work, initial and date the activity in the place provided.

Evaluation

1. Administer the quiz from this Instructor's Manual.
2. Grade the quiz, and return it with appropriate comments for the student.

Answers to Test Your Knowledge, Page 391

1. true
2. 240 V; one (single-phase)
3. generated; higher; point of use
4. magnetic induction
5. increase
6. current
7. fuses; circuit breakers
8. true
9. amperage; voltage
10. ground

Answers to Lab Manual, Pages 181–184

Activity 23-1

6. A. 240
   B. 120
   C. 120
   D. 0
   E. 120
   F. 0

Activity 23-2

1. A. ordinary (one-time blow)
   B. time-delay (dual element)
   C. Type S
2. A. ordinary (one-time blow)
   B. replaceable element
   C. time-delay (dual element)
3. Type S
4. b. wires
5. GFCI
6. 10,000; 20,000
7. Turn it to "off," then to "on" again.
8. d. any of the above

Activity 23-3

1. Volts: 600
   Amps: 70
   Fuse type: time-delay
2. A. Volts: 250
   Amps: 50
   Fuse type: one-time
   B. Volts: 250
   Amps: 30
   Fuse type: one-time
   C. Volts: 250
   Amps: 5
   Fuse type: one-time
   D. Volts: 250
   Amps: 30
   Fuse type: one-time
3. A. Volts: 250
   Amps: 150
   Fuse type: time-delay
   B. Volts: 250
   Amps: 30
   Fuse type: time-delay
   C. Volts: 250
   Amps: 50
   Fuse type: time-delay
   D. Volts: 250
   Amps: 60
   Fuse type: time-delay

Answers to Quiz

1. between any two hot wires
2. A. primary coil
   B. core (laminated soft iron plates)
   C. secondary coil
3. secondary
4. true
5. overload; short circuit (ground fault)
6. time-delay (dual-element)
7. voltage; amperage
8. a. GFCI.
9. amperage; voltage; AIC
10. d. All of the above.

Three-phase Wiring System

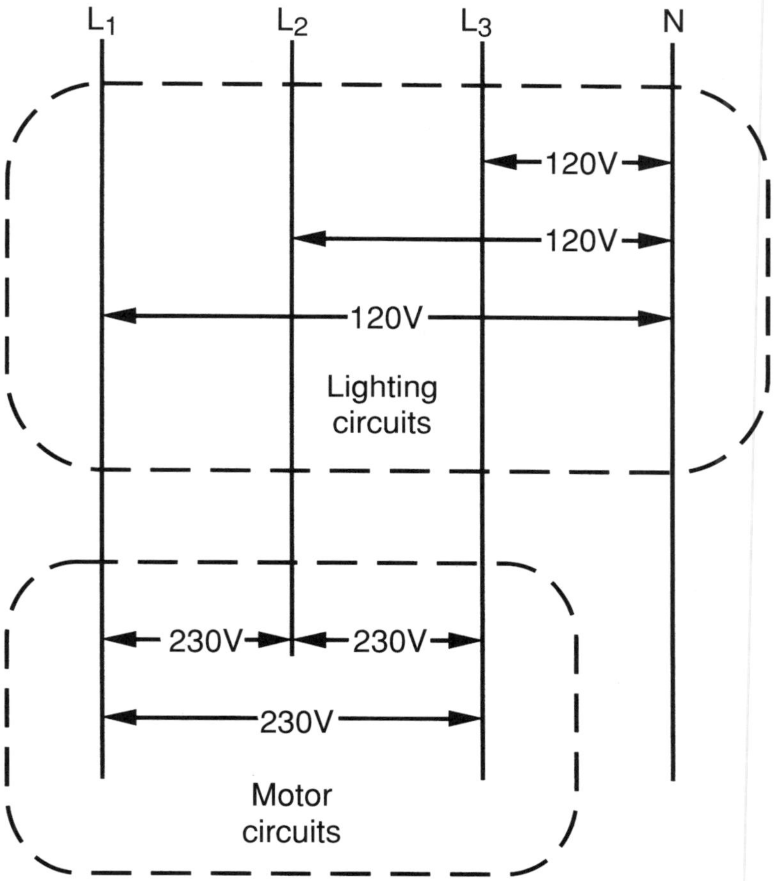

HC 23-1

Residential Service

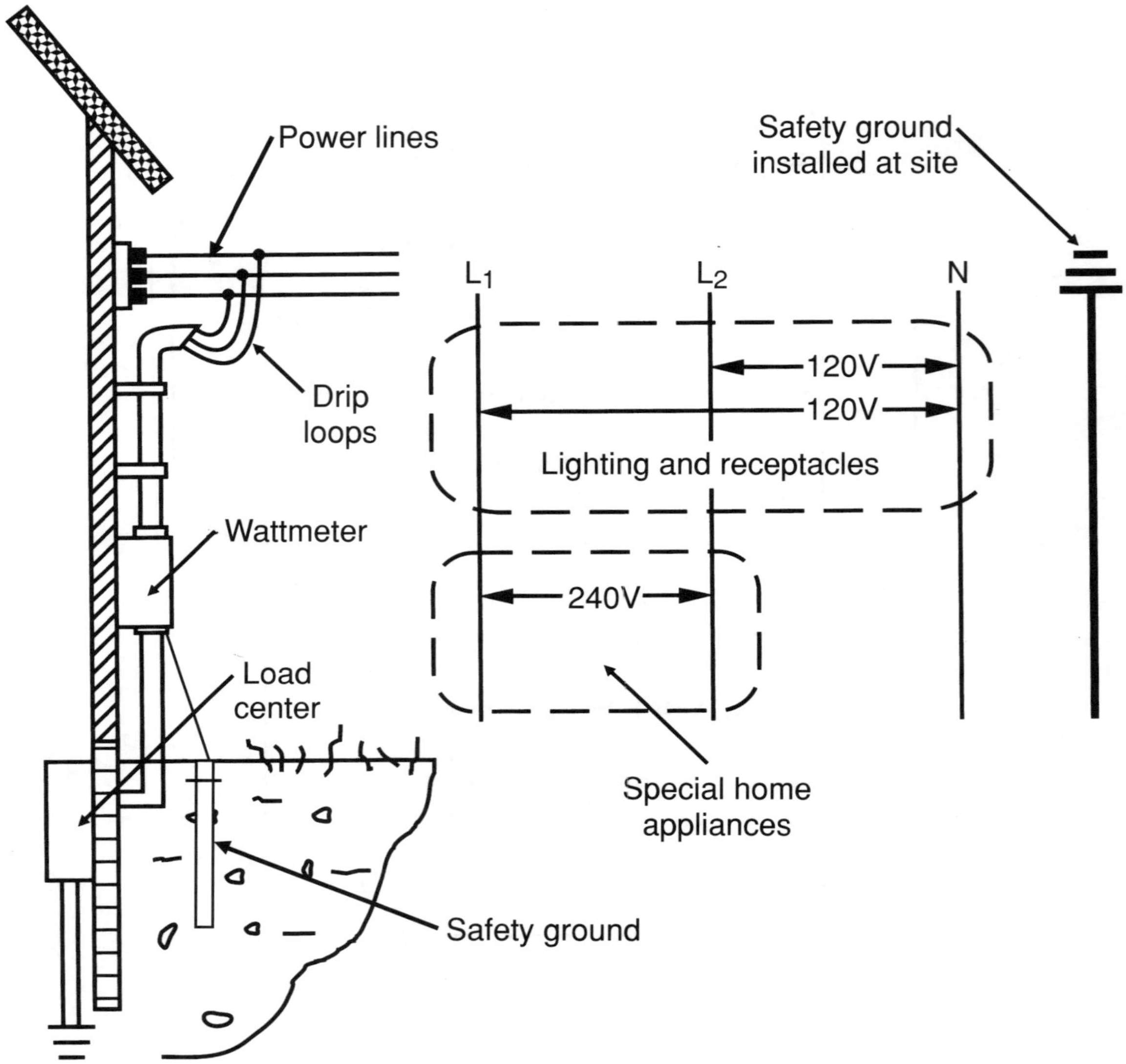

HC 23-2

Transformers

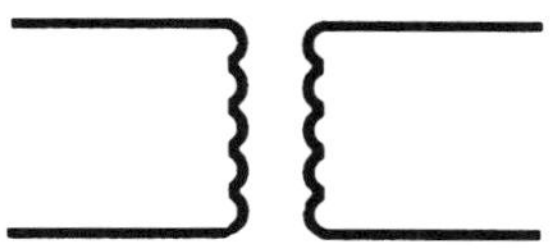

Single-
phase

3-phase
delta

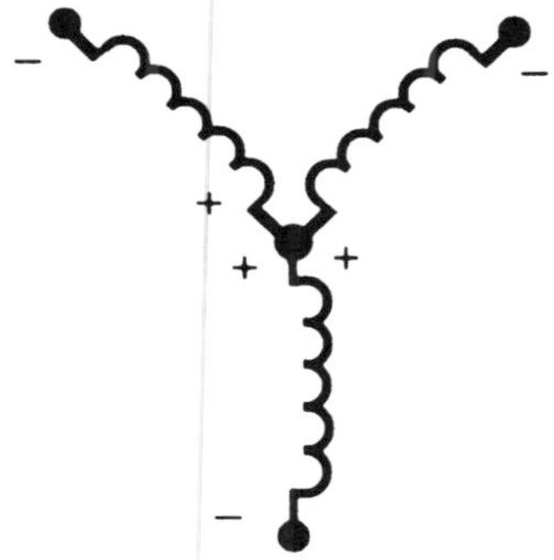

3-phase
wye

Symbols

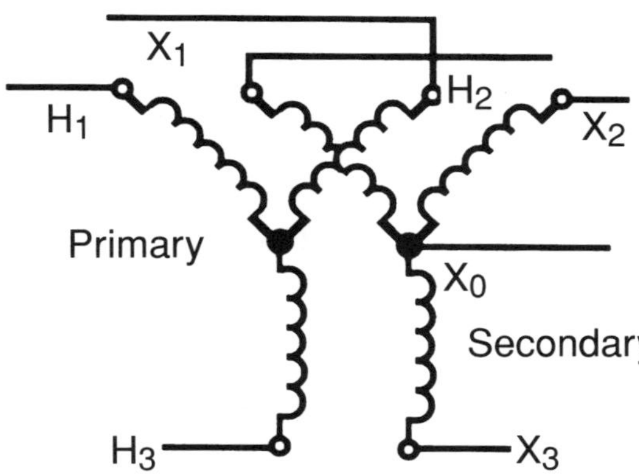

Wye

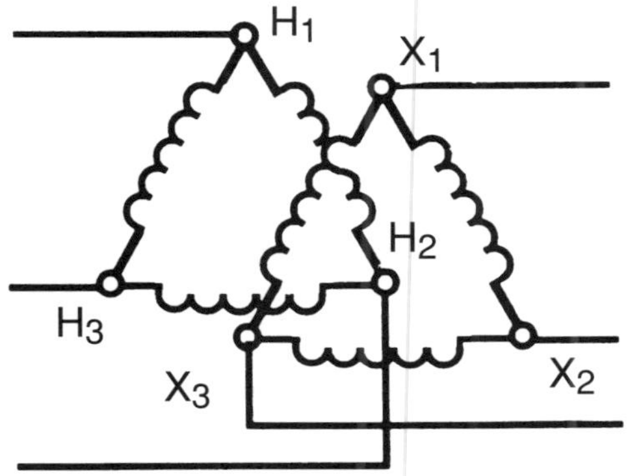

Delta

Terminals

HC 23-3

GFCI

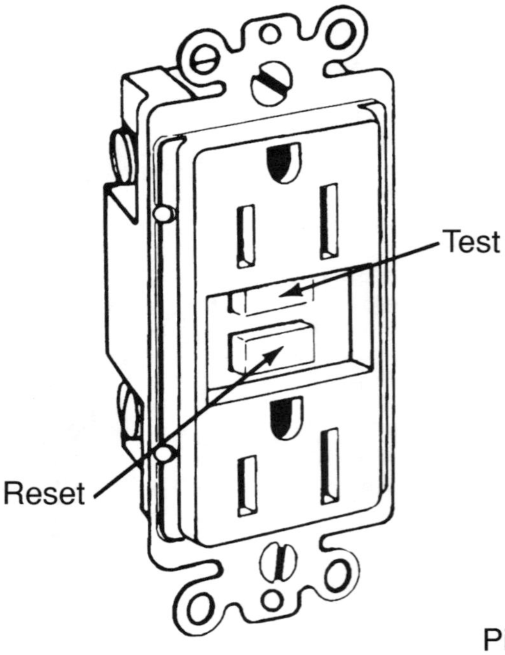

Receptacle

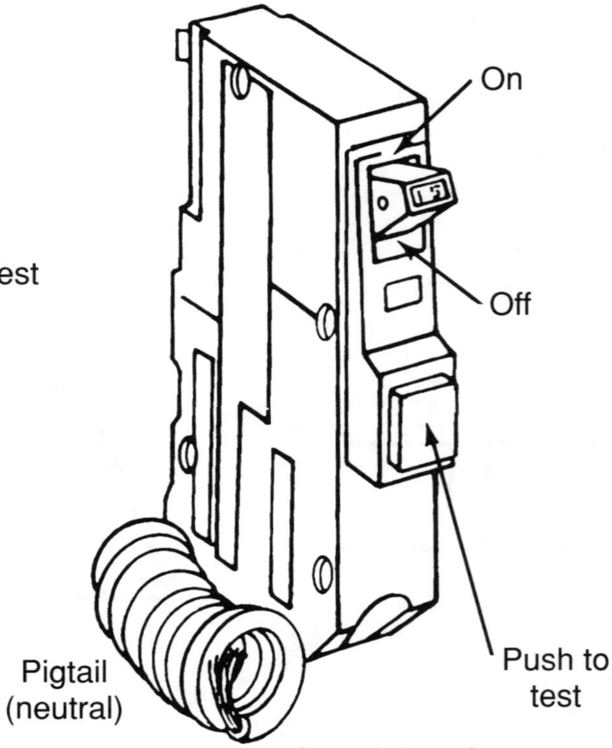

Circuit breaker

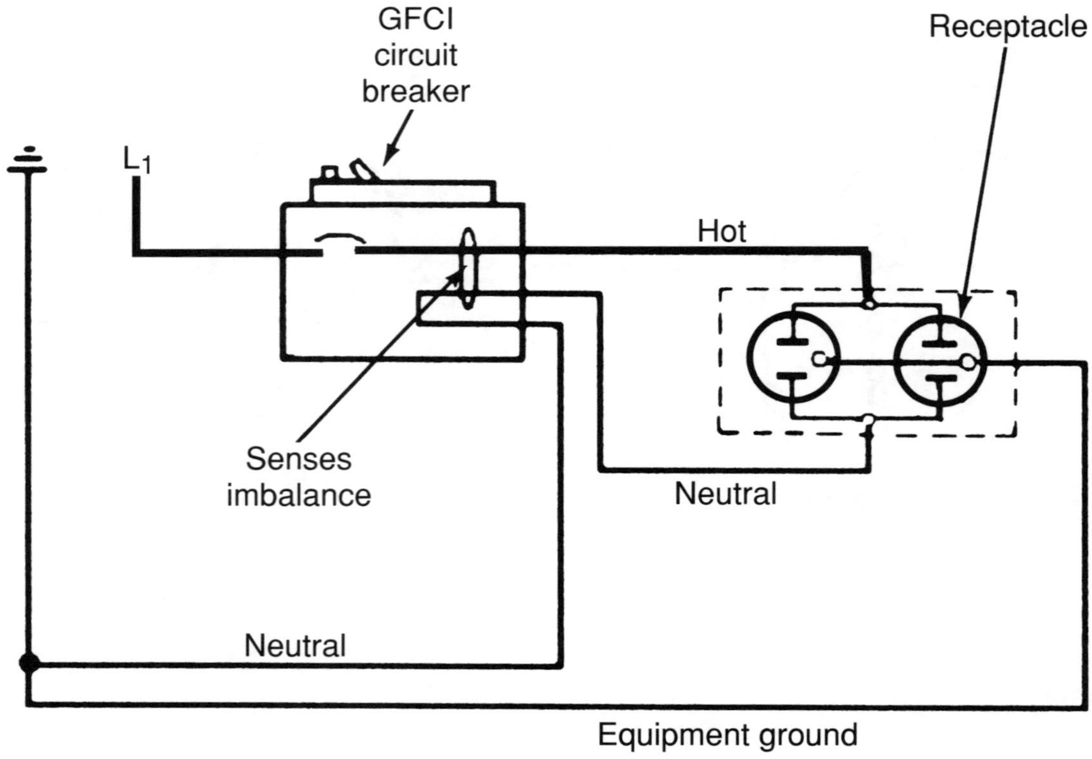

HC 23-4

Name ______________________________

Date ______________________________

Instructor ______________________________

Period ______________________________

23

Power Transmission and Circuits

Quiz

Instructions: Write your answers on the lines provided at the right-hand side of the page.

1. How do you measure voltage on a three-phase system? 1. ______________________

2. Identify the parts of this electrical transformer. 2. a. ______________________ b. ______________________ c. ______________________

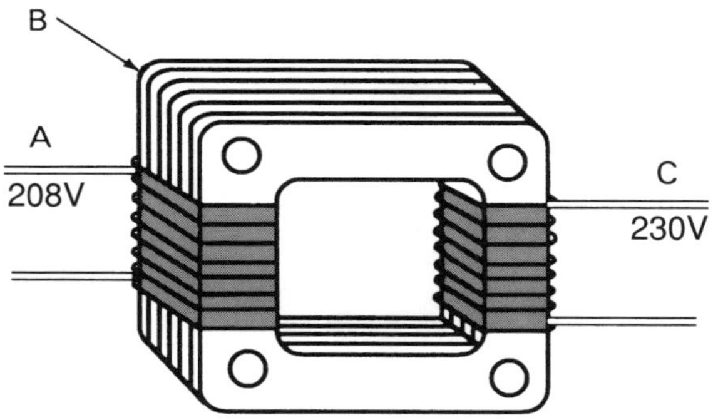

3. Is the capacity of a transformer rated at the primary or the secondary? 3. ______________________

4. *True or false?* Three-phase transformers can be of either the delta type or wye type. 4. ______________________

5. Name two circuit problems that can cause overcurrent. 5. ______________________ ______________________

6. What type of fuse is recommended to protect a circuit containing an electric motor? 6. ______________________

7. The physical size of a cartridge fuse is determined by its _____ and _____ ratings. 7. ______________________ ______________________

Continued

8. What device protects people against a ground fault problem? 8. ________________
   a. GFCI.
   b. Dual-element fuse.
   c. Type S fuse.

9. What are the three important ratings on a fuse or circuit breaker? 9. ________________
____
____

10. Which of the following can result from loose terminal connections? 10. ________________
   a. Arcing.
   b. Overheating.
   c. Voltage drop.
   d. All of the above.

24

Induction Motors

Objectives

After studying this chapter in the text, the student will be able to:

- ❑ Identify five types of single-phase induction motors.
- ❑ Describe the operation and application of induction motors.
- ❑ Demonstrate how to make proper motor connections.
- ❑ Use an ohmmeter to identify hermetic motor-compressor terminals.
- ❑ Troubleshoot induction motor problems.
- ❑ Select and connect start components.
- ❑ Identify examples of multi-speed and dual-voltage motors.
- ❑ Select and install proper fan replacements.

Instructional Materials

Text: Pages 393–420
Lab Manual: Pages 187–196
Instructor's Manual:
Visual Masters, Pages 213–217
Chapter 24 Quiz, Page 218

Teaching Strategy

Text

1. Review objectives.
2. Discuss and assign chapter contents. Use these visual masters as appropriate:
   HC 24-1: Alternating Current
   HC 24-2: Overload Protector
   HC 24-3: Motors
   HC 24-4: Identifying Compressor Terminals
   HC 24-5: Checking Motor Windings
3. Direct students to complete Test Your Knowledge questions, page 420. After completion, discuss correct answers with students.

Lab Manual

1. Provide overview of Activities. Where necessary, stress safe laboratory practices and the importance of wearing protective clothing.
2. Assign Activities to individuals or pairs of students, as appropriate.
   Activity 24-1: Examining a Shaded Pole Motor
   Activity 24-2: Examining an Open-type Split-phase Motor
   Activity 24-3: Examining a Permanent Split Capacitor (PSC) Motor
   Activity 24-4: Examining a Capacitor-start, Induction-run (CSIR) Motor
   Activity 24-5: Testing Capacitors
   Activity 24-6: Connecting Dual-voltage Three-phase Motors
   Activity 24-7: Identifying Hermetic Motor Problems
3. Observe student performance on Activities, providing guidance or correction as needed. After reviewing the student's work, initial and date the activity in the place provided.

Evaluation

1. Administer the quiz from this Instructor's Manual.
2. Grade the quiz, and return it with appropriate comments for the student.

Answers to Test Your Knowledge, Page 420

1. stator; rotor
2. stator
3. true

4. poles
5. start
6. true
7. 75
8. Because of arcing of the contacts.
9. locked rotor amps; full load amps
10. Rotor unable to turn (locked rotor); turning too slowly (overloaded).
11. d. All of the above.
12. shaded pole; split-phase; PSC; CSIR; CSCR
13. b. Run capacitor.
14. c. Start capacitor.
15. in series
16. Replace with a known good one. Use an ohmmeter. Use a capacitor tester.
17. true
18. By switching any two supply wires.
19. Common; Start; Run
20. ohmmeter

Answers to Lab Manual, Pages 187–195

Activity 24-1

Student response (answers will vary).

Activity 24-6

1.–7. Student response (answers will vary).

8. c. Operating costs are the same for either high or low-voltage connections. (The motor draws the same wattage on either connection. When the voltage is doubled the amperage is reduced by half.)

Activity 24-7

Student response (answers will vary).

Answers to Quiz

1. stator; rotor
2. start
3. centrifugal switch
4. six
5. in series
6. shaded pole; split phases; PSC; CSIR; CSCR
7. run; start
8. true
9. Common; Start; Run
10. d. all of the above.

Alternating Current

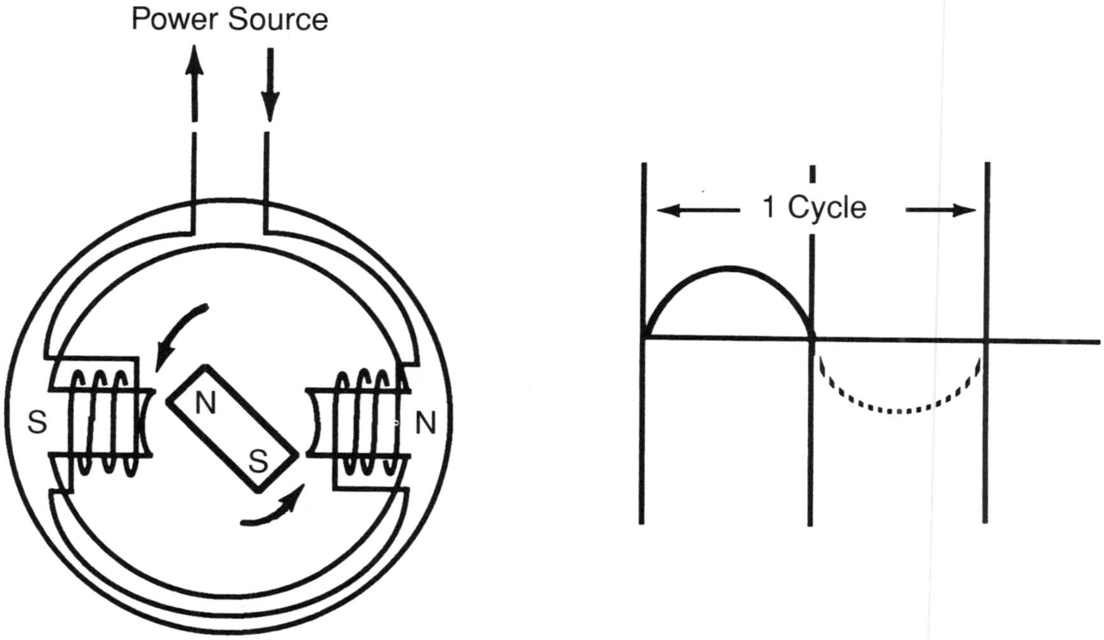

A

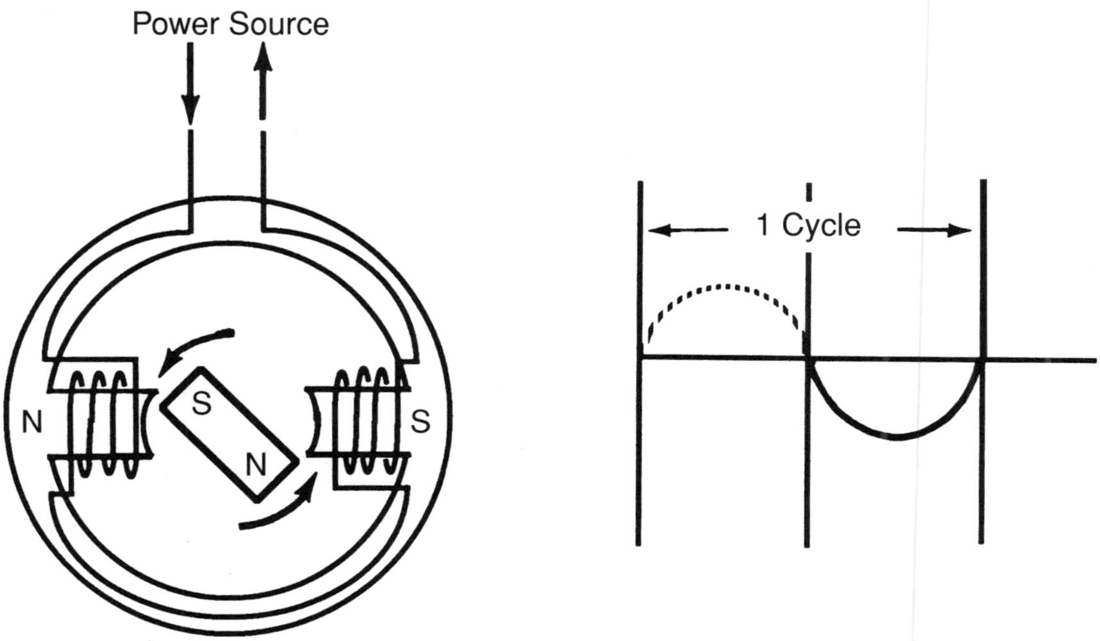

B

HC 24-1

Overload Protector

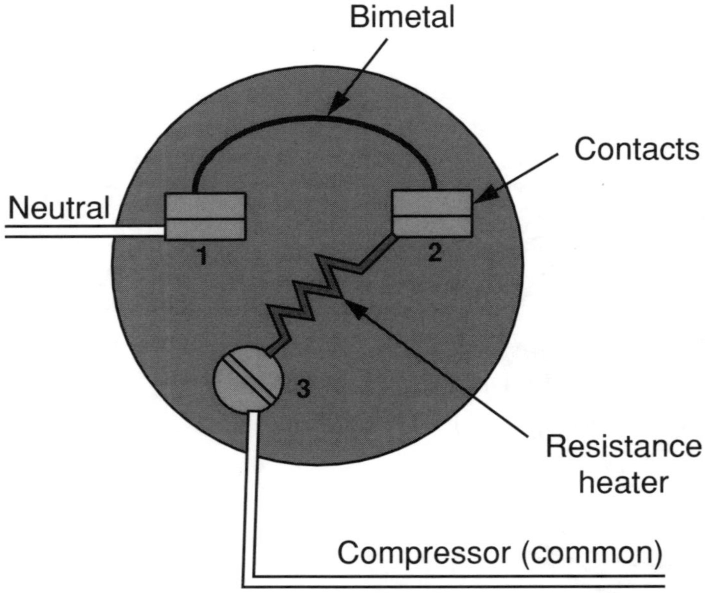

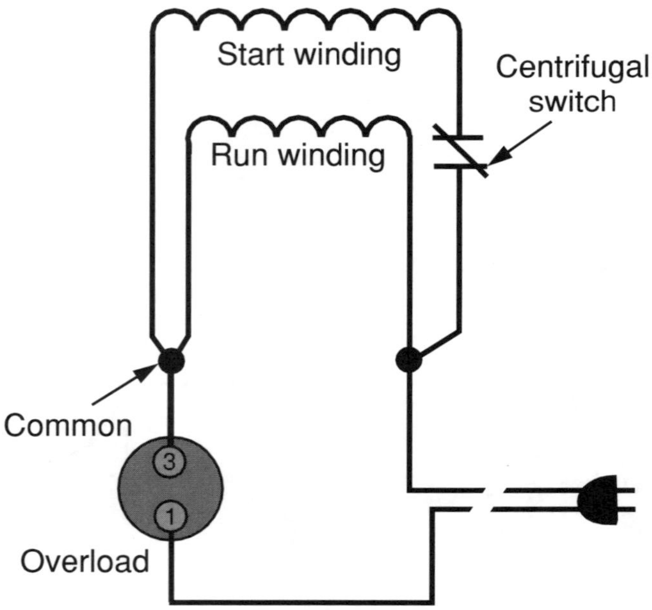

HC 24-2

Motors

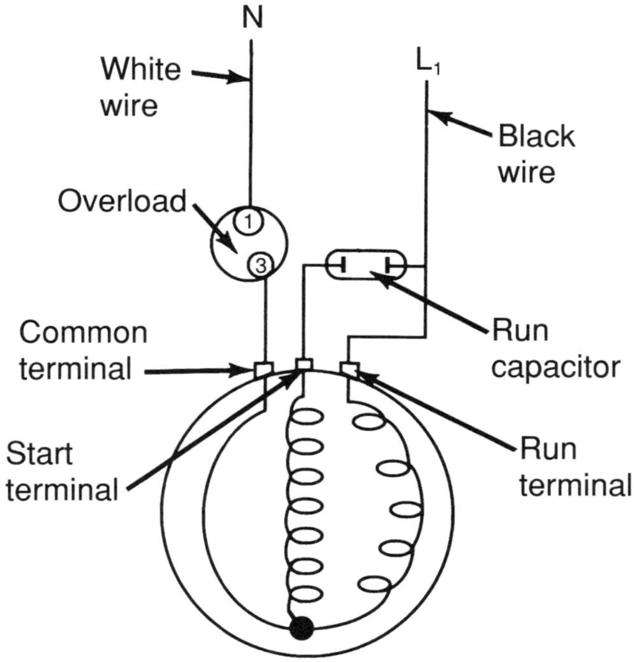

PSC

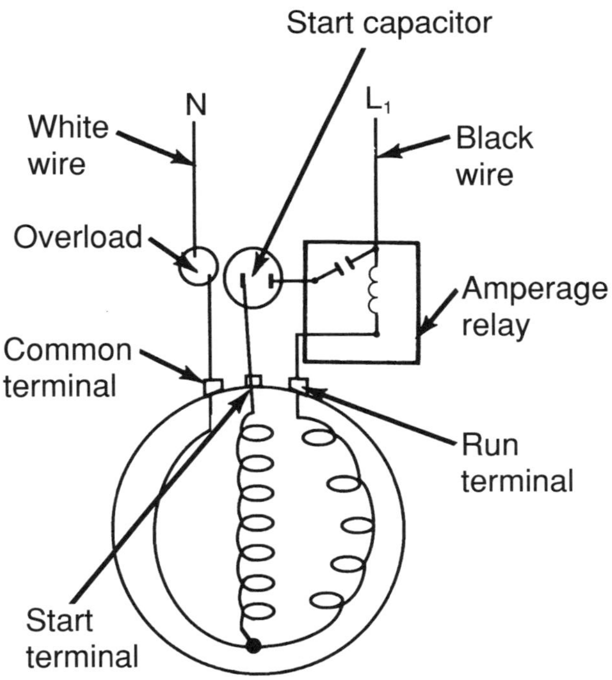

CSIR

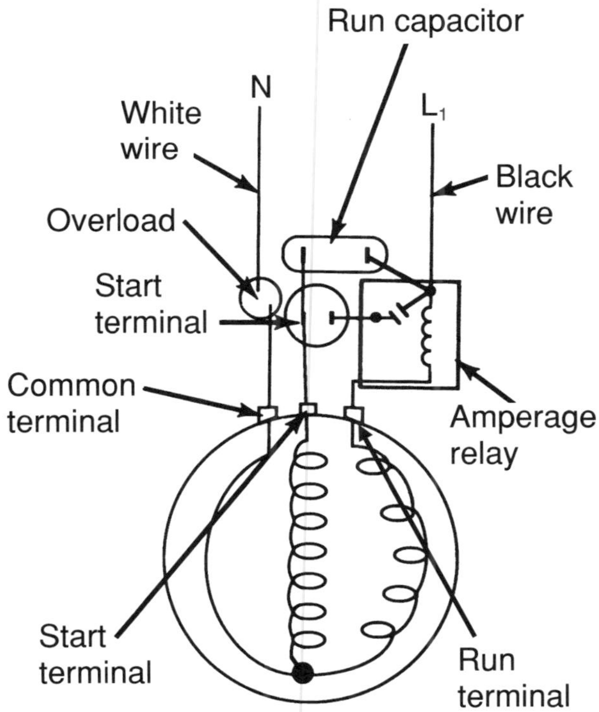

CSCR

HC 24-3

Identifying Compressor Terminals

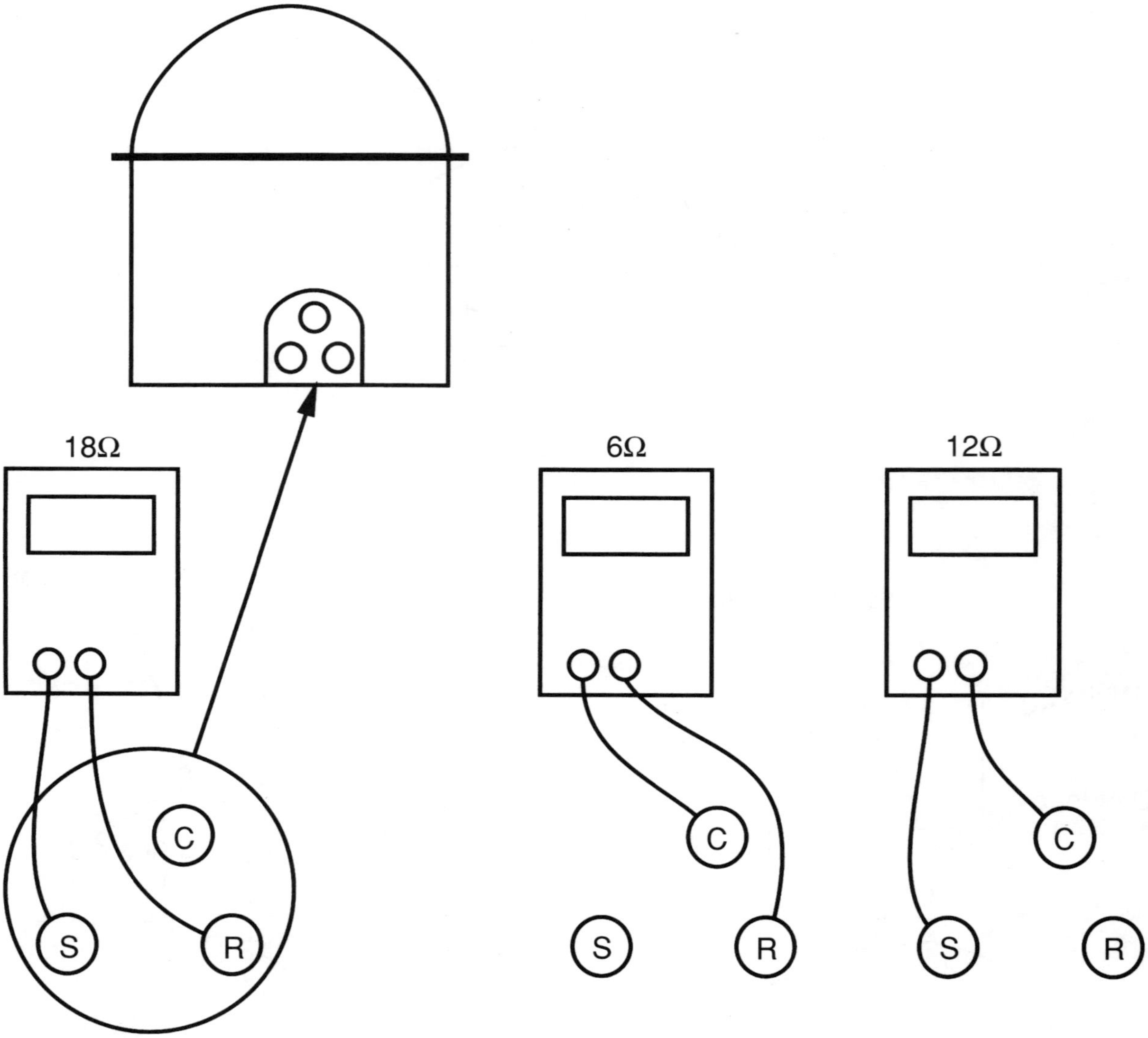

HC 24-4

Checking Motor Windings

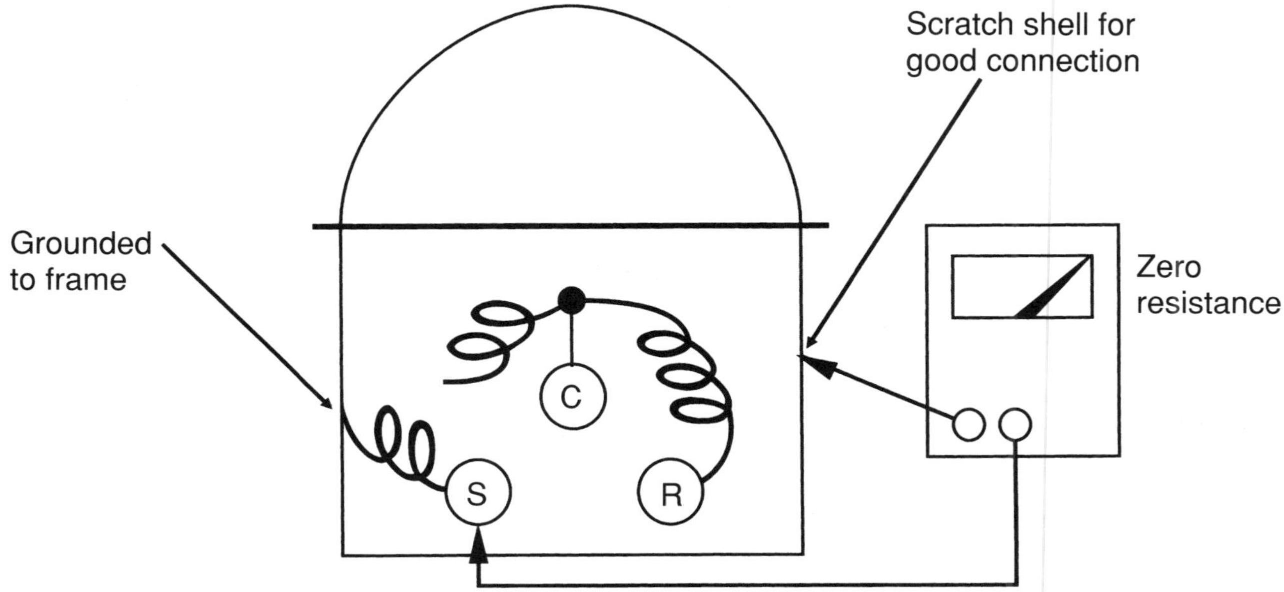

Grounded

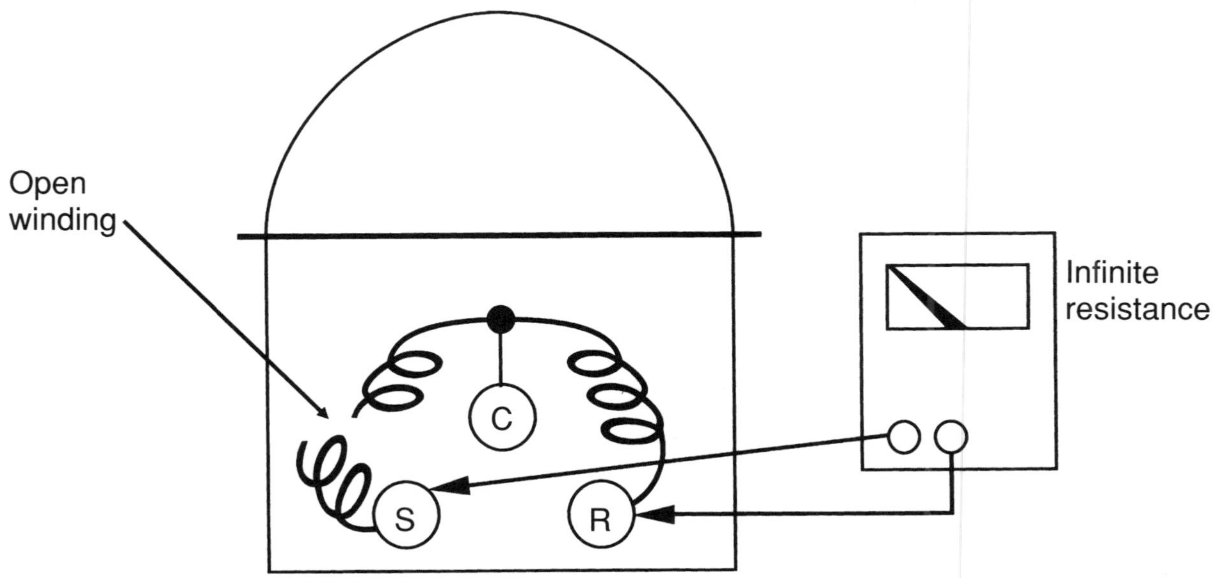

Open

HC 24-5

Name ______________________________

Date ______________________________

Instructor ______________________________

Period______________________________

24 Induction Motors

Quiz

Instructions: Write your answers on the lines provided at the right-hand side of the page.

1. What are the two main parts of an induction motor? 1. ______________

2. In a single-phase motor, which winding determines direction of rotation? 2. ______________

3. What device is used to disconnect the start winding on an open-type motor? 3. ______________

4. LRA is about _____ times higher than FLA. 4. ______________

5. Is the overload normally connected in series or in parallel with the common terminal? 5. ______________

6. List five types of single-phase induction motors. 6. ______________

7. The PSC motor uses a _____ capacitor connected in series with the _____ winding. 7. ______________

8. *True or false?* Dual voltage motors have two separate run windings. 8. ______________

9. Name the three electrical motor connections on a hermetic compressor. 9. ______________

10. Cycling on the overload can be caused by: 10. ______________
    a. excess amperage.
    b. excessive motor heat.
    c. excessive compressor heat.
    d. all of the above.

25

Electromagnetic Control Devices

Objectives

After studying this chapter in the text, the student will be able to:

- ❑ Use and troubleshoot solenoid valves.
- ❑ Recognize and troubleshoot relay circuits.
- ❑ Connect and troubleshoot contactors and line starters.
- ❑ Read and use both pictorial and schematic diagrams.

Instructional Materials

Text: Pages 421-446
Lab Manual: Pages 197–204
Instructor's Manual:
Visual Masters, Pages 221–226
Chapter 25 Quiz, Page 227

Teaching Strategy

Text

1. Review objectives.
2. Discuss and assign chapter contents. Use these visual masters as appropriate:
   HC 25-1: Solenoid Valve
   HC 25-2: Relay Operation
   HC 25-3: Condenser Fan Connections
   HC 25-4: Connecting Semihermetic Compressors
   HC 25-5: Residential System with Contactor
   HC 25-6: Complex Control Circuit
3. Direct students to complete Test Your Knowledge questions, page 446. After completion, discuss correct answers with students.

Lab Manual

1. Provide overview of Activities. Where necessary, stress safe laboratory practices and the importance of wearing protective clothing.
2. Assign Activities to individuals or pairs of students, as appropriate.
   Activity 25-1: Examining a Solenoid Valve
   Activity 25-2: Checking Amperage Relays
   Activity 25-3: Checking a Potential Relay
   Activity 25-4: Wiring a Potential Relay
   Activity 25-5: Reconnecting a Potential Relay
   Activity 25-6: Wiring a Contactor
   Activity 25-7: Checking a Line Starter
3. Observe student performance on Activities, providing guidance or correction as needed. After reviewing the student's work, initial and date the activity in the place provided.

Evaluation

1. Administer the quiz from this Instructor's Manual.
2. Grade the quiz, and return it with appropriate comments for the student.

Answers to Test Your Knowledge, Page 446

1. electromagnetism
2. small
3. large
4. defective
5. false
6. L_1; L_2; L_3
7. control
8. unlimited
9. The line starter has built-in overloads.
10. series

Answers to Lab Manual, Pages 197–204

Activities 25-1 to 25-3

Student response (answers will vary).

Activity 25-4

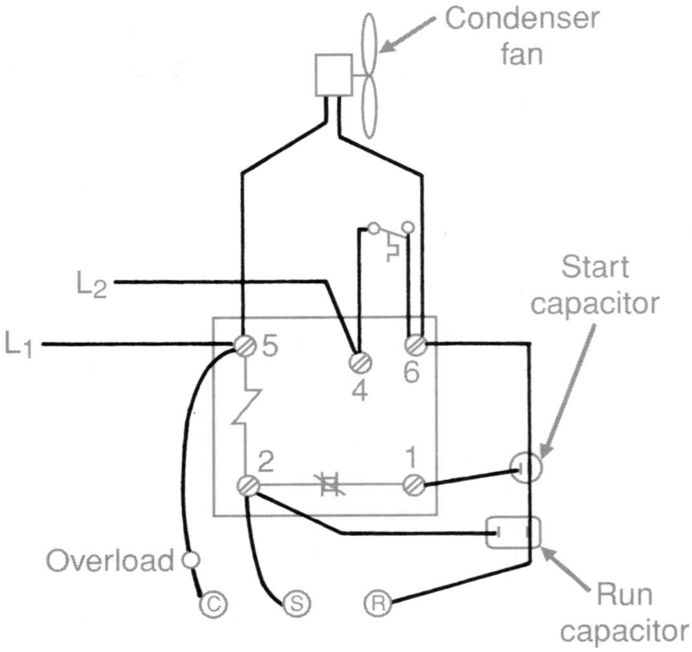

2. A. 1
   B. 2
   C. 2
   D. Common

Activity 25-5

Student response (answers will vary).

Activity 25-7

Student response (answers will vary).

Answers to Quiz

1. ferrous
2. d. All of the above.
3. false
4. amperage; potential
5. Reversing rotation. Using a high-MFD start capacitor. Using higher voltage.
6. true
7. two-pole
8. yes
9. an unlimited number
10. pictorial; schematic (or ladder)

Solenoid Valve

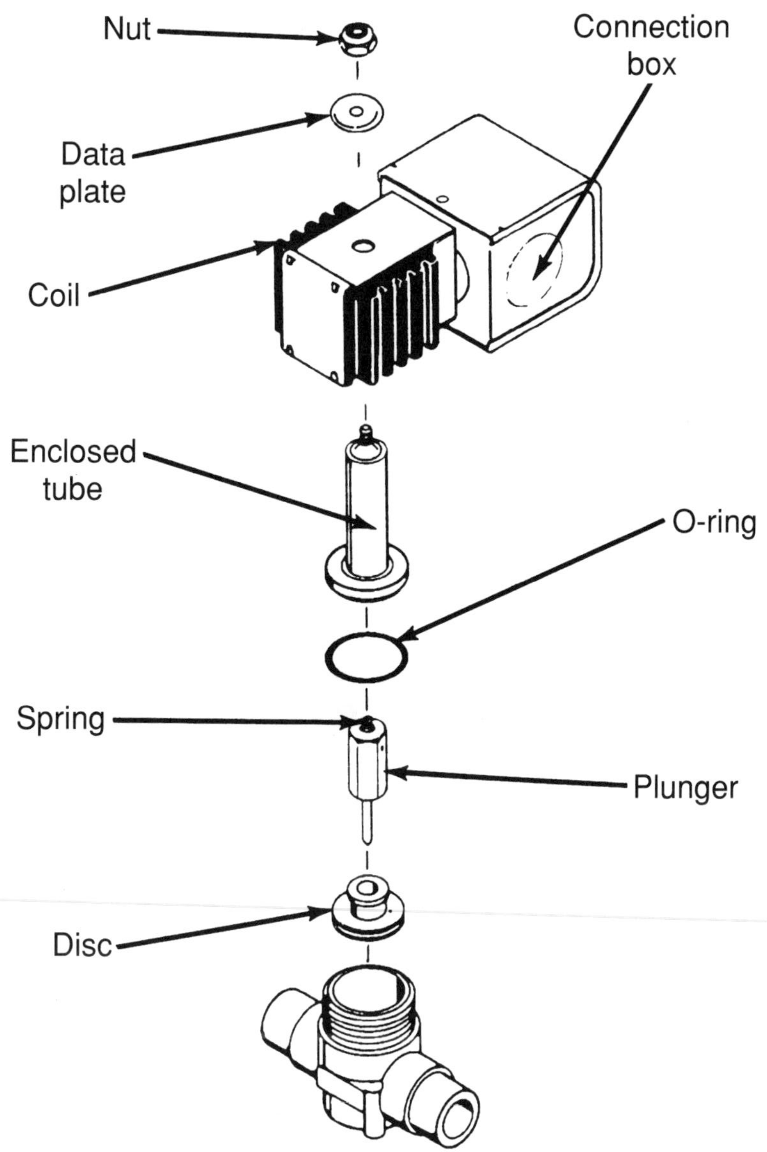

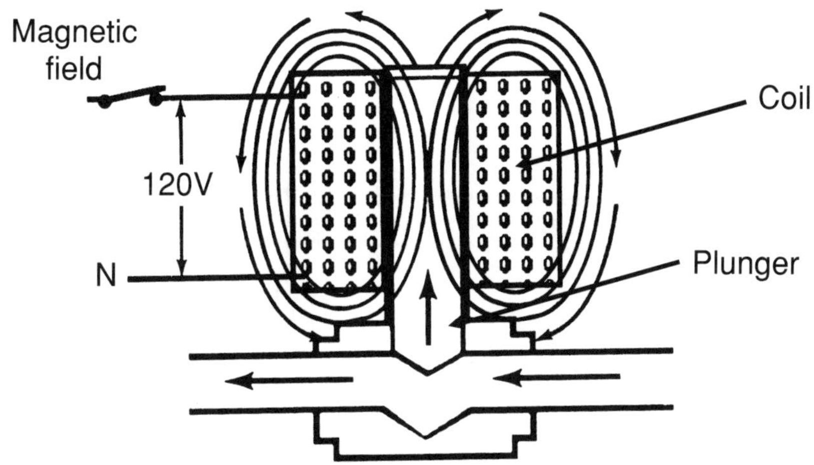

Energized (Valve Open)

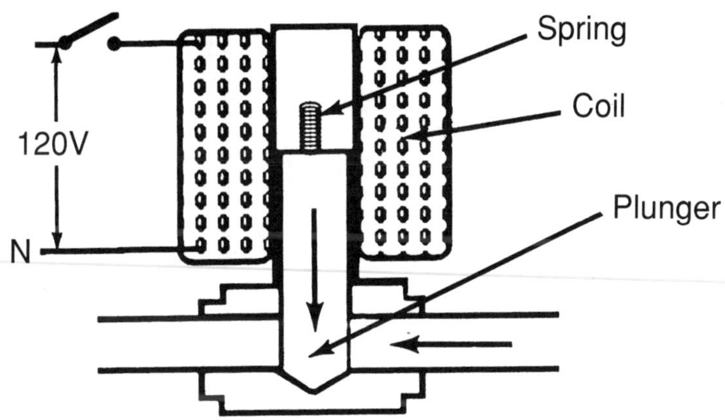

Deenergized (Valve Closed)

HC 25-1

Relay Operation

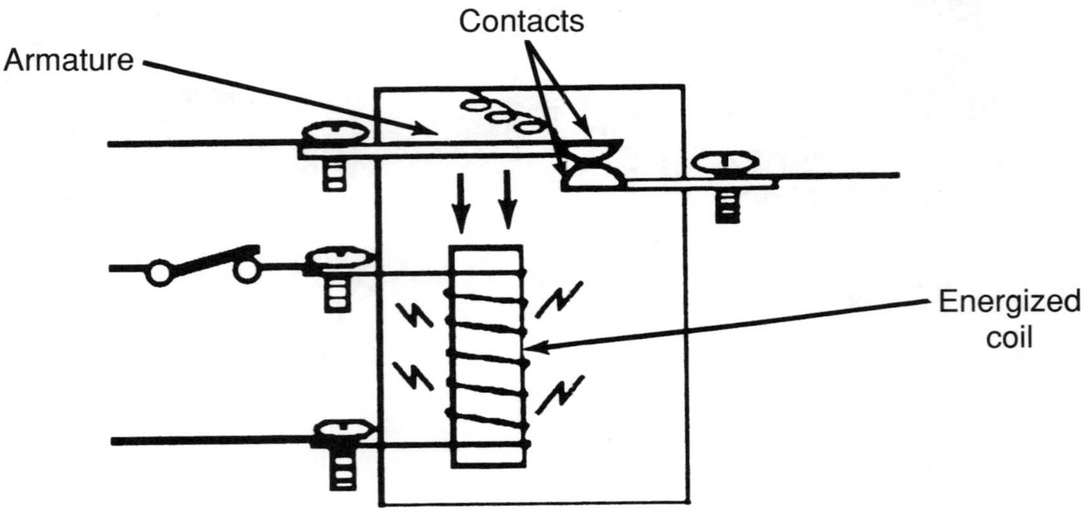

Energized

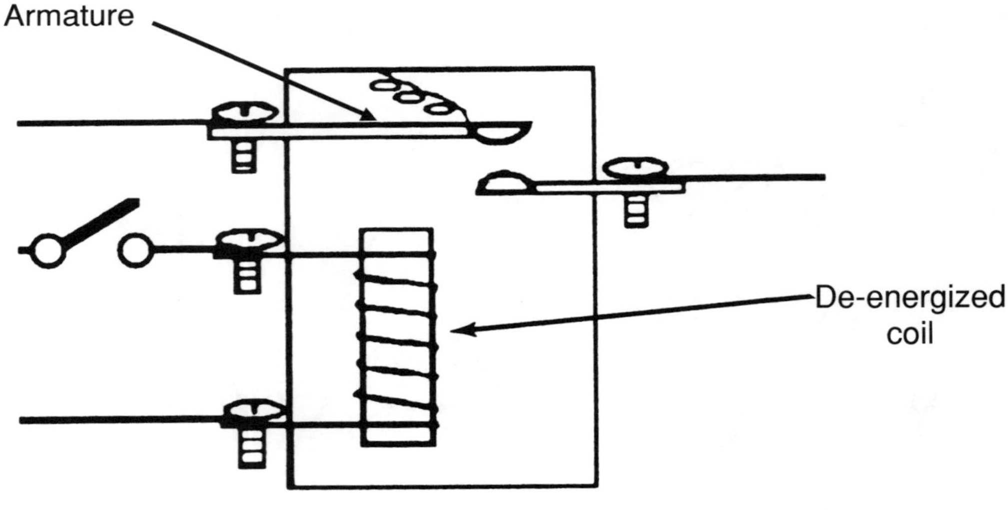

De-energized

HC 25-2

Condenser Fan Connections

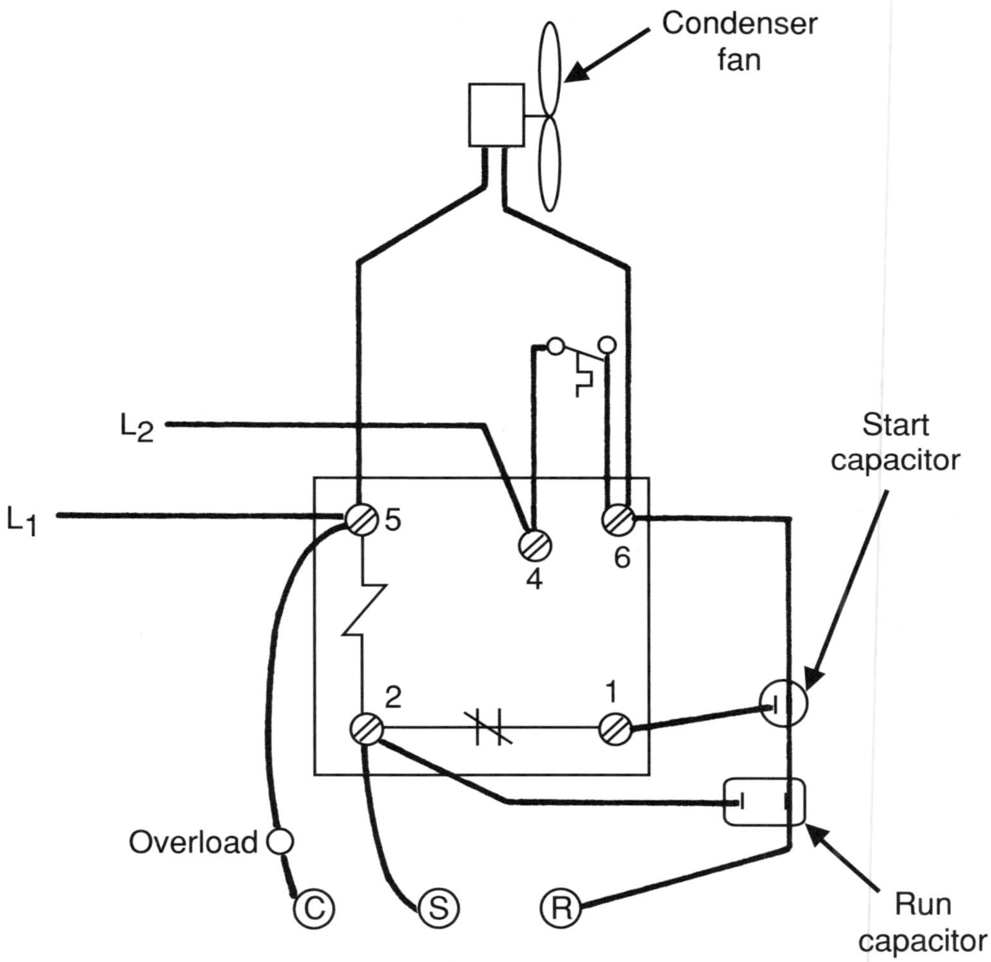

HC 25-3

Connecting Semihermetic Compressors

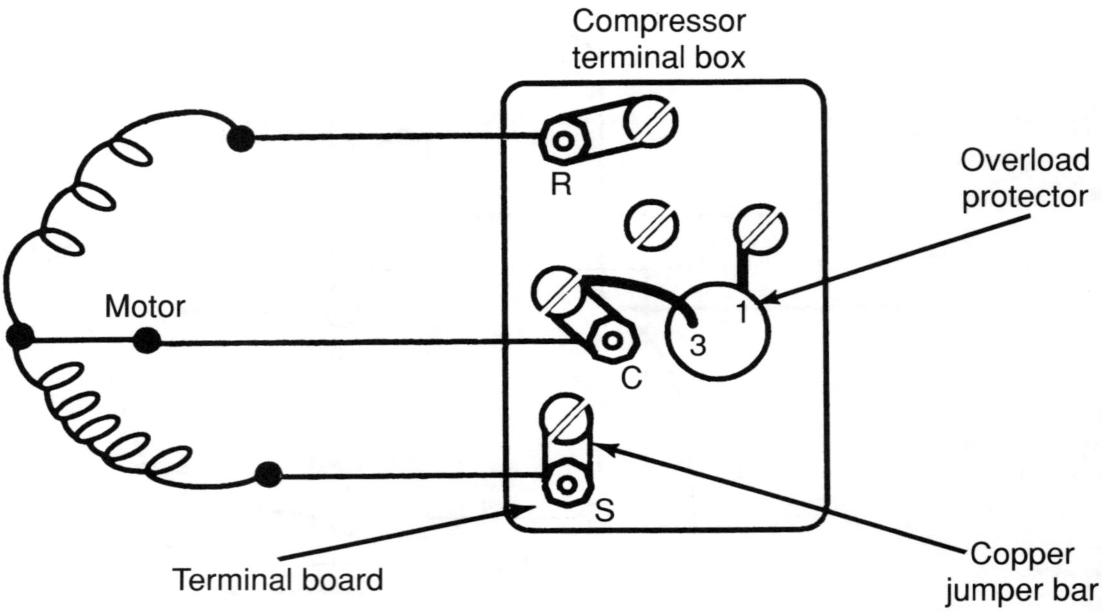

HC 25-4

Residential System with Contactor

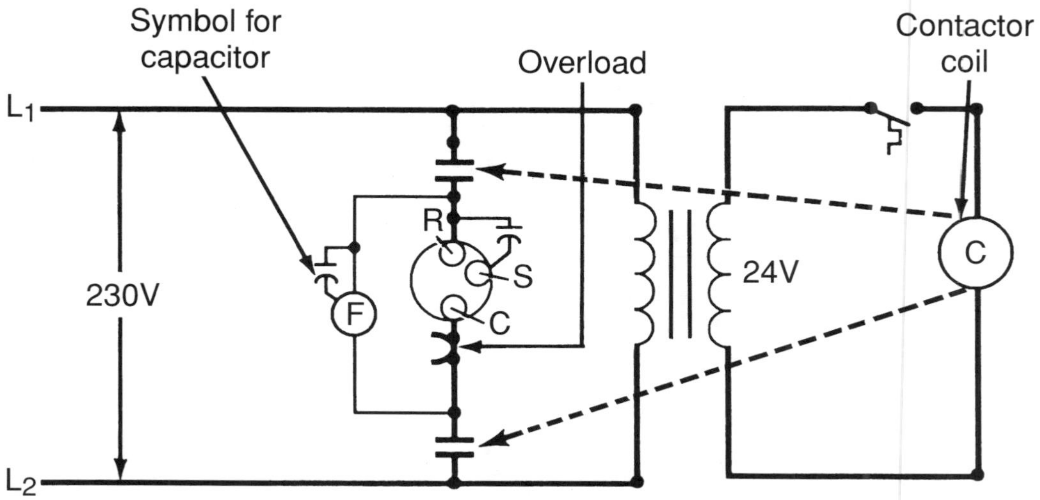

Schematic

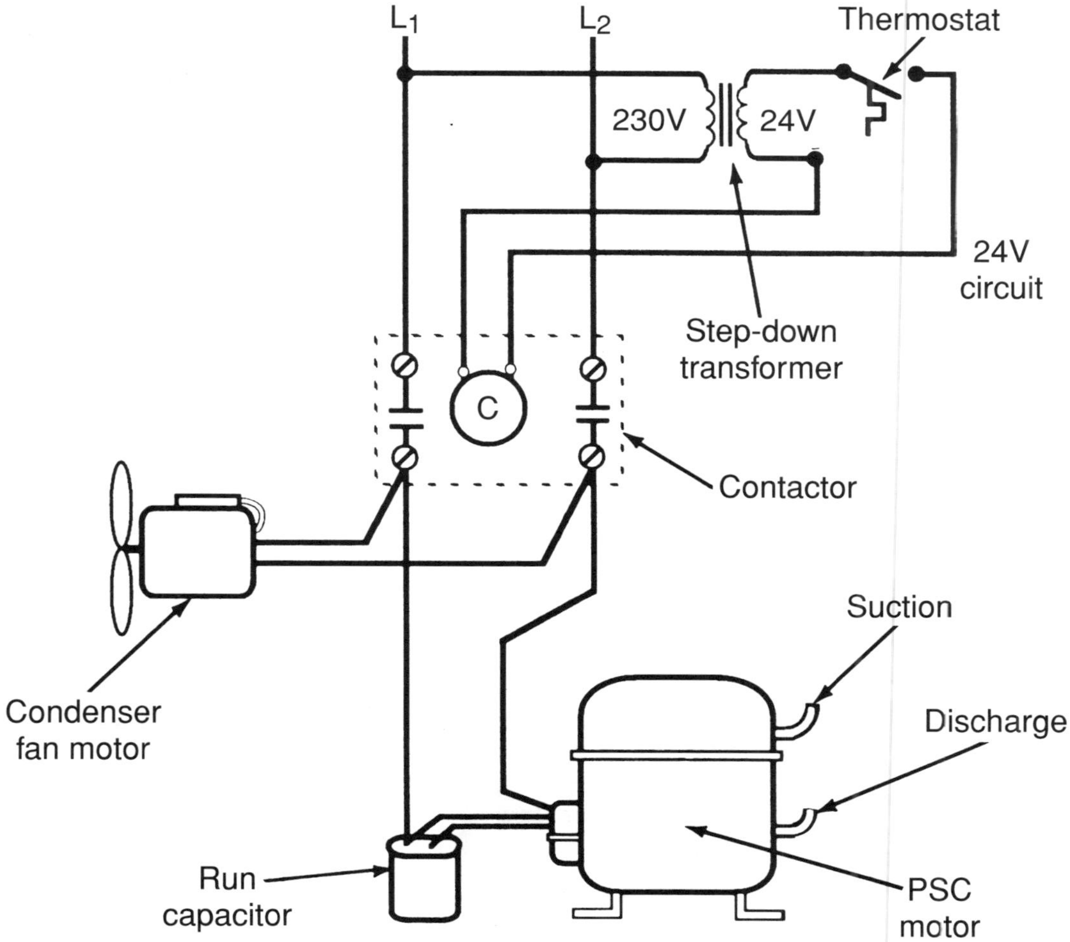

Pictorial

HC 25-5

Complex Control Circuit

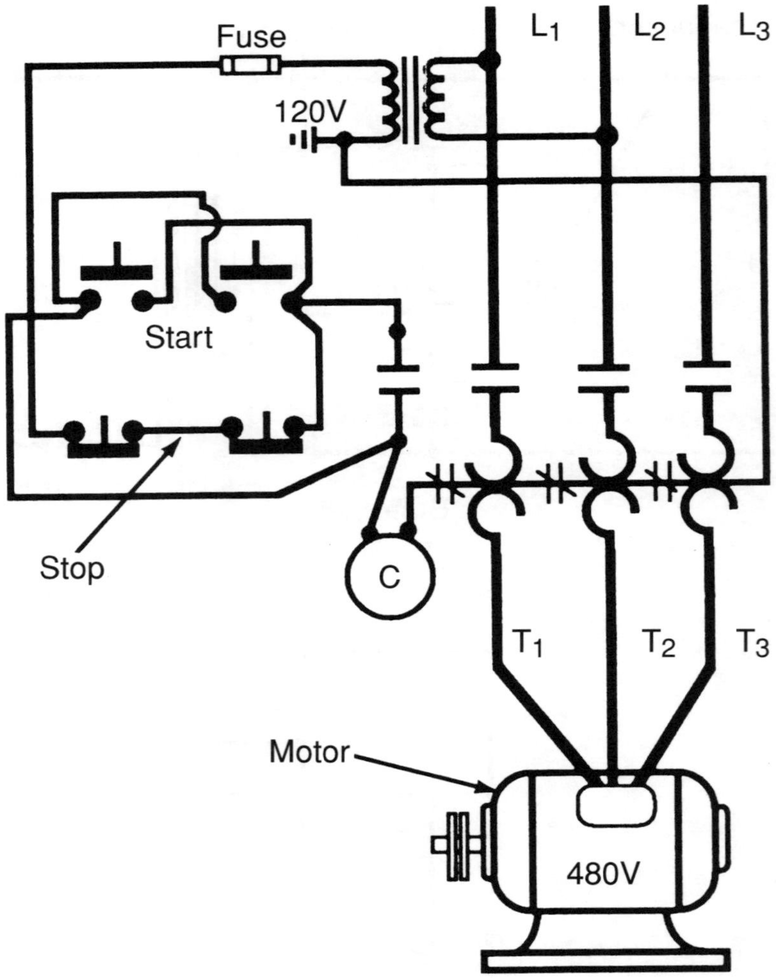

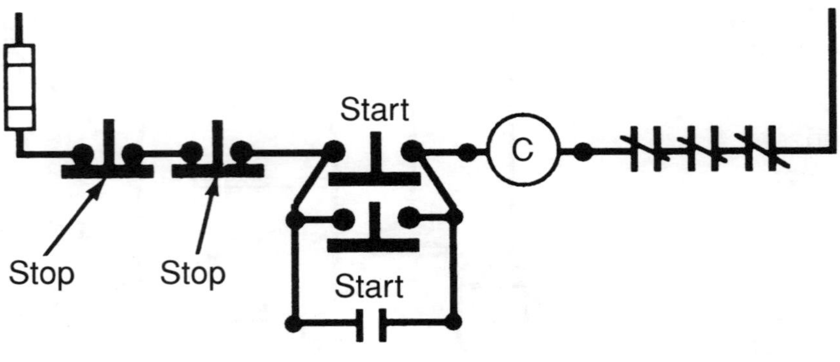

Motor Control from a Separate Location

HC 25-6

Name ______________________

Date ______________________

Instructor ______________________

Period______________________

25

Electromagnetic Control Devices

Quiz

Instructions: Write your answers on the lines provided at the right-hand side of the page.

1. What type of metal can be easily magnetized? 1. ______________________

2. Electromagnetism is used to operate which of the devices listed below? 2. ______________________
   a. Solenoid valves.
   b. Relays.
   c. Contactors.
   d. All of the above.

3. *True or false?* Relay contacts are normally limited to 20 A. 3. ______________________

4. What are the two most common types of relays used on single-phase compressors to disconnect the start winding and/or capacitor? 4. ______________________ ______________________

5. Name three methods for freeing a locked-up hermetic compressor. 5. ______________________ ______________________ ______________________

6. *True or false?* The major difference between a relay and a contactor is the size of the contacts. 6. ______________________

7. Does a residential air conditioning system normally use a two-pole or a three-pole contactor? 7. ______________________

8. Is the contactor coil electrically separated from the main contacts? 8. ______________________

9. How many switches can be located in the control circuit of a contactor or line-starter? 9. ______________________

10. Name two types of electrical wiring diagrams. 10. ______________________ ______________________

26

Motor Controls

Objectives

After studying this chapter in the text, the student will be able to:

- ❑ Identify and adjust primary motor controls.
- ❑ Troubleshoot thermostatic and pressure-type motor controls.
- ❑ Install and properly adjust safety controls.
- ❑ Troubleshoot safety controls.
- ❑ Recognize and properly adjust pumpdown cycles.
- ❑ Install and troubleshoot oil pressure safety controls.

Instructional Materials

Text: Pages 447–468
Lab Manual: Pages 205–212
Instructor's Manual:
Visual Masters, Pages 231–234
Chapter 26 Quiz, Page 235

Teaching Strategy

Text

1. Review objectives.
2. Discuss and assign chapter contents. Use these visual masters as appropriate:
   HC 26-1: Pressure Settings
   HC 26-2: Remote (Split) System
   HC 26-3: Pumpdown Cycle
   HC 26-4: Oil Pressure Safety Control
3. Direct students to complete Test Your Knowledge questions, page 468. After completion, discuss correct answers with students.

Lab Manual

1. Provide overview of Activities. Where necessary, stress safe laboratory practices and the importance of wearing protective clothing.
2. Assign Activities to individuals or pairs of students, as appropriate.
   Activity 26-1: Adjusting Commercial Thermostats
   Activity 26-2: Adjusting Pressure Controls
   Activity 26-3: Adjusting a Low-pressure Safety Control
   Activity 26-4: Adjusting a High-pressure Safety Control
   Activity 26-5: Checking an Oil Pressure Safety Control
3. Observe student performance on Activities, providing guidance or correction as needed. After reviewing the student's work, initial and date the activity in the place provided.

Evaluation

1. Administer the quiz from this Instructor's Manual.
2. Grade the quiz, and return it with appropriate comments for the student.

Answers to Test Your Knowledge, Page 468

1. temperature-sensitive; pressure-sensitive
2. cut-in
3. differential
4. power supply
5. range adjustment
6. cut-out
7. in series
8. 2 psig
9. thermostat
10. To prevent evaporator freeze-up.

11. 58 to 60
12. evaporator
13. abnormally low suction pressure
14. false
15. control low head pressure during low ambient
16. liquid line solenoid
17. To control the condensing unit (or compressor).
18. net
19. T_1 and T_2
20. manually

Answers to Lab Manual, Pages 207–212

Activity 26-2

Note: The following are the cut-in/cut-out values the student should be using at the designated steps in this activity.

2. 15 psig cut-in
3. 2 psig cut-out
4. 15 psig cut-in
5. 13 psig differential
6. 33 psig cut-in; 13 psig cut-out
7. 33 psig cut-in
8. 11 psig cut-out
9. 33 psig cut-in
10. 22 psig differential

Activity 26-5

3. no
4. yes

Answers to Quiz

1. thermostatic; pressure-type
2. differential
3. It can call for system operation when the system is malfunctioning.
4. 20 degrees Fahrenheit (11 degrees Celsius)
5. d. all of the above.
6. 2 psig (14 kPa)
7. true
8. 40 psig to 45 psig (276 kPa to 310 kPa)
9. head pressure safety control
10. lubrication

Pressure Settings

Example:

Cut-in – Differential = Out
36 20 16

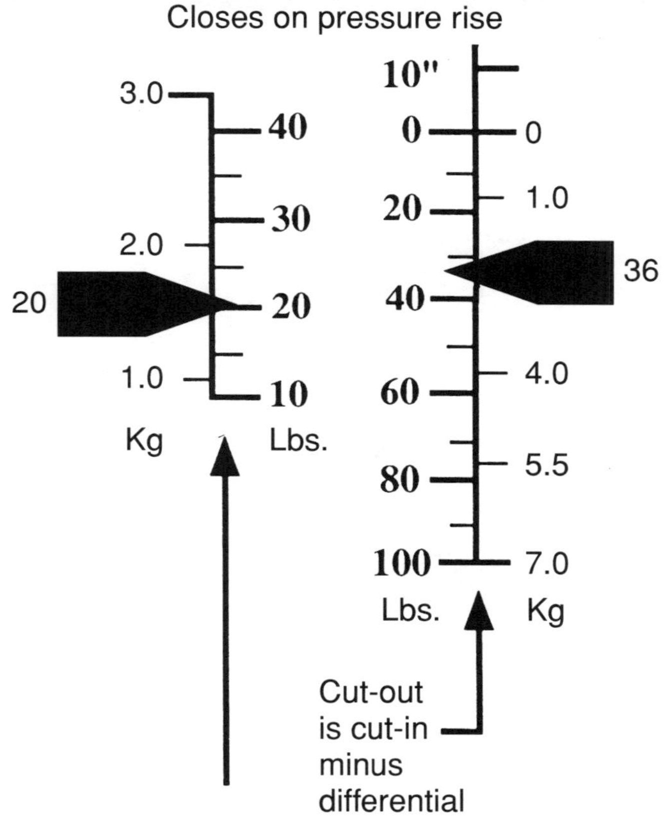

HC 26-1

Remote (Split) System

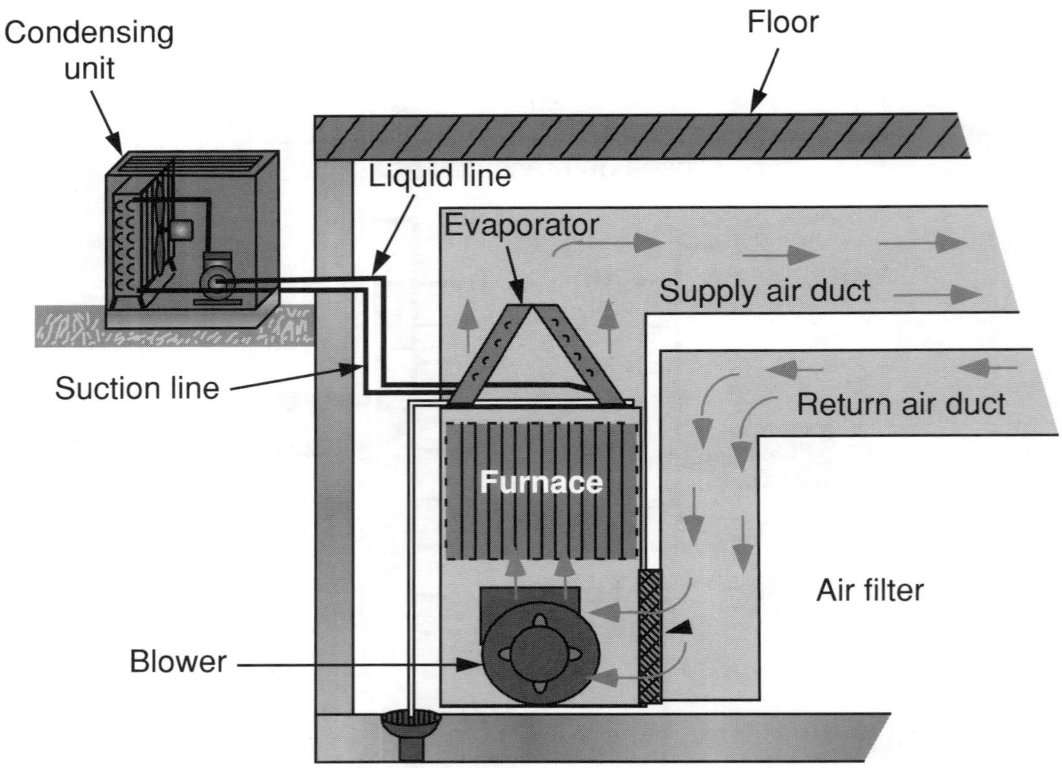

HC 26-2

Pumpdown Cycle

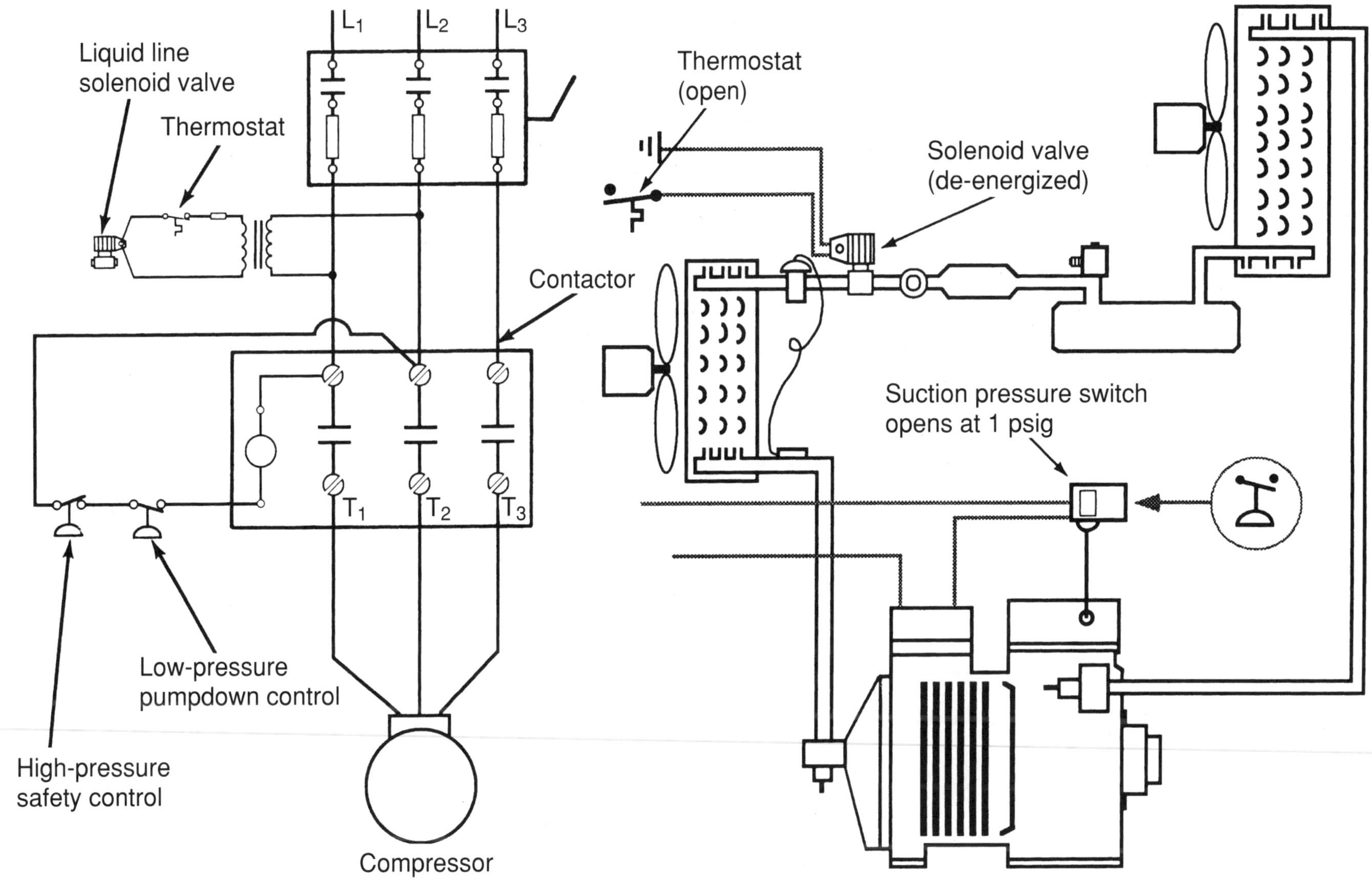

HC 26-3

Oil Pressure Safety Control

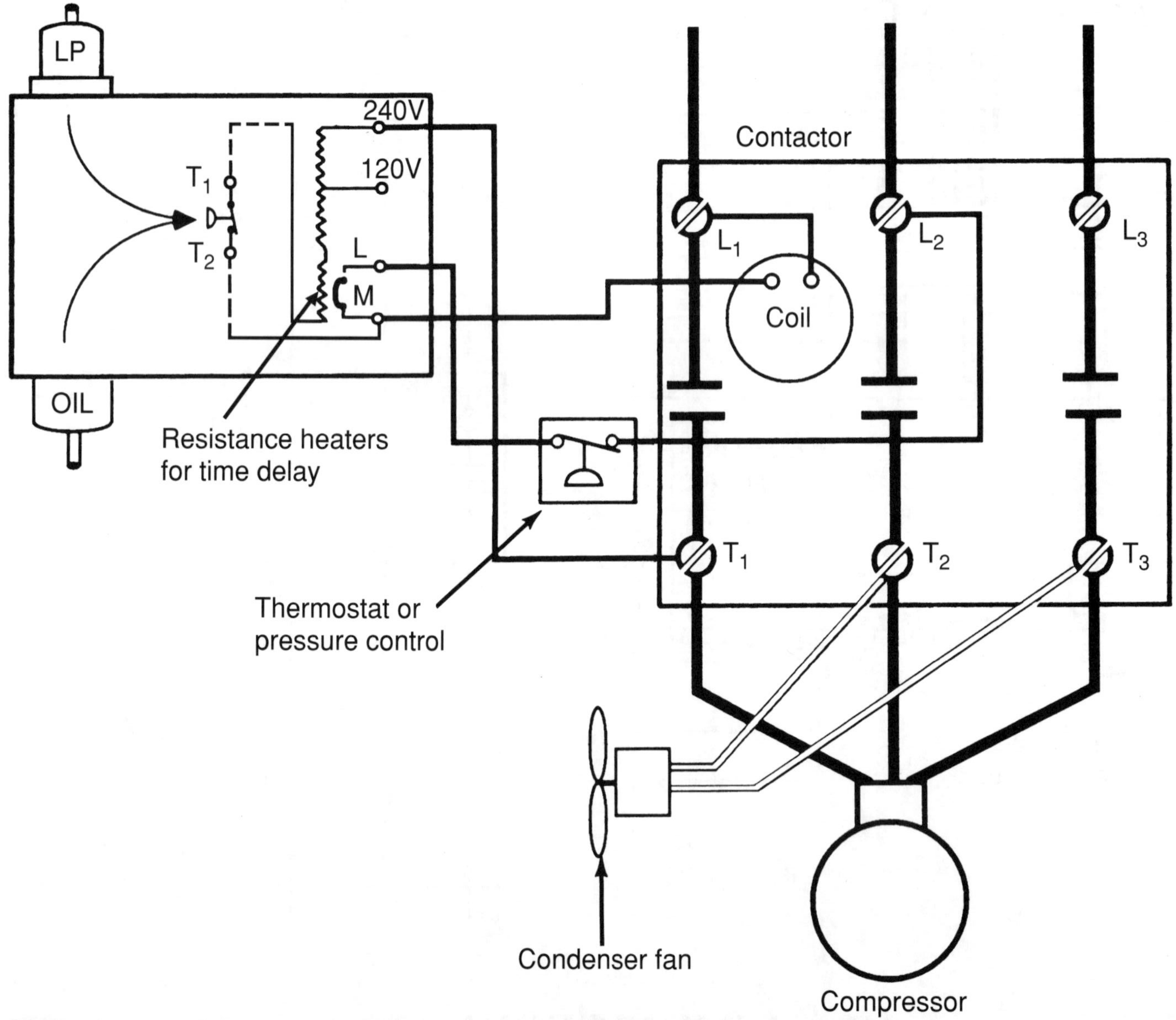

HC 26-4

Name ______________________________

Date ______________________________

Instructor ______________________________

Period______________________________

26 Motor Controls

Quiz

Instructions: Write your answers on the lines provided at the right-hand side of the page.

1. Name two types of primary motor controls. 1. ______________________ ______________________

2. The difference between the control device's cut-in and cut-out is called the _____. 2. ______________________

3. What is the disadvantage of a thermostatic motor control? 3. ______________________

4. During the run cycle on a walk-in freezer, what is the temperature difference between the air temperature and the vaporizing refrigerant? 4. ______________________

5. Pressure controls can be used for: 5. ______________________
   a. primary control.
   b. low-pressure safety control.
   c. pumpdown cycle.
   d. all of the above.

6. When setting the cut-out on a pressure control, how much allowance must be made for pressure drop through the suction line? 6. ______________________

7. *True or false?* A low-pressure safety control will turn the system off when low-side pressure is reduced to unsafe levels. 7. ______________________

8. The low-pressure safety control on an air conditioning system is set to cut out at ____. 8. ______________________

9. What control is used to shut off the system if head pressure exceeds proper limits? 9. ______________________

10. The oil pressure safety control protects the compressor from lack of _____. 10. ______________________

27

Defrost Cycles

Objectives

After studying this chapter in the text, the student will be able to:

- ❑ Identify various domestic and commercial defrost timeclocks.
- ❑ Describe the five types of defrost cycles.
- ❑ Install and properly adjust defrost timeclocks.
- ❑ Connect and troubleshoot defrost terminator thermostats.
- ❑ Explain the operation of, and make necessary adjustments to, domestic and commercial defrost systems.
- ❑ Adjust temperature and pressure termination.

Instructional Materials

Text: Pages 469–494
Lab Manual: Pages 213–218
Instructor's Manual:
Visual Masters, Pages 239–243
Chapter 27 Quiz, Page 244

Teaching Strategy

Text

1. Review objectives.
2. Discuss and assign chapter contents. Use these visual masters as appropriate:
   HC 27-1: Defrost Heater
   HC 27-2: Testing a Defrost Timeclock
   HC 27-3: Fan Time-Delay Thermostat
   HC 27-4: Multiplexed System
   HC 27-5: Defrost Timeclock Controls
3. Direct students to complete Test Your Knowledge questions, page 494. After completion, discuss correct answers with students.

Lab Manual

1. Provide overview of Activities. Where necessary, stress safe laboratory practices and the importance of wearing protective clothing.
2. Assign Activities to individuals or pairs of students, as appropriate.
   Activity 27-1: Checking a Domestic Electric Defrost Timer
   Activity 27-2: Checking a Commercial Off-time Defrost Timer
   Activity 27-3: Checking a Commercial Defrost Timer
   Activity 27-4: Wiring Defrost Systems
3. Observe student performance on Activities, providing guidance or correction as needed. After reviewing the student's work, initial and date the activity in the place provided.

Evaluation

1. Administer the quiz from this Instructor's Manual.
2. Grade the quiz, and return it with appropriate comments for the student.

Answers to Test Your Knowledge, Page 494

1. 32°F (0°C)
2. off-cycle; off-time; electric; hot gas; reverse-air
3. compressor; evaporator fan
4. solenoid valve and evaporator fan
5. condensing unit
6. commercial
7. time-initiated/time-terminated; time-initiated/temperature-terminated; time-initiated/pressure-terminated
8. initiate
9. true
10 false

11. open
12. close
13. clock solenoid
14. three
15. To maintain head pressure.
16. To carry and control high amperage flow to the heaters.
17. Behind the kickrail.
18. series
19. condensing unit and fan reversing relay
20. They have two sets of main windings.

Answers to Lab Manual, Pages 213–218

Activity 27-1

Student response (answers will vary).

Activity 27-2

5. Compressor and condenser fan.

Activity 27-3

6. Student response (answers will vary).
7. normally closed; zero ohms
8. normally open; infinity
9. 4
10. 4

Activity 27-4

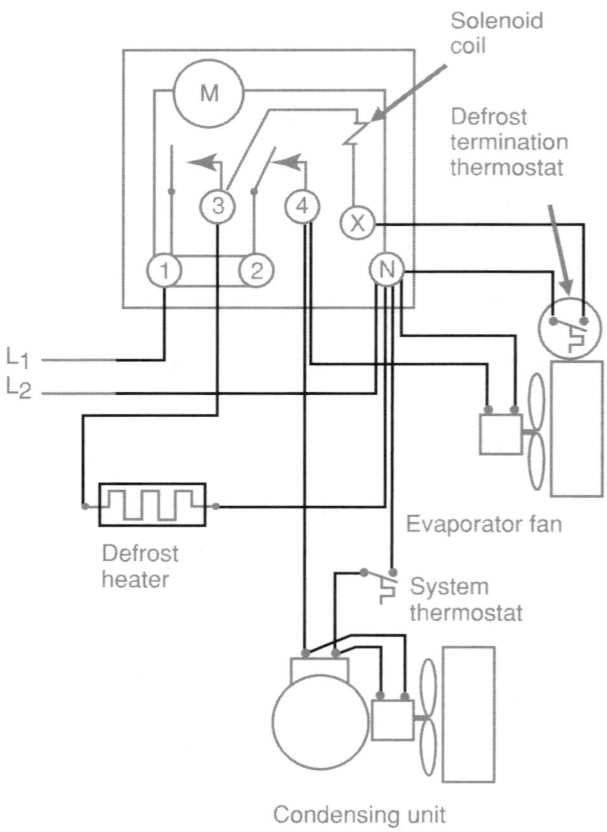

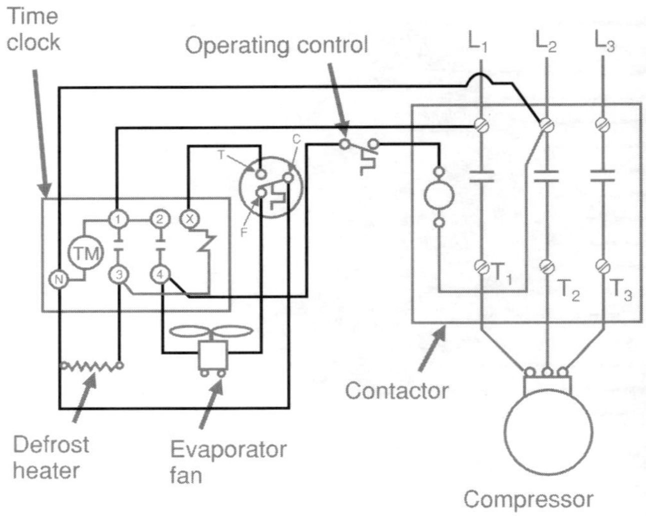

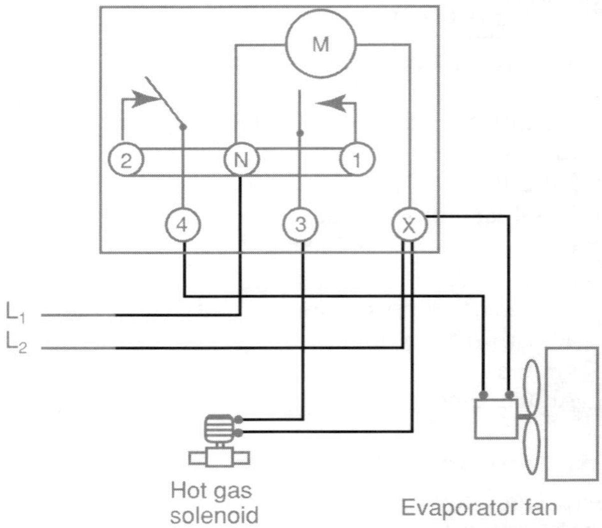

Answers to Quiz

1. b. 100
2. It removes moisture.
3. off-cycle; off-time; electric; hot gas; reverse-air
4. in series
5. time-initiated/time-terminated; time-initiated/temperature-terminated; time-initiated/pressure-terminated
6. open
7. close
8. condensing unit (compressor and condenser)
9. condensing unit; evaporator fans
10. true

Defrost Heater

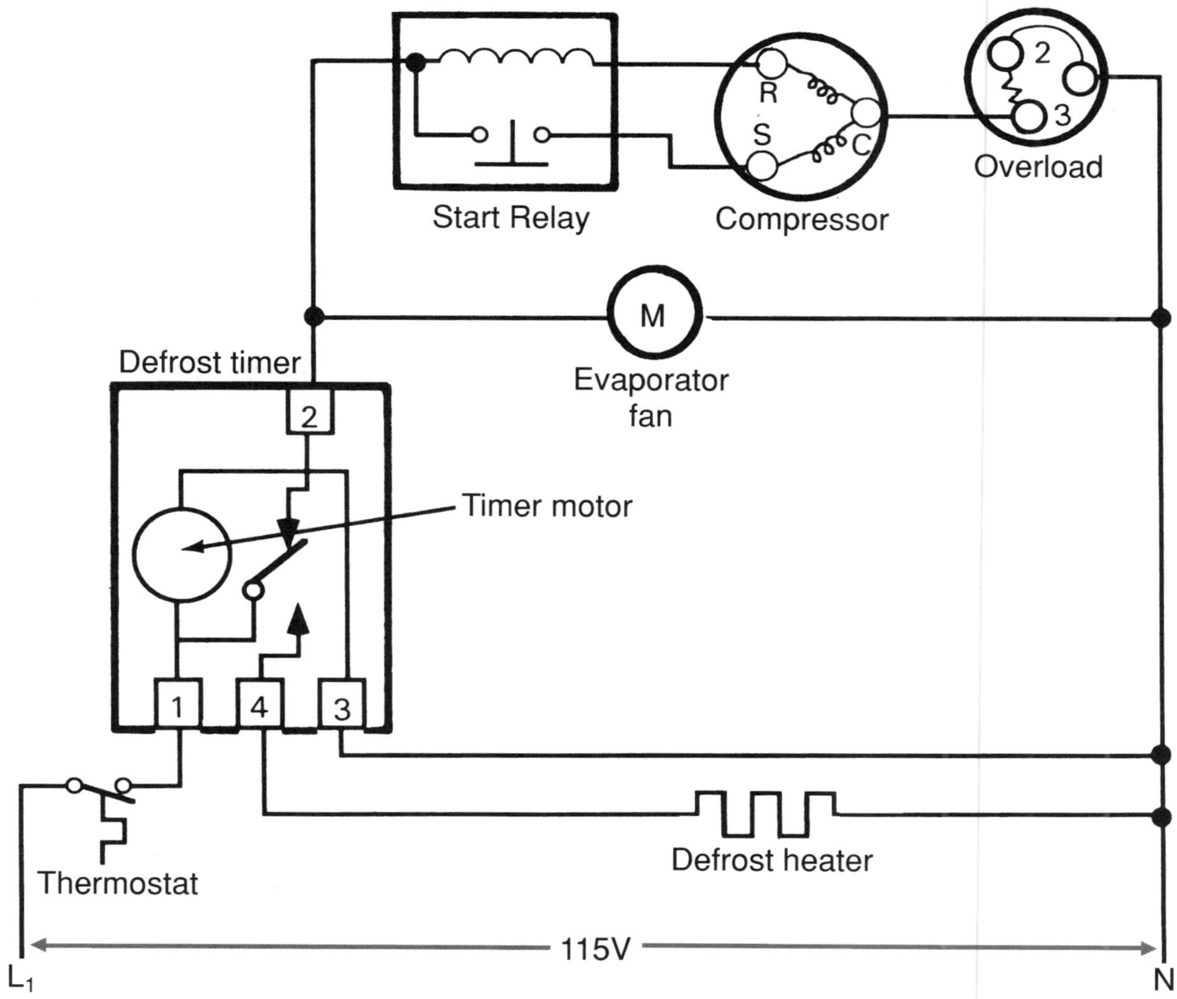

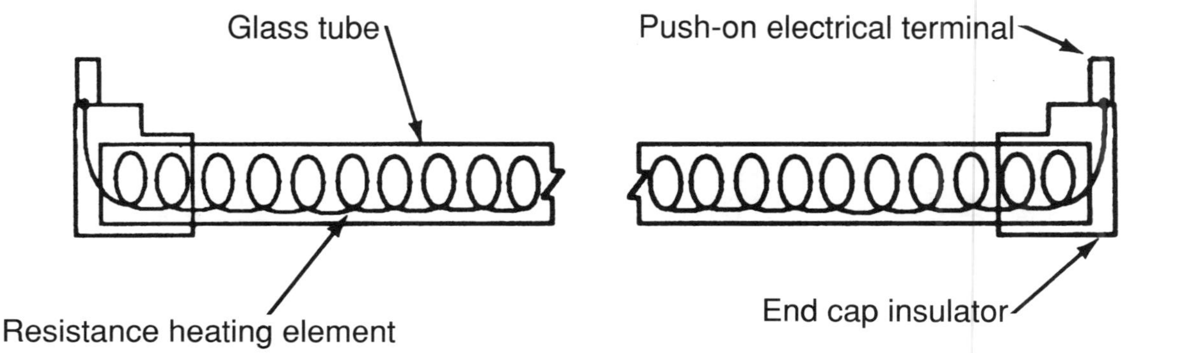

HC 27-1

Testing a Defrost Timeclock

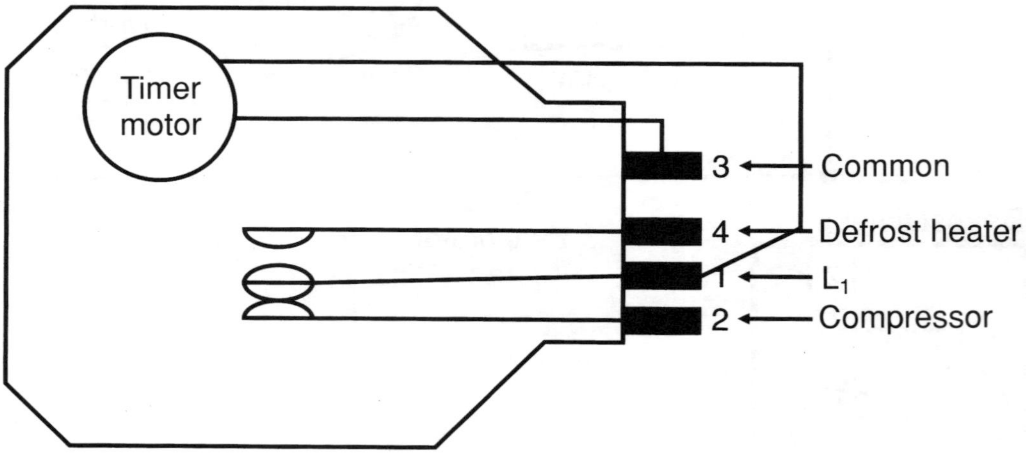

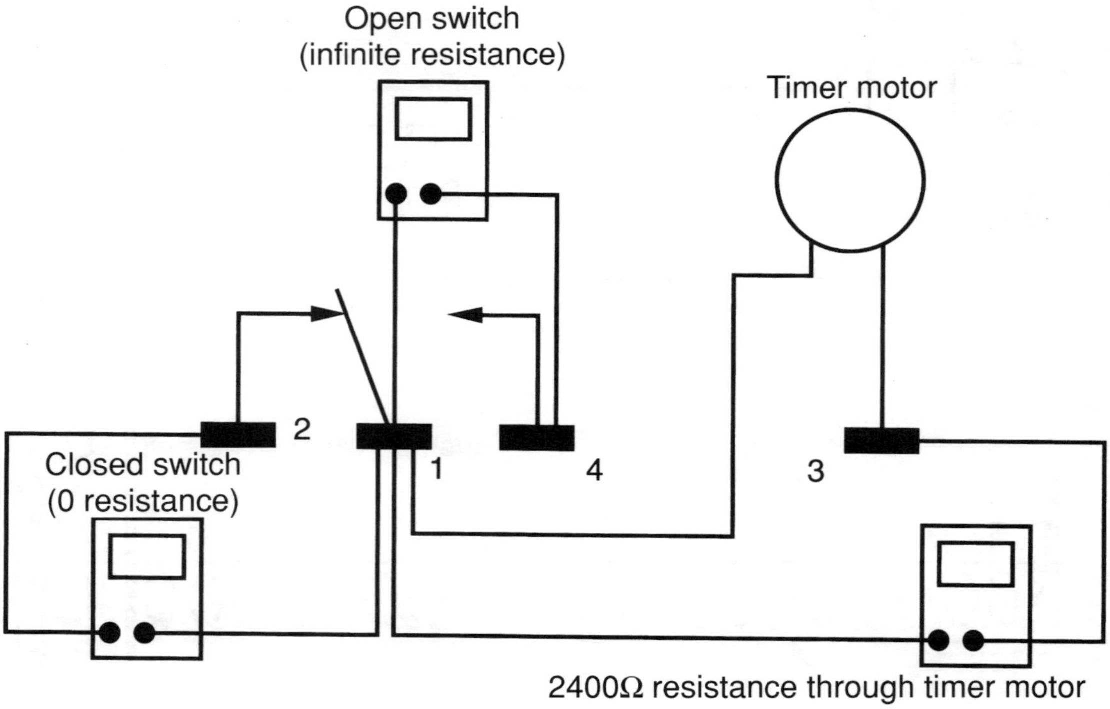

HC 27-2

Fan Time-Delay Thermostat

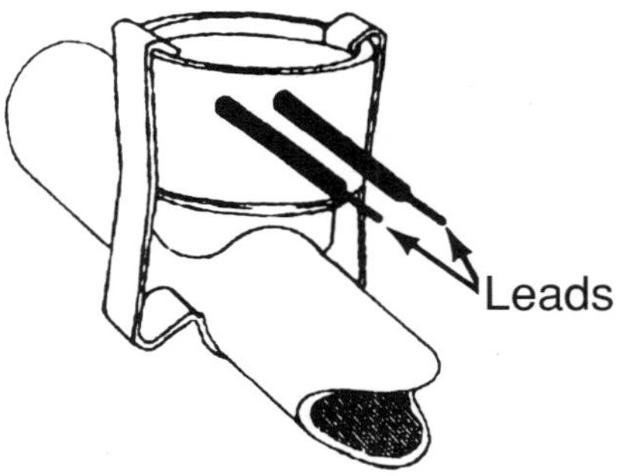

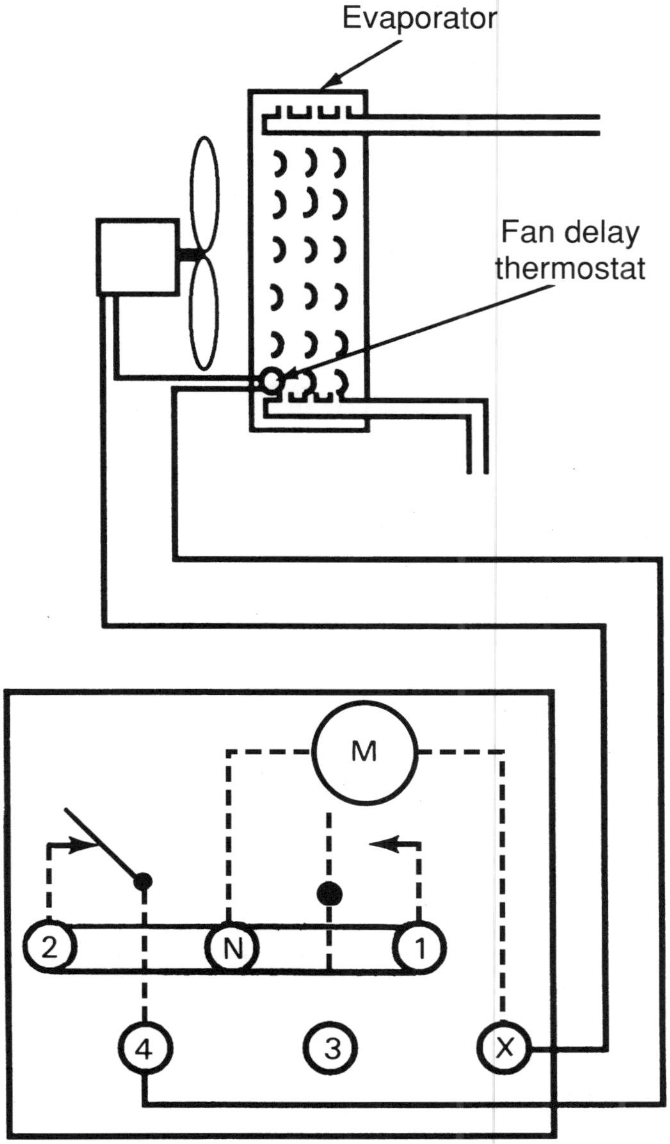

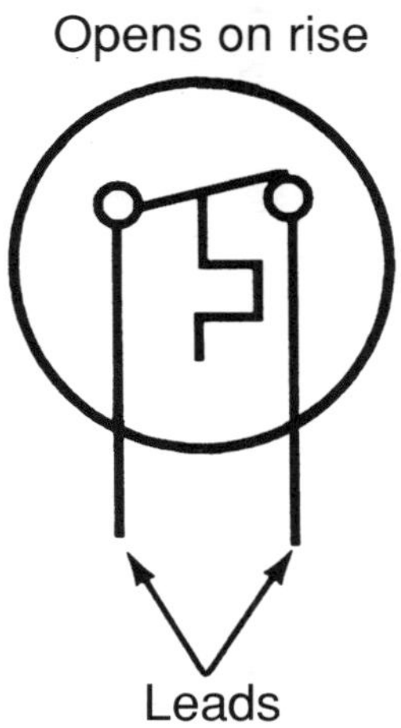

HC 27-3

Multiplexed System

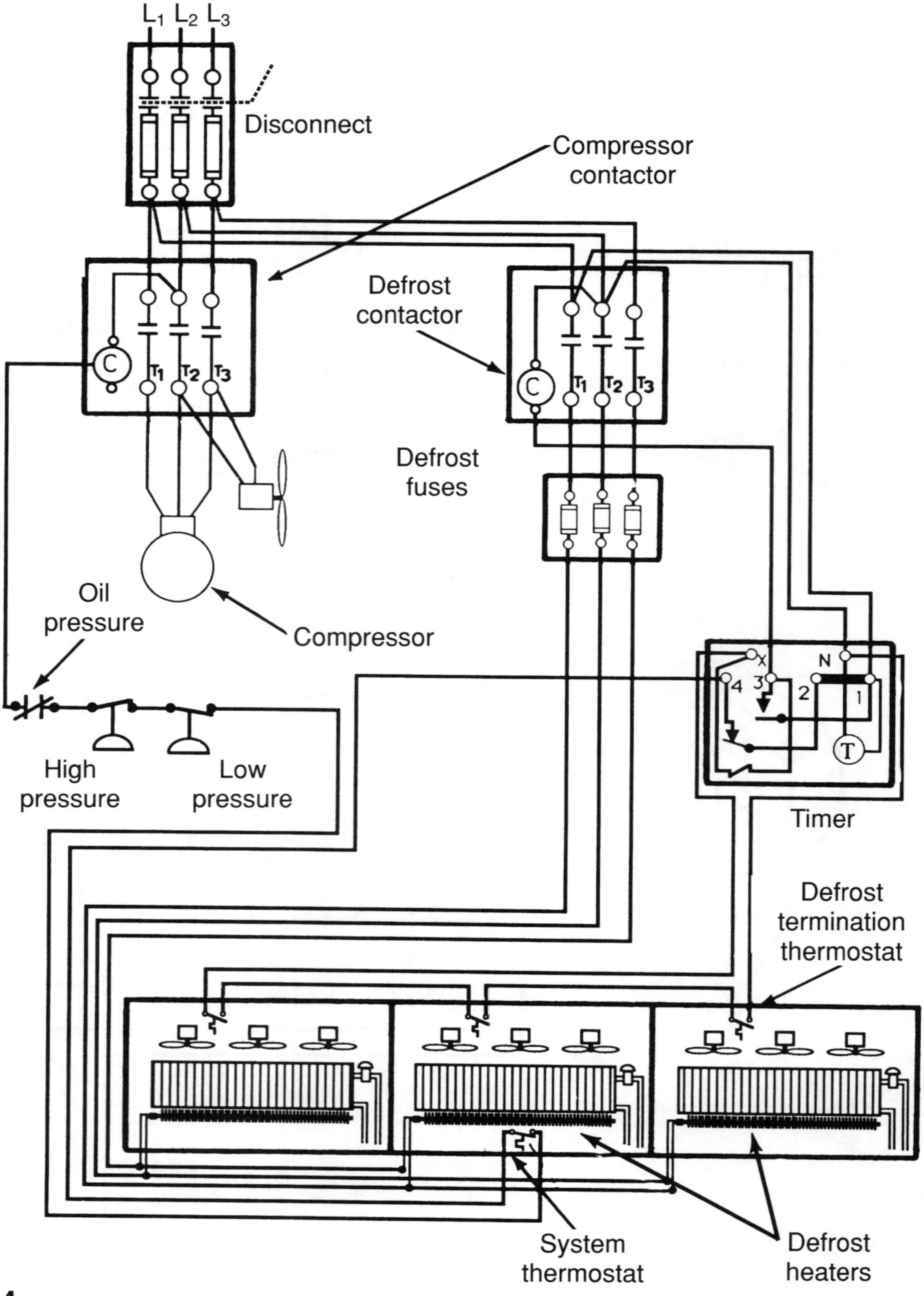

HC 27-4

Defrost Timeclock Controls

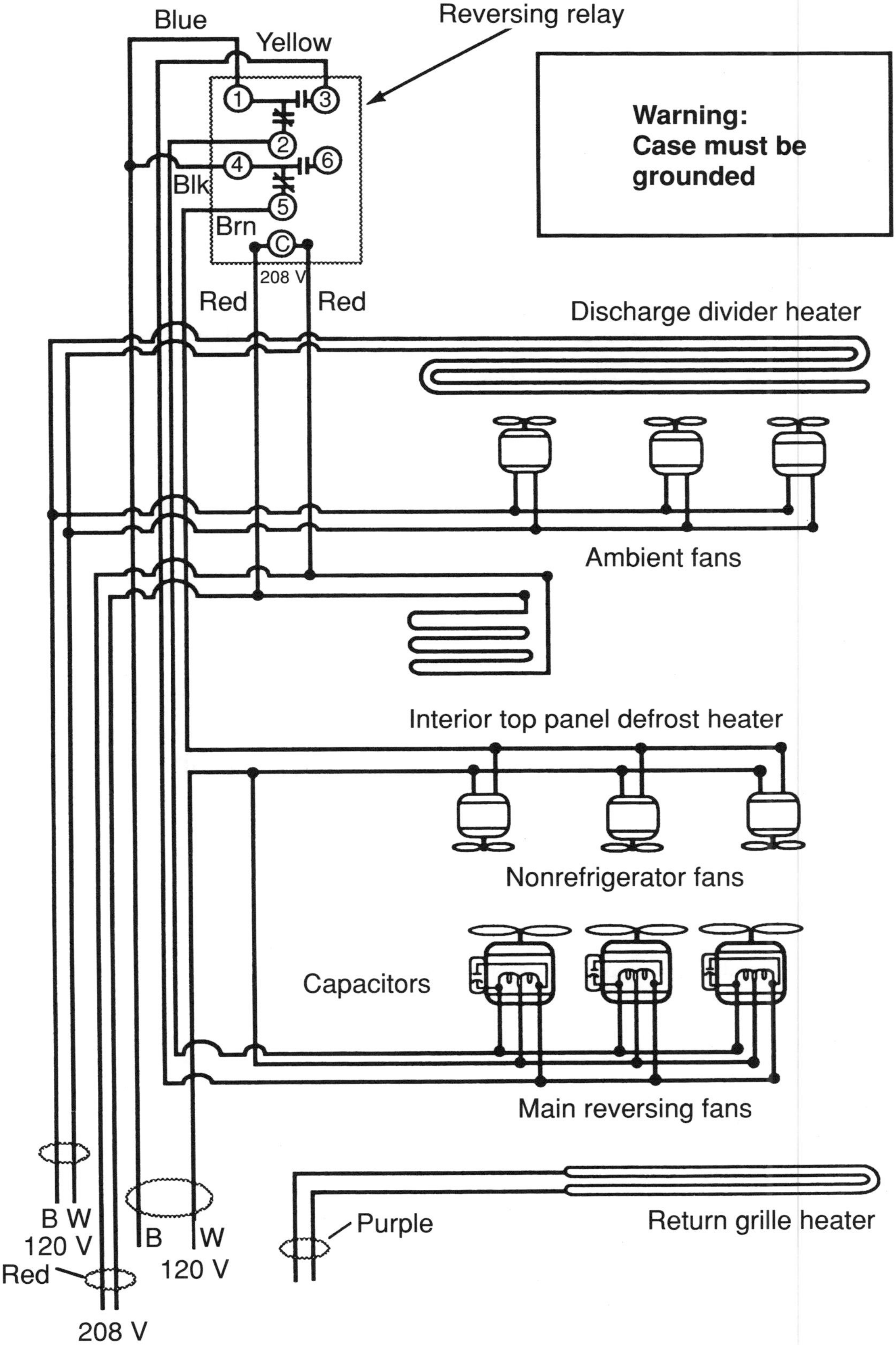

HC 27-5

Name ______________________________

Date ______________________________

Instructor ______________________________

Period______________________________

27

Defrost Cycles

Quiz

Instructions: Write your answers on the lines provided at the right-hand side of the page.

1. Dewpoint is another way of describing _____% relative humidity. 1. ______________
   a. 50
   b. 100
   c. 125
2. Does a dehumidifier add or remove moisture? 2. ______________
3. Name five types of automatic defrost systems. 3. ______________
   ____
   ____
   ____
   ____
4. On a domestic electric defrost, is the defrost terminator thermostat connected in series or in parallel with the defrost heater? 4. ______________
5. Name three types of commercial defrost timeclocks. 5. ______________
   ____
   ____
6. Will contacts on the fan delay thermostat open or close on a rise in temperature? 6. ______________
7. Will contacts on the defrost terminator thermostat open or close on a rise in temperature? 7. ______________
8. During an off-time defrost, which components are controlled by the timeclock? 8. ______________
9. Which components are controlled by the timeclock on a reverse air defrost system? 9. ______________
10. *True or false?* Short, frequent defrosts are better then long, few defrosts. 10. ______________

28

Ductwork

Objectives

After studying this chapter in the text, the student will be able to:

- ❑ Explain operation of the supply and return air duct systems.
- ❑ Assemble, join, install, and repair all types of duct.
- ❑ Install registers, diffusers, and grilles.
- ❑ Insulate and properly support piping.
- ❑ Explain why and how to use mastic and fiberglass membrane.
- ❑ Name and describe the use of many types of fittings.

Instructional Materials

Text: Pages 495-518
Lab Manual: Pages 219-230
Instructor's Manual:
Visual Masters, Pages 247-251
Chapter 28 Quiz, Page 252

Teaching Strategy

Text

1. Review objectives.
2. Discuss and assign chapter contents. Use these visual masters as appropriate:
   HC 28-1: Radial Piping System
   HC 28-2: Extended Plenum System
   HC 28-3: Elbows for Rectangular Duct
   HC 28-4: Installing a Floor Diffuser
   HC 28-5: Sheeted Joists for Return Air
3. Direct students to complete Test Your Knowledge Questions, page 518. After completion, discuss correct answers with the students.

Lab Manual

1. Provide overview of Activities. Where necessary, stress safe laboratory practices and the importance of wearing protective clothing.
2. Assign Activities to individuals or pairs of students, as appropriate.
   Activity 28-1: Identifying Ductwork Tools and Supplies
   Activity 28-2: Joining Round Metal Duct
   Activity 28-3: Identifying Evaporators
   Activity 28-4: Cutting Openings in Ductwork
3. Observe student performance on Activities, providing guidance or correction as needed. After reviewing the student's work, initial and date the Activity in the place provided.

Evaluation

1. Administer the quiz from this Instructor's Manual.
2. Grade the quiz, and return it with appropriate comments to the student.

Answers to Test Your Knowledge, Page 518

1. outside
2. upflow
3. c. both heating and air conditioning modes
4. c. mobile home
5. a. A-coil
6. plenum
7. rectangular
8. 6-inch
9. takeoffs
10. crimped
11. rectangular
12. c. woven polypropylene webbing
13. insulated flexible duct

14. registers
15. unheated

Answers to Lab Manual, Pages 219-230

Activity 28-1

A. aviation snips, straight cut
B. hand seamer
C. scratch awl
D. magnetic screw holder
E. drive screw
F. mastic
G. fiberlass reinforced membrane

Activity 28-3

A.

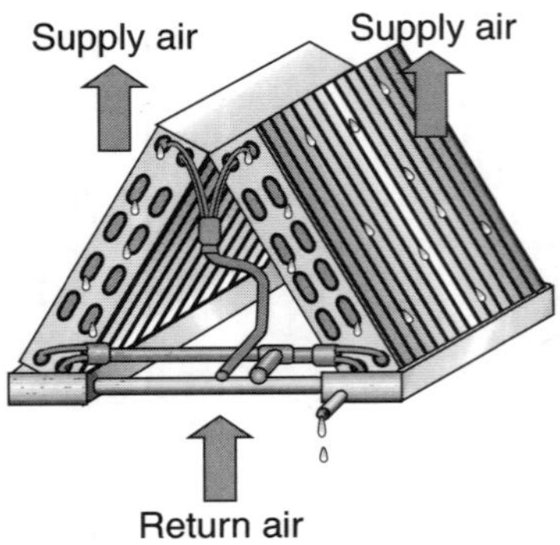

B.

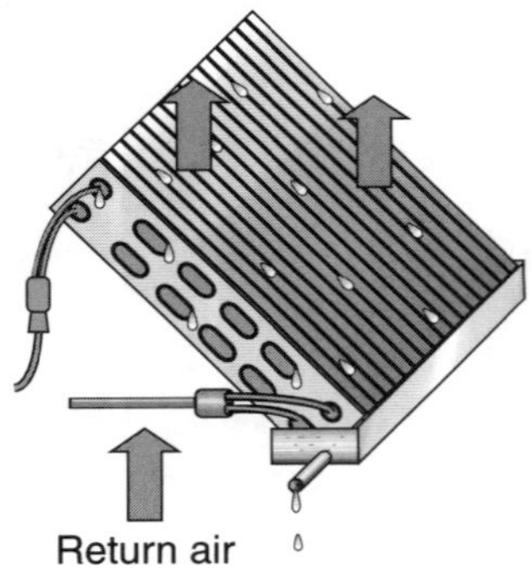

C.

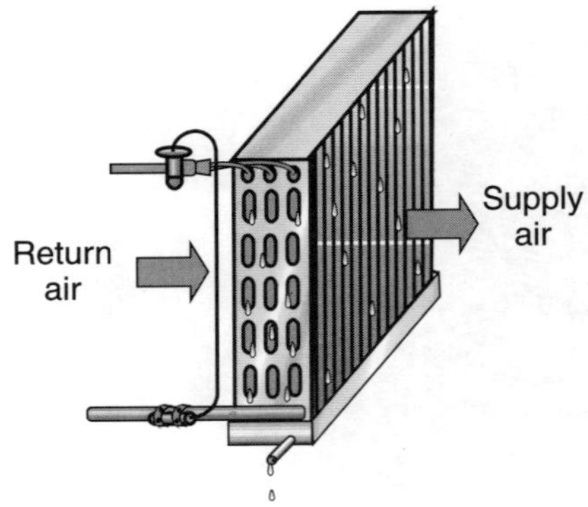

Answers to Quiz

1. forced
2. downflow
3. c. extended plenum
4. vapor barrier
5. wire
6. takeoff
7. False
8. diffuser
9. outside
10. mastic

Radial Piping System

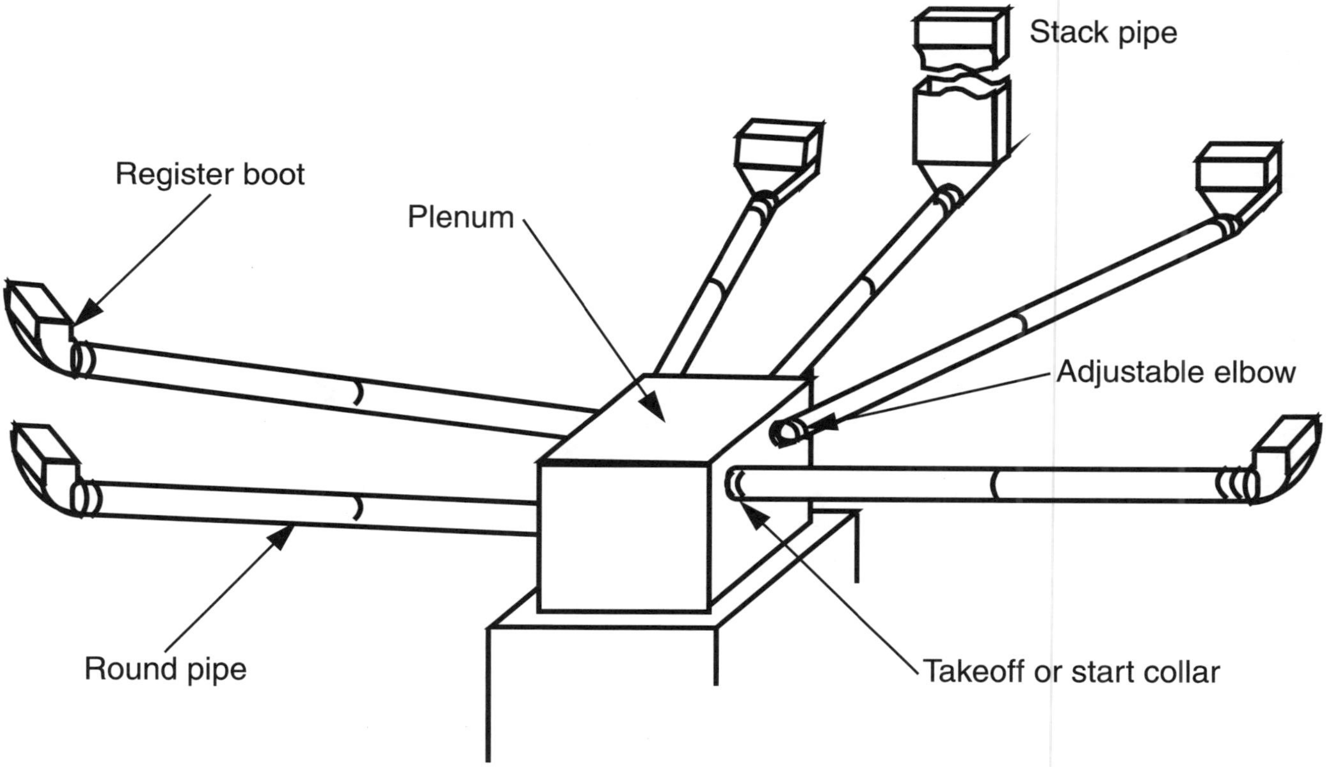

HC 28-1

Extended Plenum System

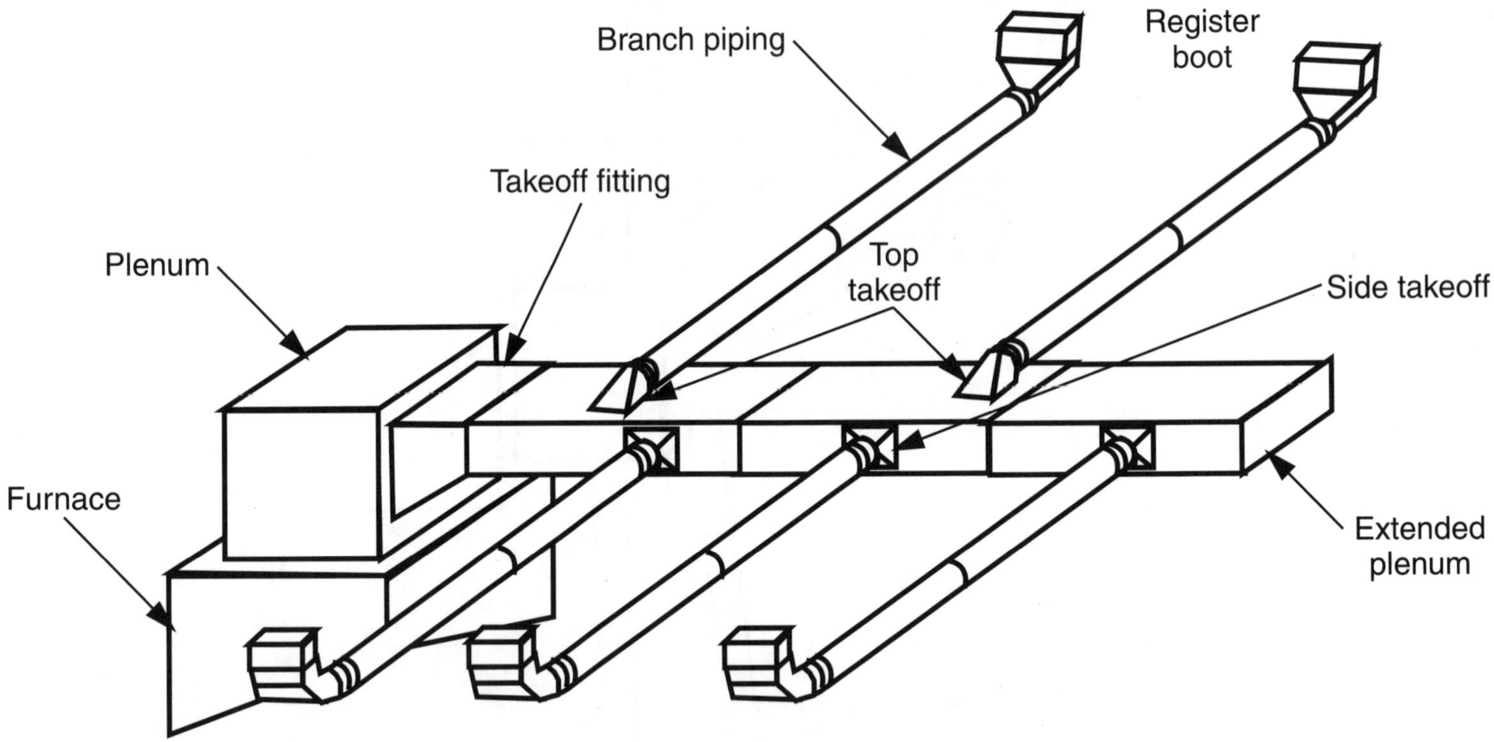

HC 28-2

Elbows for Rectangular Duct

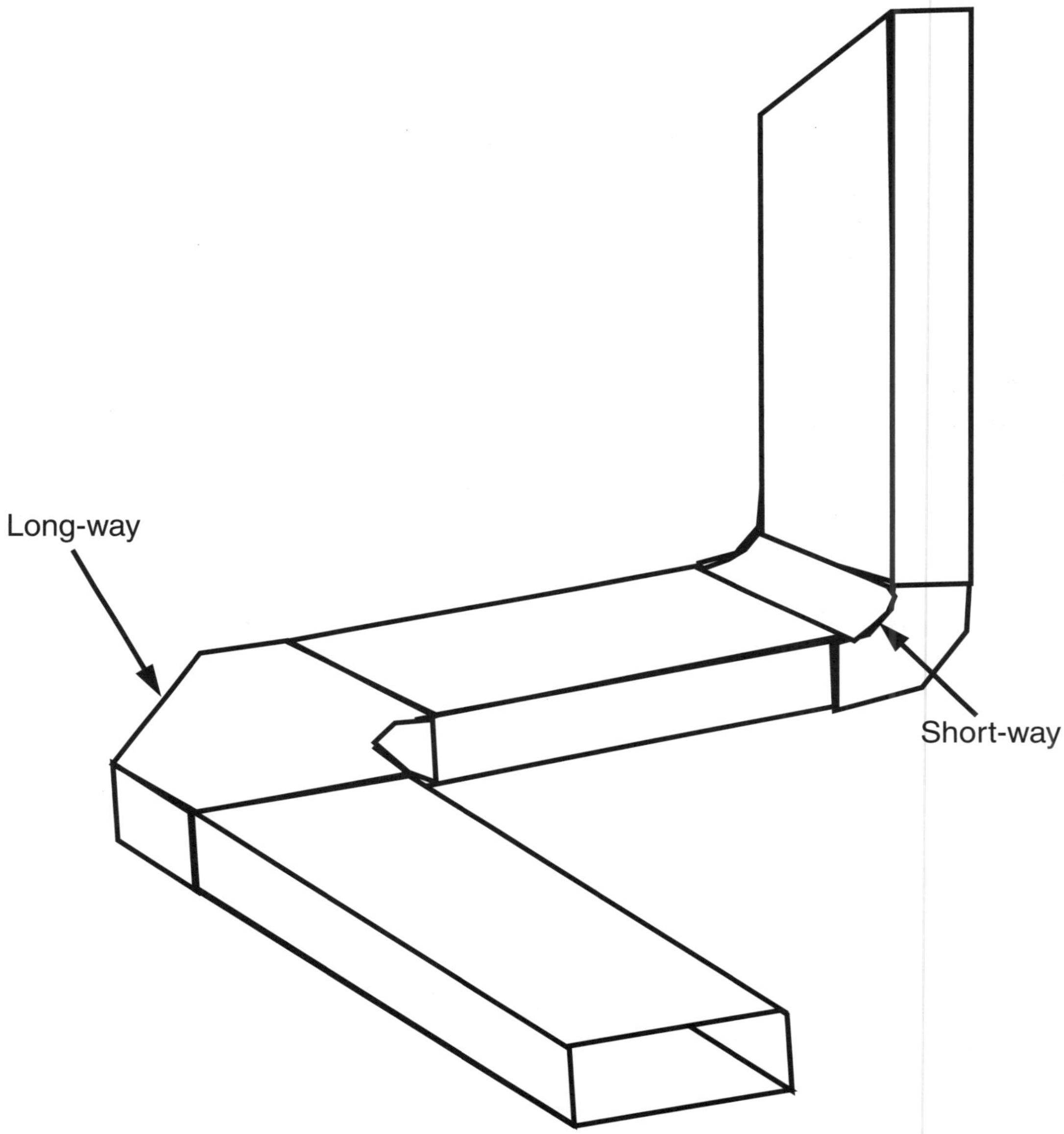

HC 28-3

Installing a Floor Diffuser

A

Cut Opening in Floor

B

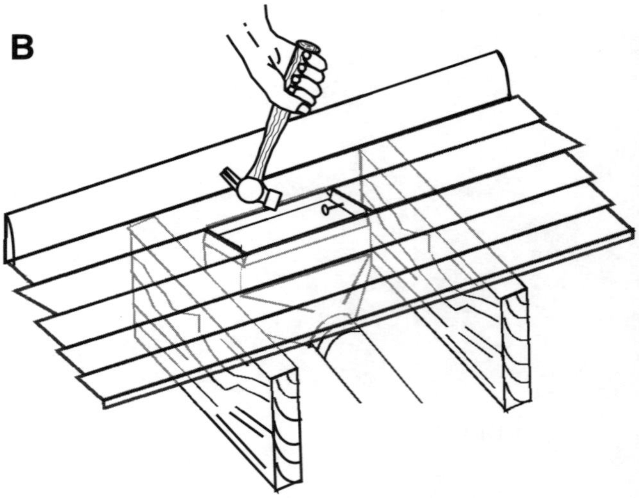

Nail Register Boot to Floor

C

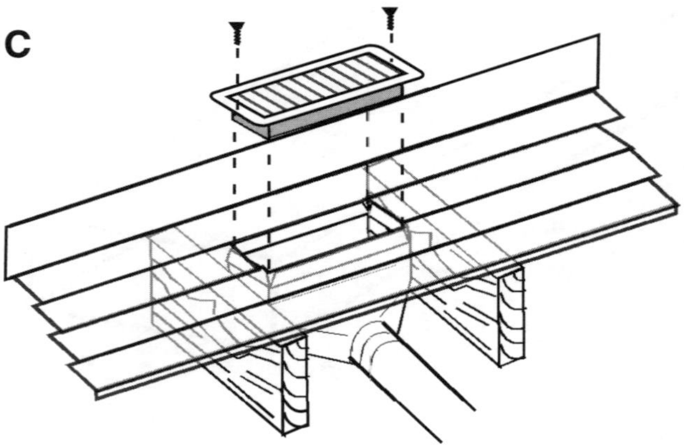

Install and Secure Diffuser

HC 28-4

Sheeted Joists for Return Air

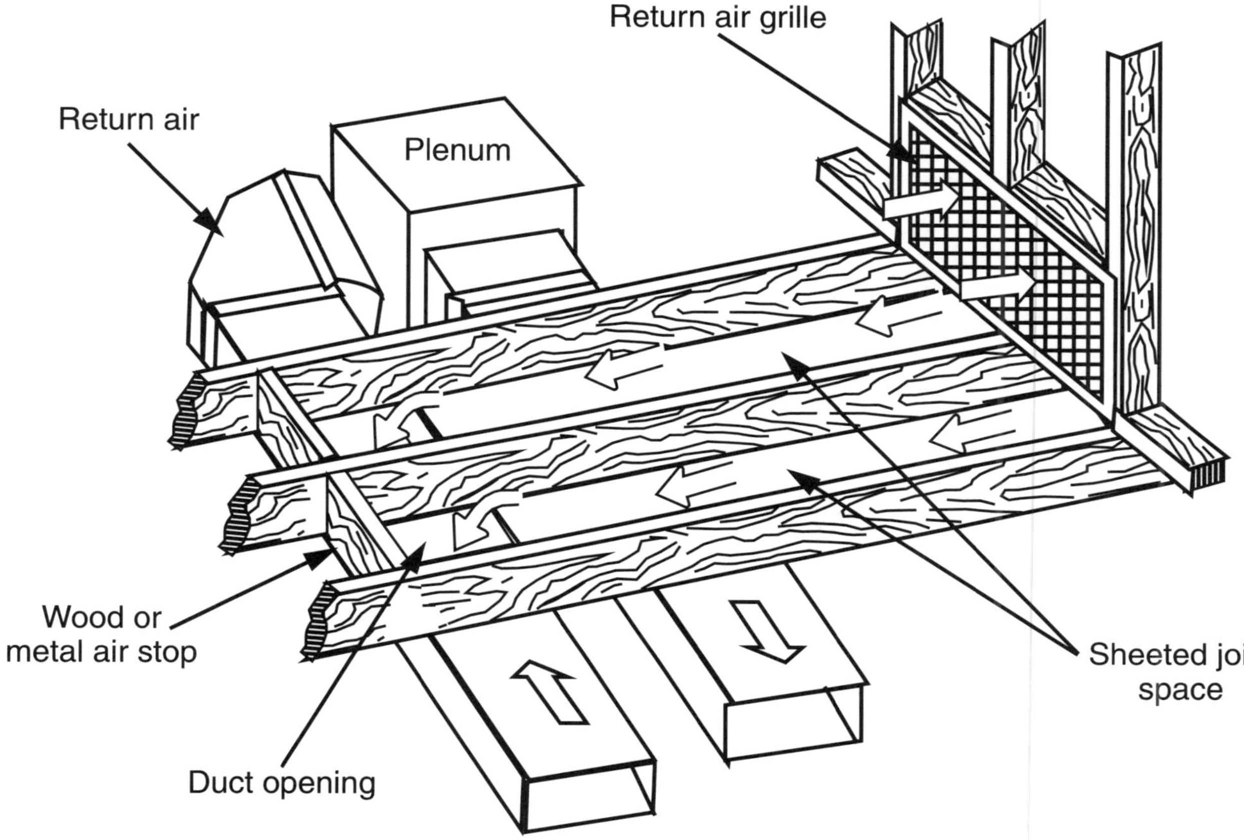

HC 28-5

Name ______________________________

Date ______________________________

Instructor ______________________________

Period______________________________

28 Ductwork

Quiz

Instructions: Write your answers on the lines provided at the right-hand side of the page.

1. Most heating and cooling systems in use today are of the _____-air type. 1. ______________
2. Upflow and _____ furnaces are most common, but horizontal-flow units must be used for some applications. 2. ______________
3. The _____ distribution system combines rectangular ducts with round pipe. 3. a. ______________
   a. centrifugal b. ______________
   b. expanded ducting c. ______________
   c. extended plenum d. ______________
   d. radial piping
4. The outermost layer of insulated flexible duct is called the _____. 4. ______________
5. What material should aviation snips never be used to cut? 5. ______________
6. A starting collar is installed in the furnace plenum to begin a run of round pipe; the fitting used to begin a run of rectangular duct is called a _____. 6. ______________
7. *True or false?* To cut a length of round pipe, you should first snap together the seam. 7. ______________
8. Supply air outlets are called registers or _____. 8. ______________
9. In a perimiter system, return air grilles are usually positioned on an _____ wall. 9. ______________
10. To prevent air leaks, joints in ductwork should be sealed with _____ and fiberglass membrane. 10. ______________

29

Gas Heat with Air Conditioning

Objectives

After studying this chapter in the text, the student will be able to:

- ❑ Identify each component of a gas heating system and describe its purpose.
- ❑ Discuss proper adjustment techniques for burner flames, fan/limit controls, and heat anticipators in gas heating systems.
- ❑ Demonstrate how to wire heating and cooling system components.
- ❑ Describe methods for properly connecting two-speed motors.

Instructional Materials

Text: Pages 519–538
Lab Manual: Pages 223–230
Instructor's Manual:
Visual Masters, Pages 255–259
Chapter 29 Quiz, Page 260

Teaching Strategy

Text

1. Review objectives.
2. Discuss and assign chapter contents. Use these visual masters as appropriate:
   HC 29-1: Heat Exchanger
   HC 29-2: Ignition Systems
   HC 29-3: Limit Switch
   HC 29-4: Fan Switch
   HC 29-5: Gas Heat with Air Conditioning
3. Direct students to complete Test Your Knowledge questions, page 538. After completion, discuss correct answers with students.

Lab Manual

1. Provide overview of Activities. Where necessary, stress safe laboratory practices and the importance of wearing protective clothing.
2. Assign Activities to individuals or pairs of students, as appropriate.
   Activity 29-1: Disassembling and Reassembling a Gas Furnace Blower Assembly
   Activity 29-2: Adjusting the Pilot Light and Thermocouple
   Activity 29-3: Wiring a Limit Switch
   Activity 29-4: Wiring a Furnace Circuit
   Activity 29-5: Wiring a Two-speed Blower Motor Circuit
   Activity 29-6: Wiring a Circuit for a Furnace with Air Conditioning
3. Observe student performance on Activities, providing guidance or correction as needed. After reviewing the student's work, initial and date the activity in the place provided.

Evaluation

1. Administer the quiz from this Instructor's Manual.
2. Grade the quiz, and return it with appropriate comments for the student.

Answers to Test Your Knowledge, Page 538

1. air ducts
2. supply air
3. plenum
4. false
5. They reduce airflow.
6. burners
7. It proves that the pilot flame is functioning.
8. heating
9. 200

10. the transformer secondary
11 the gas valve
12. Y (yellow)
13. blower relay coil
14. heating
15. blower relay

Answers to Lab Manual, Pages 226–229

Activity 29-3

1.

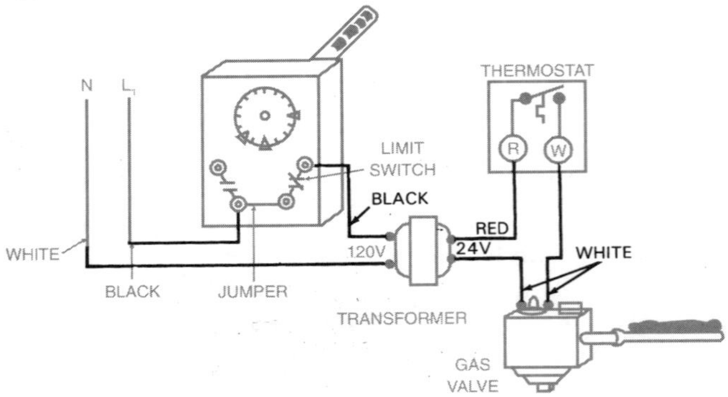

2.

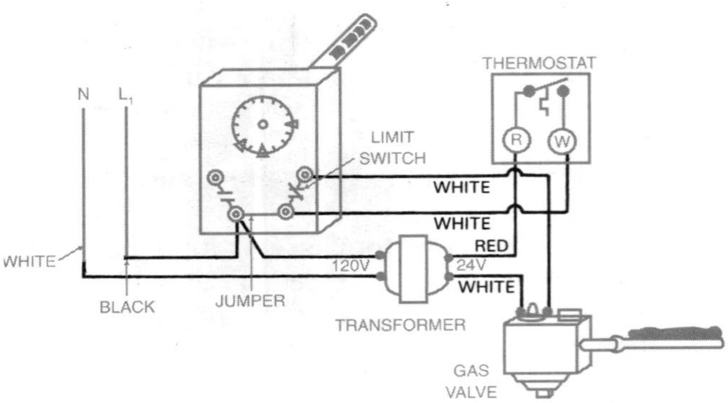

Activity 29-4

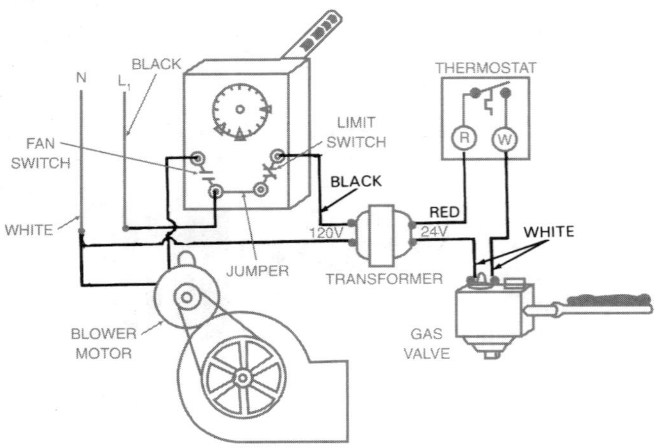

Activity 29-5

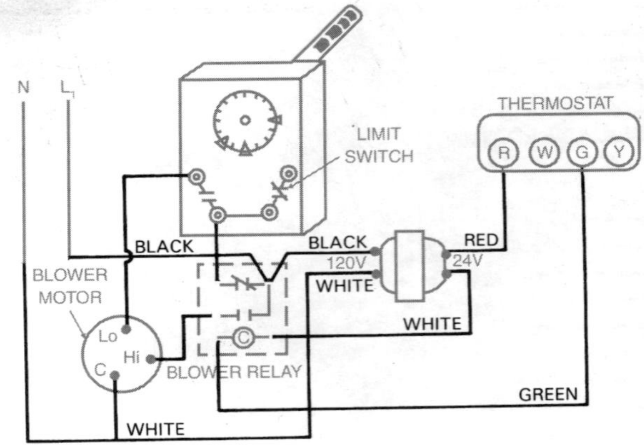

Activity 29-6

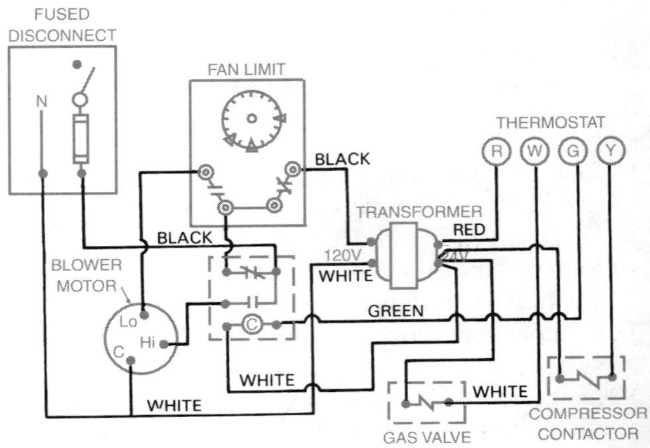

Answers to Quiz

1. supply air
2. direct
3. combustion
4. Primary
5. a. burner body
   b. secondary air
   c. manifold
   d. orifice
   e. primary air
6. combination fan/limit control
7. b. gas valve
8. blower relay
9. b. 200
10. false

Heat Exchanger

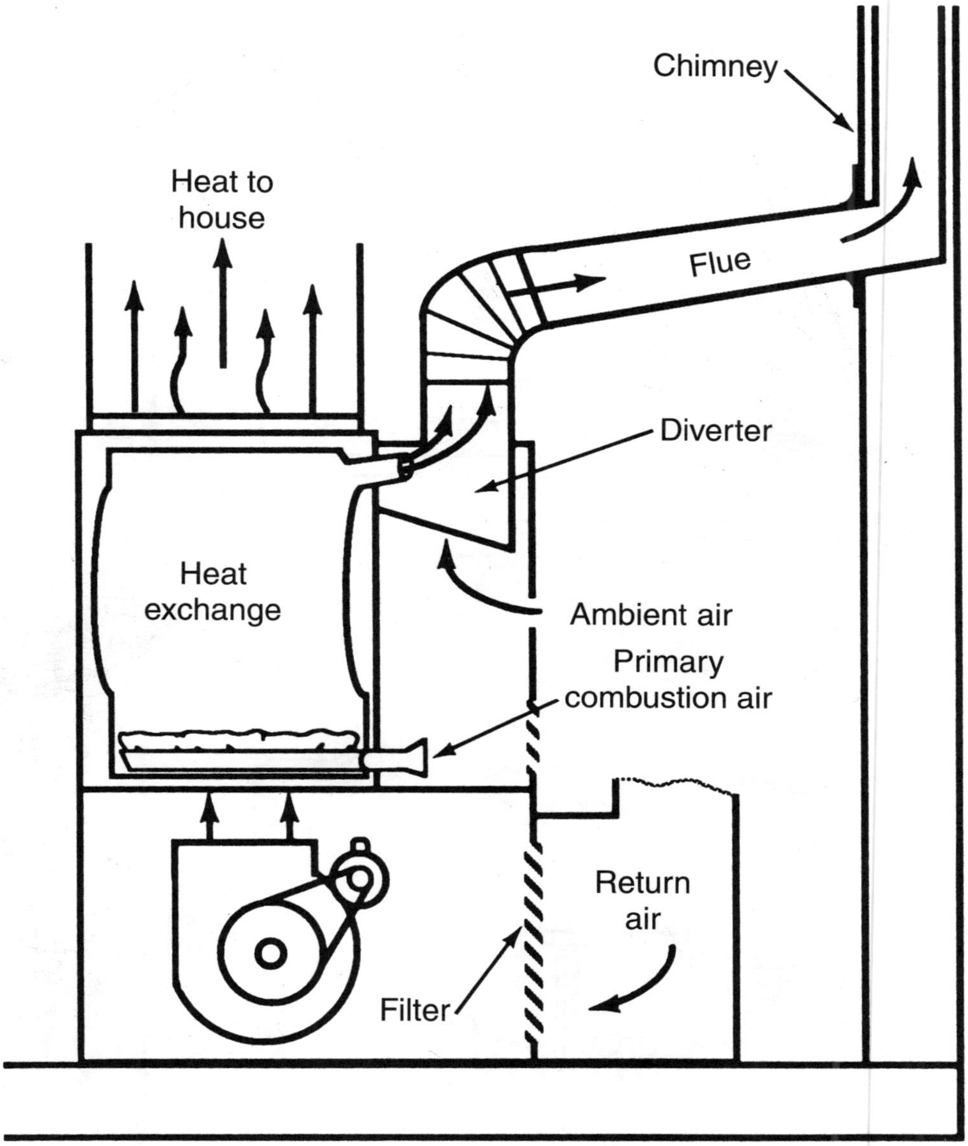

HC 29-1

Ignition Systems

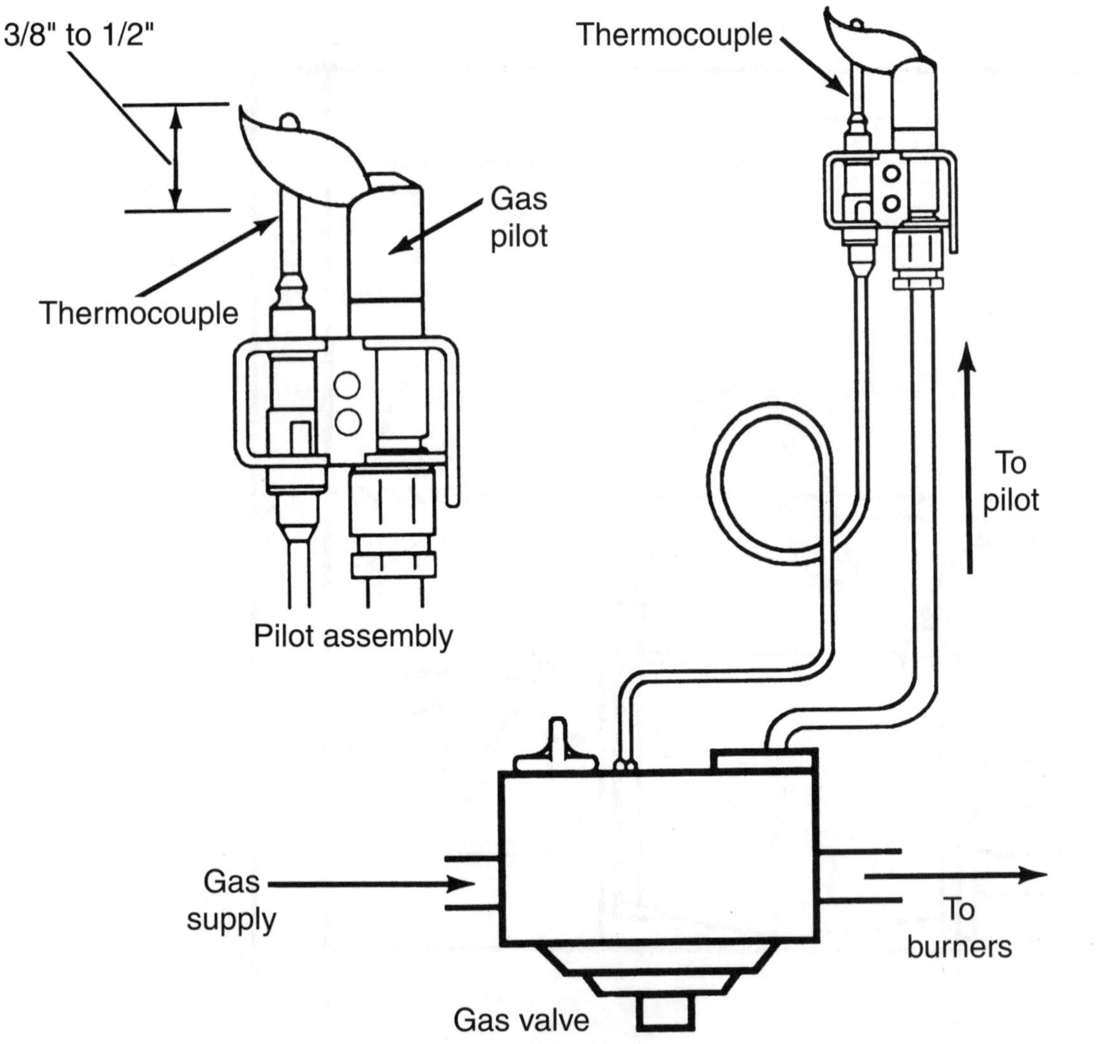

Pilot Flame

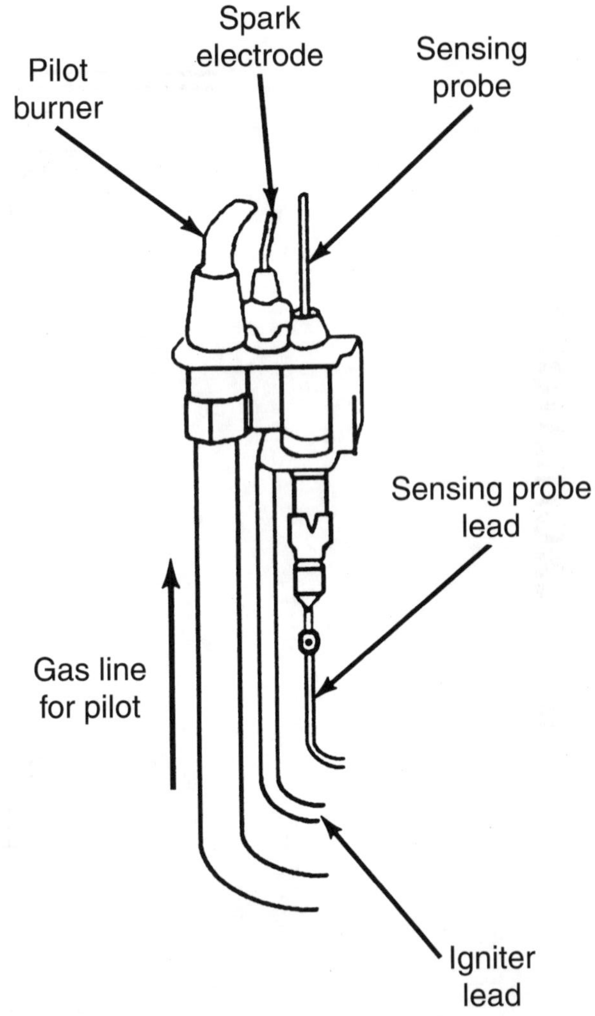

Electronic (Pilotless)

HC 29-2

Limit Switch

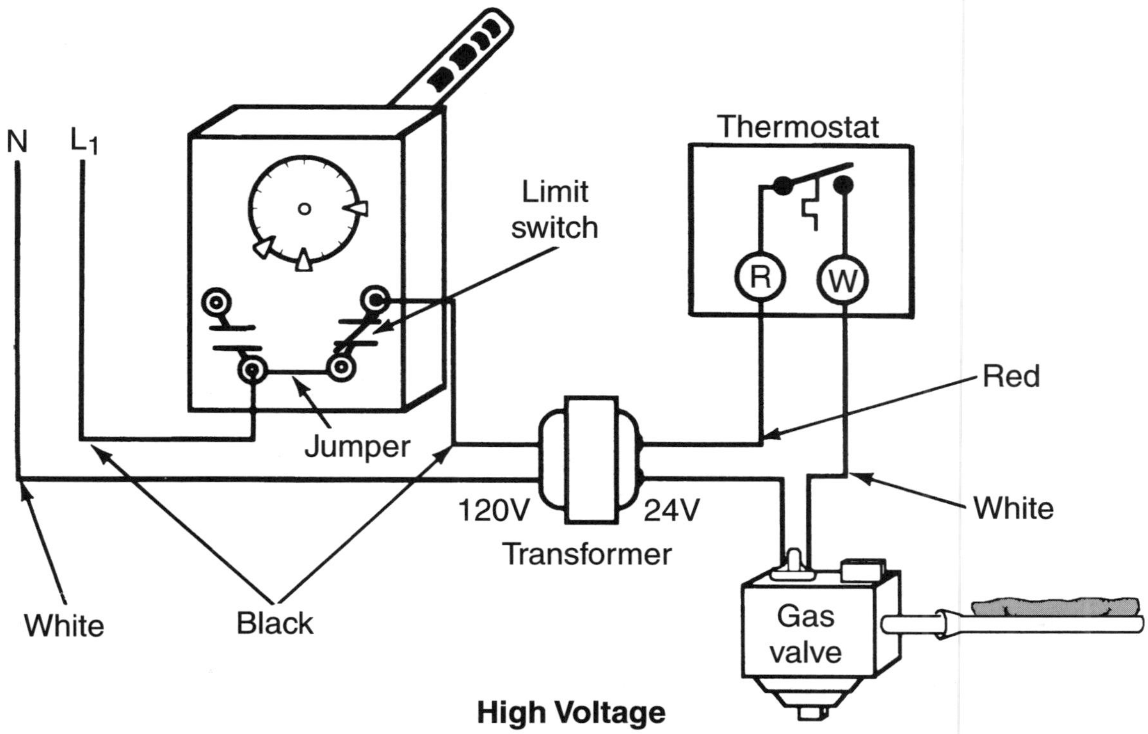

High Voltage

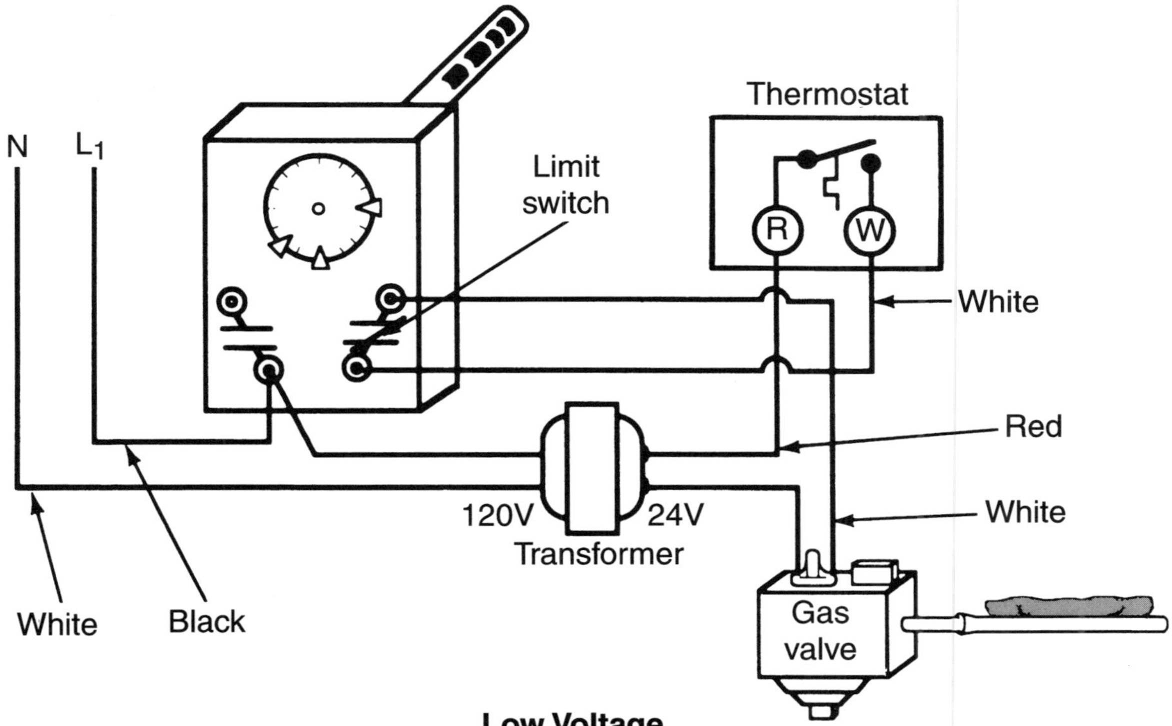

Low Voltage

HC 29-3

Fan Switch

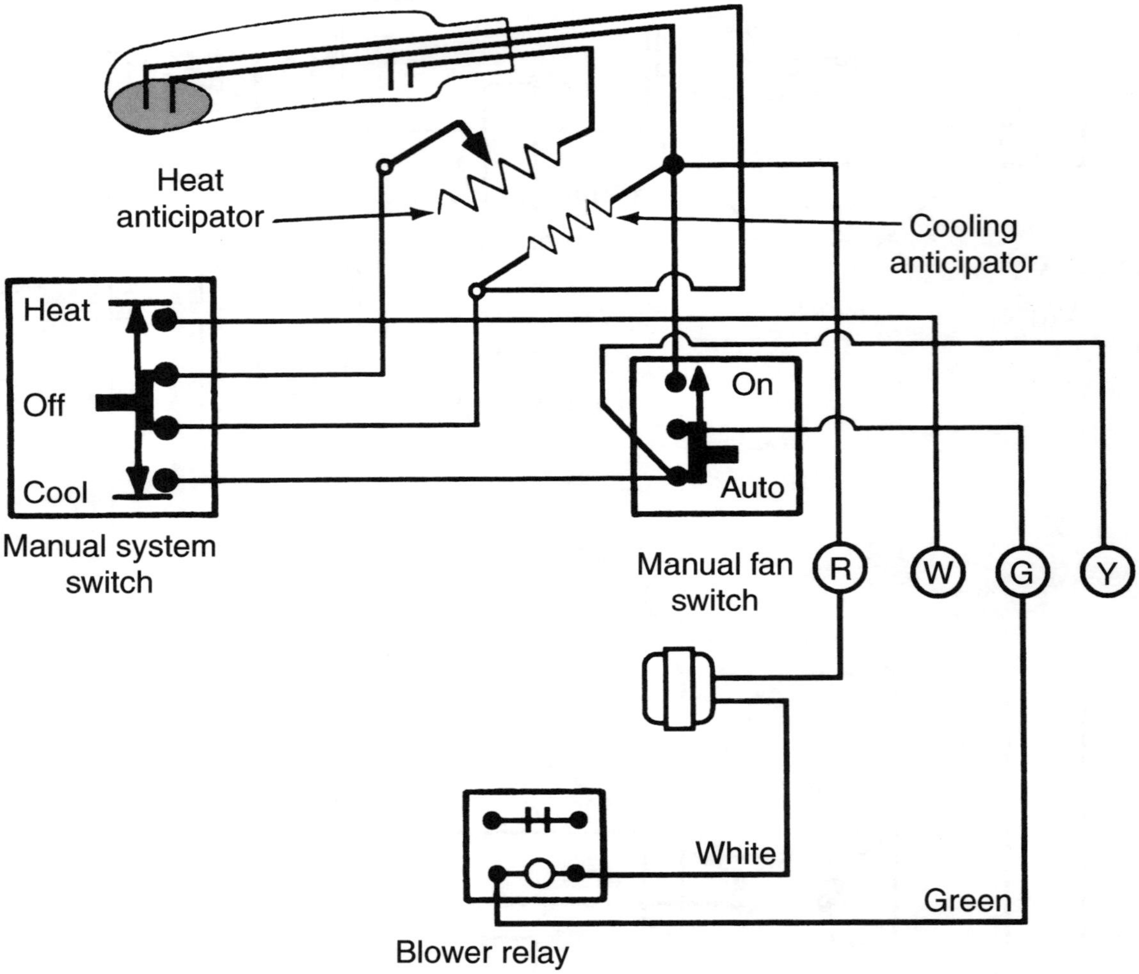

HC 29-4

Gas Heat with Air Conditioning

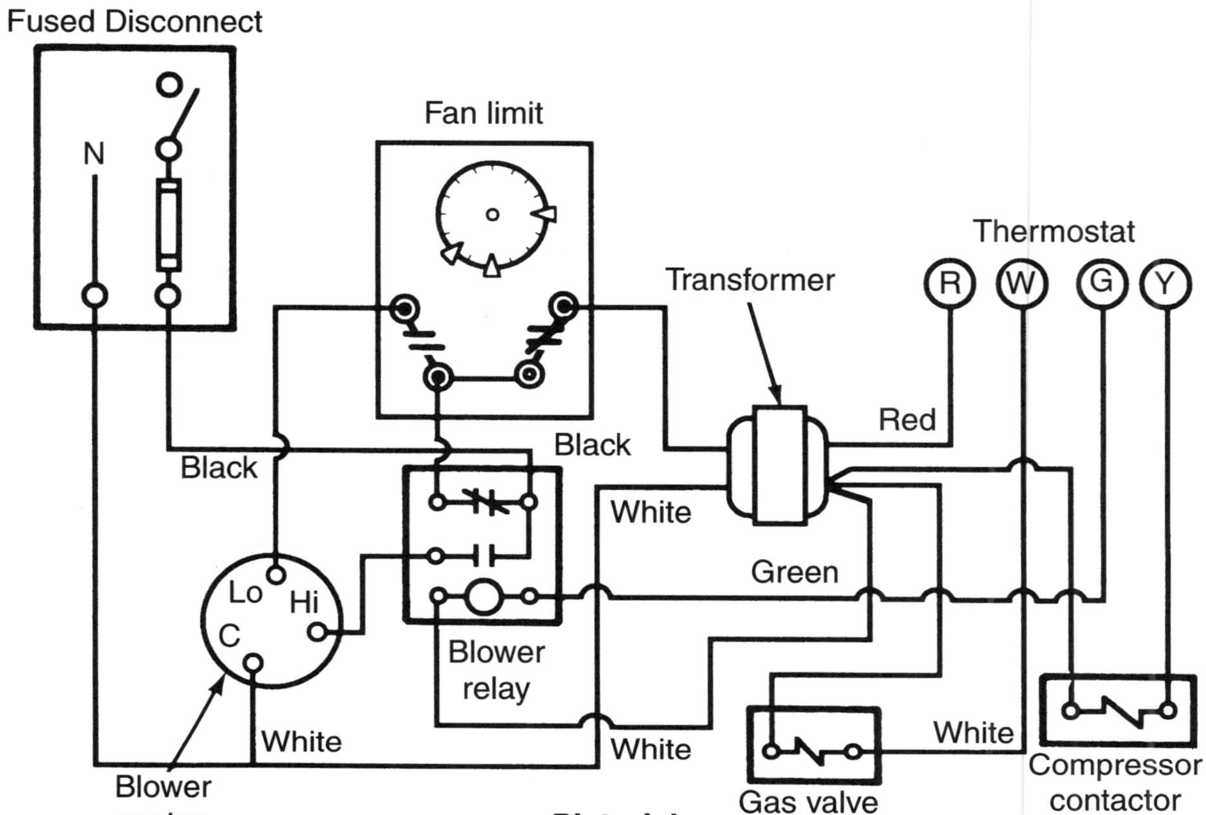

Pictorial

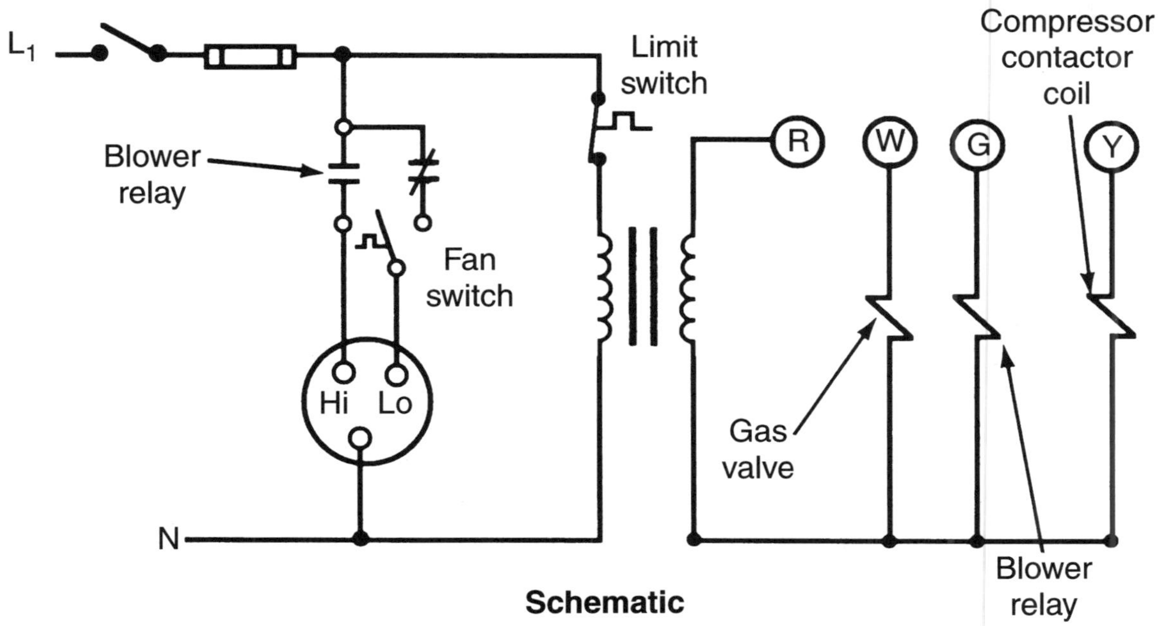

Schematic

HC 29-5

Name ______________________________

Date ______________________________

Instructor ______________________________

Period______________________________

29

Gas Heat with Air Conditioning

Quiz

Instructions: Write your answers on the lines provided at the right-hand side of the page.

1. Ducts carrying air to the conditioned space are called _____ ducts. 1. ____________________

2. The two types of furnace blower are belt-drive and _____-drive. 2. ____________________

3. Gas-fueled furnaces produce heat by burning fuel, a process called _____. 3. ____________________

4. _____ air is drawn into the burner by the jet of fuel gas. 4. ____________________

5. Identify the components and airflows through this burner body. 5. a. ____________________ b. ____________________ c. ____________________ d. ____________________ e. ____________________

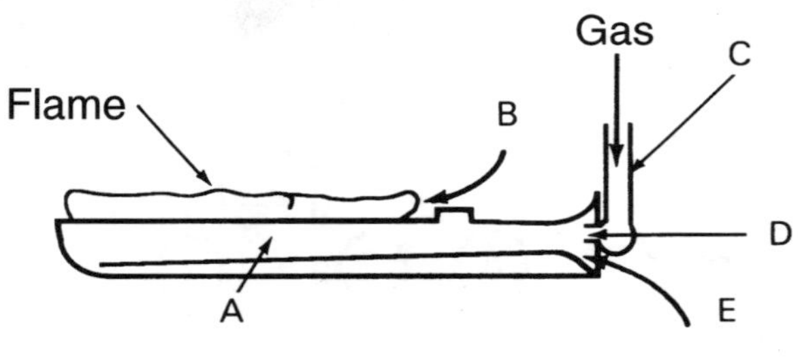

6. What control is normally used to turn the furnace blower on and off? 6. ____________________

7. The thermostat controls power supply to the _____. 7. ____________________
   a. transformer
   b. gas valve
   c. thermocouple

8. What device is used to control the blower motor during the cooling cycle? 8. ____________________

9. The limit switch will open its contacts when the furnace temperature reaches approximately _____°F. 9. ____________________
   a. 100
   b. 200
   c. 300

10. *True or false?* The R (red) terminal on the thermostat is connected to the compressor contactor coil. 10. ____________________

30

Oil Heat with Air Conditioning

Objectives

After studying this chapter in the text, the student will be able to:

- ❑ Describe how fuel oil and air are mixed in the combustion chamber for proper combustion.
- ❑ Explain how the mixture of oil and air is ignited.
- ❑ State dimensions to be used for proper electrode placement.
- ❑ Define the purpose of primary air adjustments.
- ❑ Explain why draft control is used and how it works.
- ❑ Describe the purpose and location of flame safeguards.
- ❑ Describe the electrical connections needed for heating and air conditioning circuits.
- ❑ Describe the operation and purpose of safety controls.

Instructional Materials

Text: Pages 539–552
Lab Manual: Pages 231–238
Instructor's Manual:
Visual Masters, Pages 265–268
Chapter 30 Quiz, Page 269

Teaching Strategy

Text

1. Review objectives.
2. Discuss and assign chapter contents. Use these visual masters as appropriate:
   HC 30-1: Oil Supply System
   HC 30-2: Ignition System
   HC 30-3: Stack Control
   HC 30-4: Oil Heat with Air Conditioning
3. Direct students to complete Test Your Knowledge questions, page 552. After completion, discuss correct answers with students.

Lab Manual

1. Provide overview of Activities. Where necessary, stress safe laboratory practices and the importance of wearing protective clothing.
2. Assign Activities to individuals or pairs of students, as appropriate.
   Activity 30-1: Examining a Gun-type Oil Burner Assembly
   Activity 30-2: Checking and Adjusting an Oil-burning Furnace
   Activity 30-3: Wiring an Oil Furnace Circuit
   Activity 30-4: Wiring a Furnace Circuit Including an Isolation Relay
   Activity 30-5: Wiring a Circuit for a Furnace with Air Conditioning
3. Observe student performance on Activities, providing guidance or correction as needed. After reviewing the student's work, initial and date the activity in the place provided.

Evaluation

1. Administer the quiz from this Instructor's Manual.
2. Grade the quiz, and return it with appropriate comments for the student.

Answers to Test Your Knowledge, Page 552

1. No. 2
2. oil pump and burner fan
3. To supply primary air for mixing with oil.
4. true
5. 100
6. To bleed air from the system.

7. degree of spray angle; gallons per hour
8. 1/2″ above the nozzle opening; 1/2″ in front of the nozzle opening; and 1/8″ to 3/16″ gap
9. 120 V primary; 10,000 V secondary
10. safe
11. T
12. combination fan-limit control (or fan switch)
13. cad cell
14. Any three of the following: heating-cooling thermostat; 40 VA transformer; isolation relay; blower relay
15. To isolate the heating control circuits from the cooling control circuits.

Answers to Lab Manual, Pages 232–237

Activity 30-2

Student response (answers will vary).

Activity 30-3

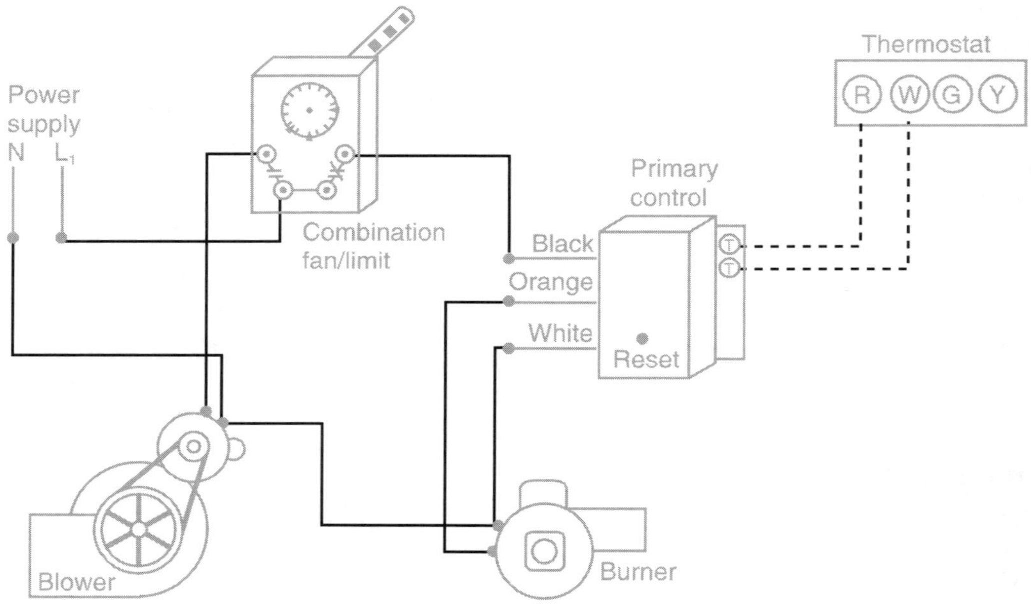

Activity 30-4

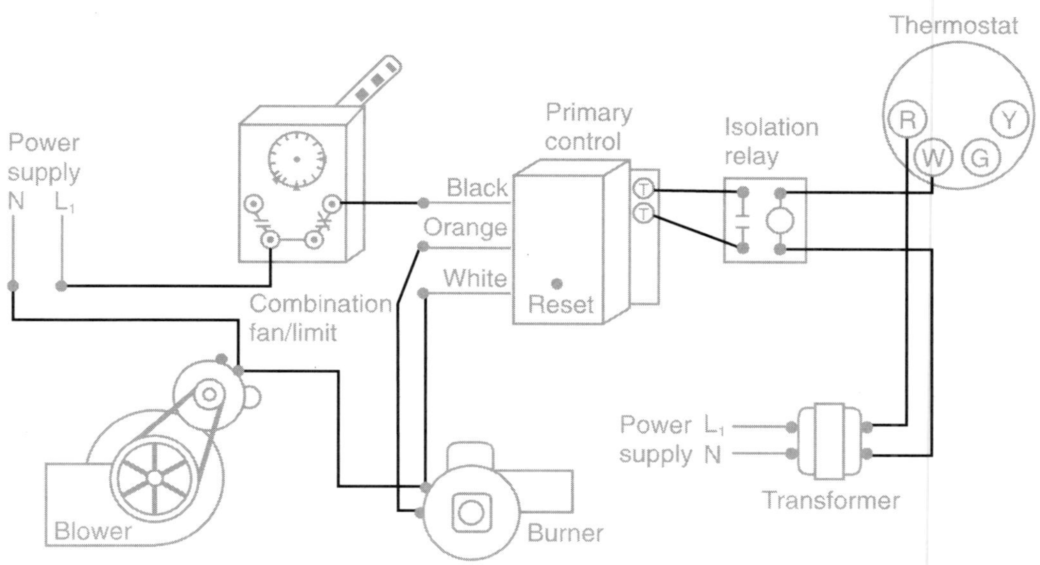

Activity 30-5

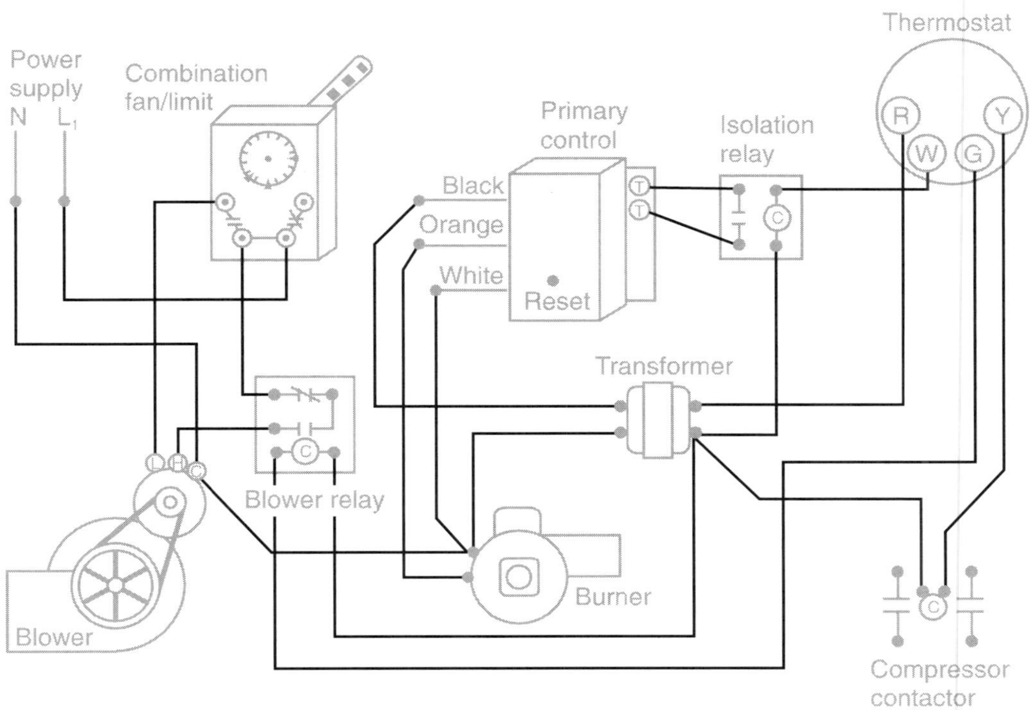

Answers to Quiz

1. 140,000
2. true
3. a. 100
4. To bleed air from the system.
5. flow rate (gph); spray angle (degrees)
6. 1/8″ to 3/16″
7. 10,000
8. false
9. b. cadmium cell
10. primary control

Oil Supply System

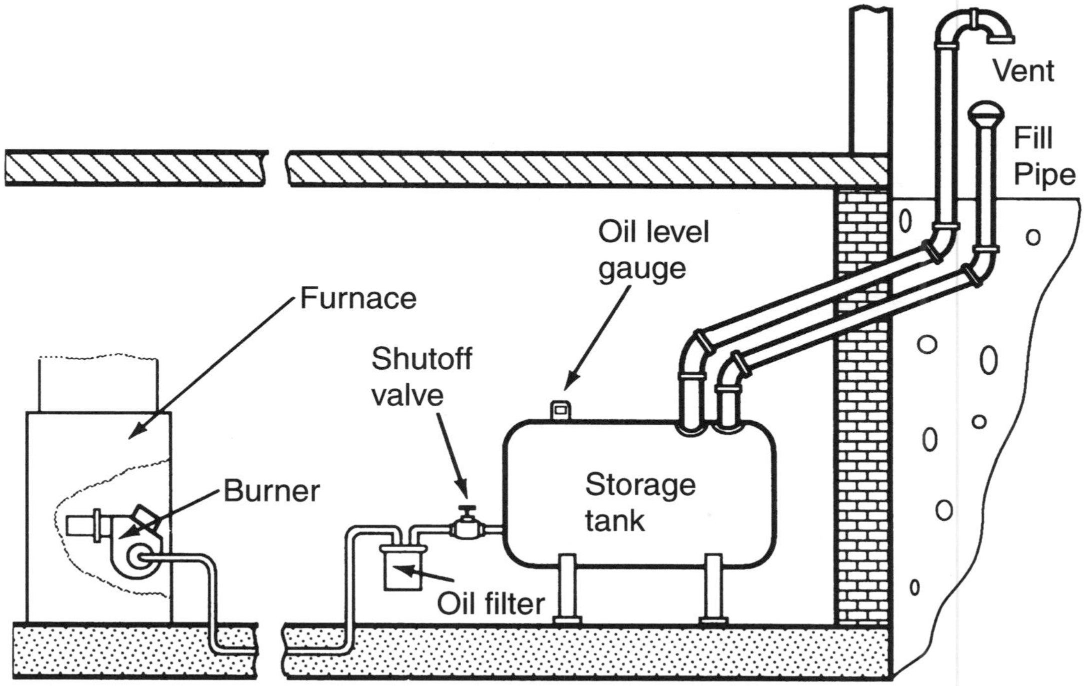

One-pipe System

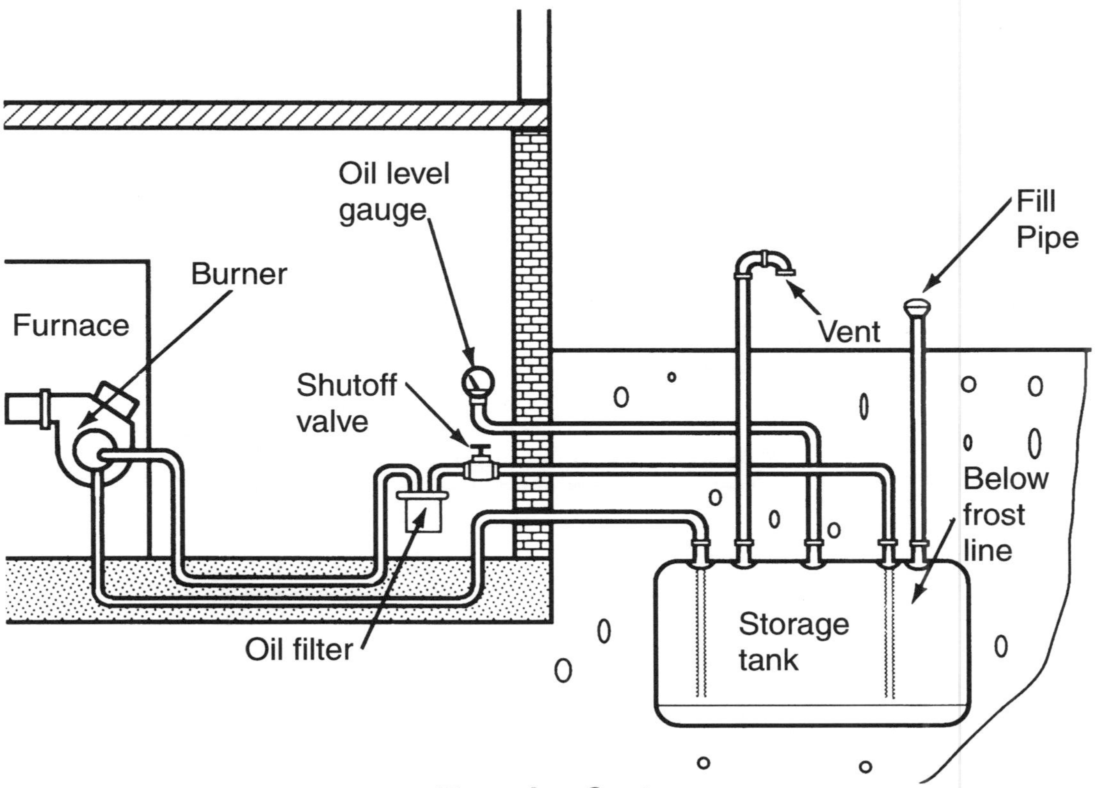

Two-pipe System

HC 30-1

Ignition System

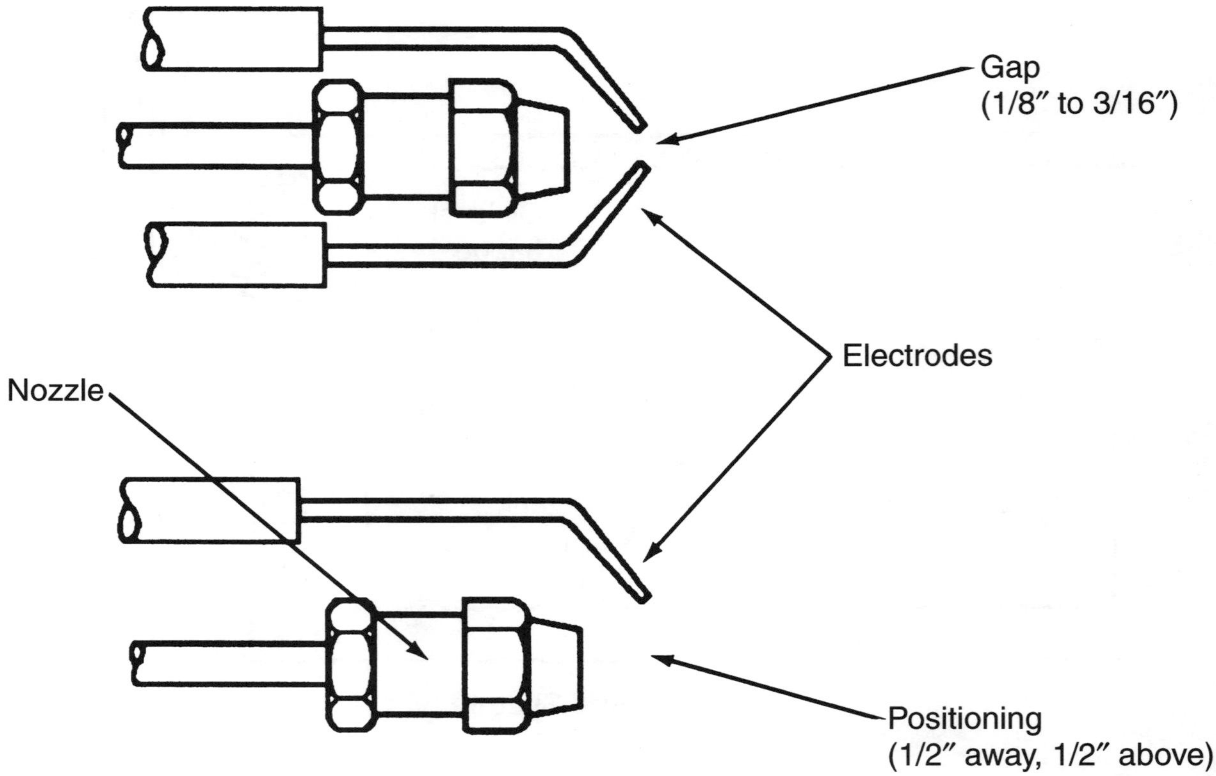

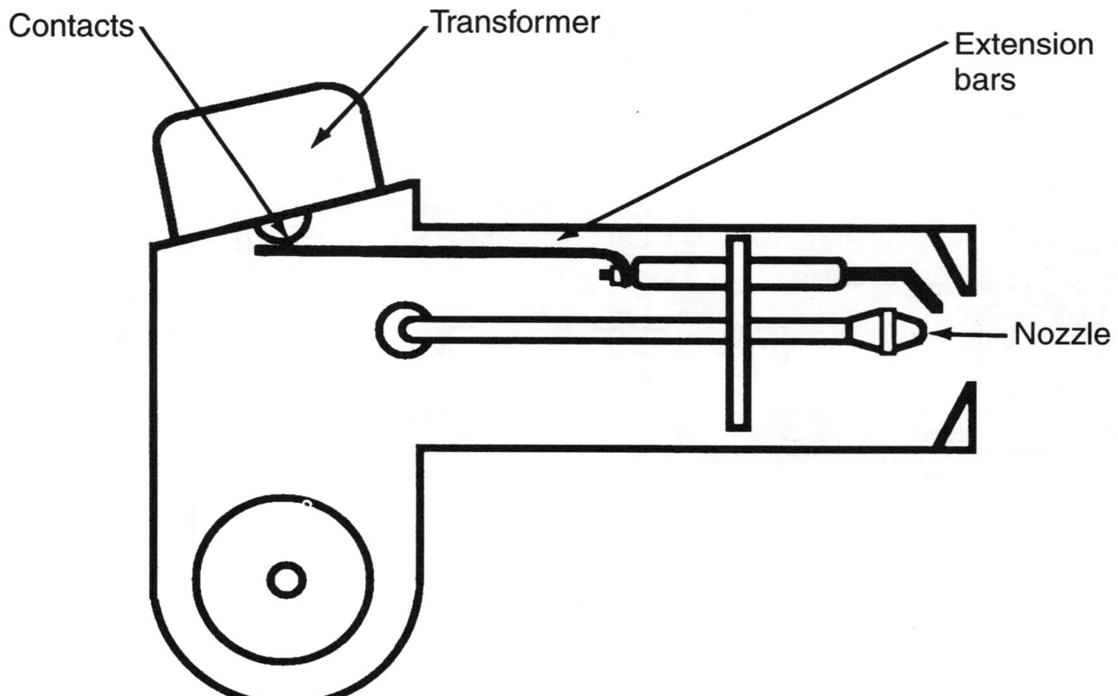

HC 30-2

Stack Control

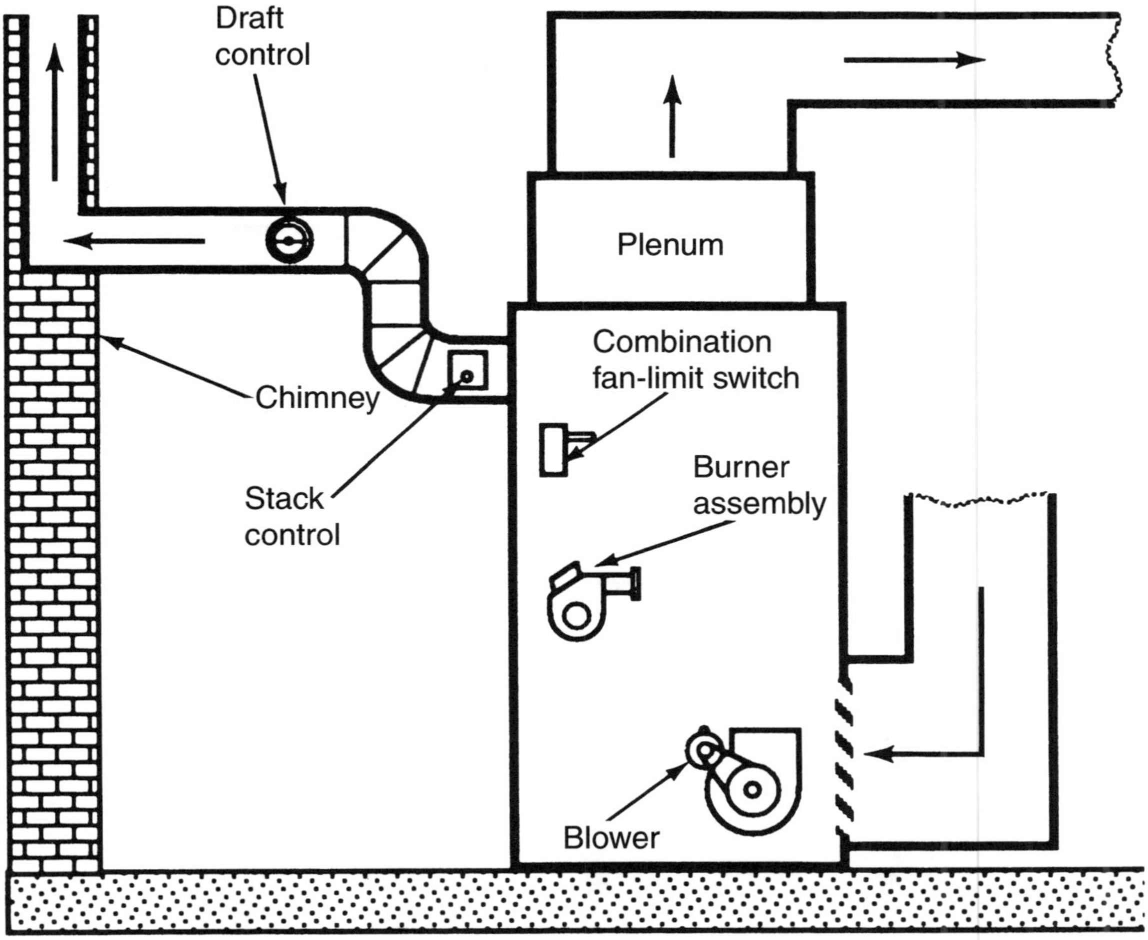

HC 30-3

Oil Heat with Air Conditioning

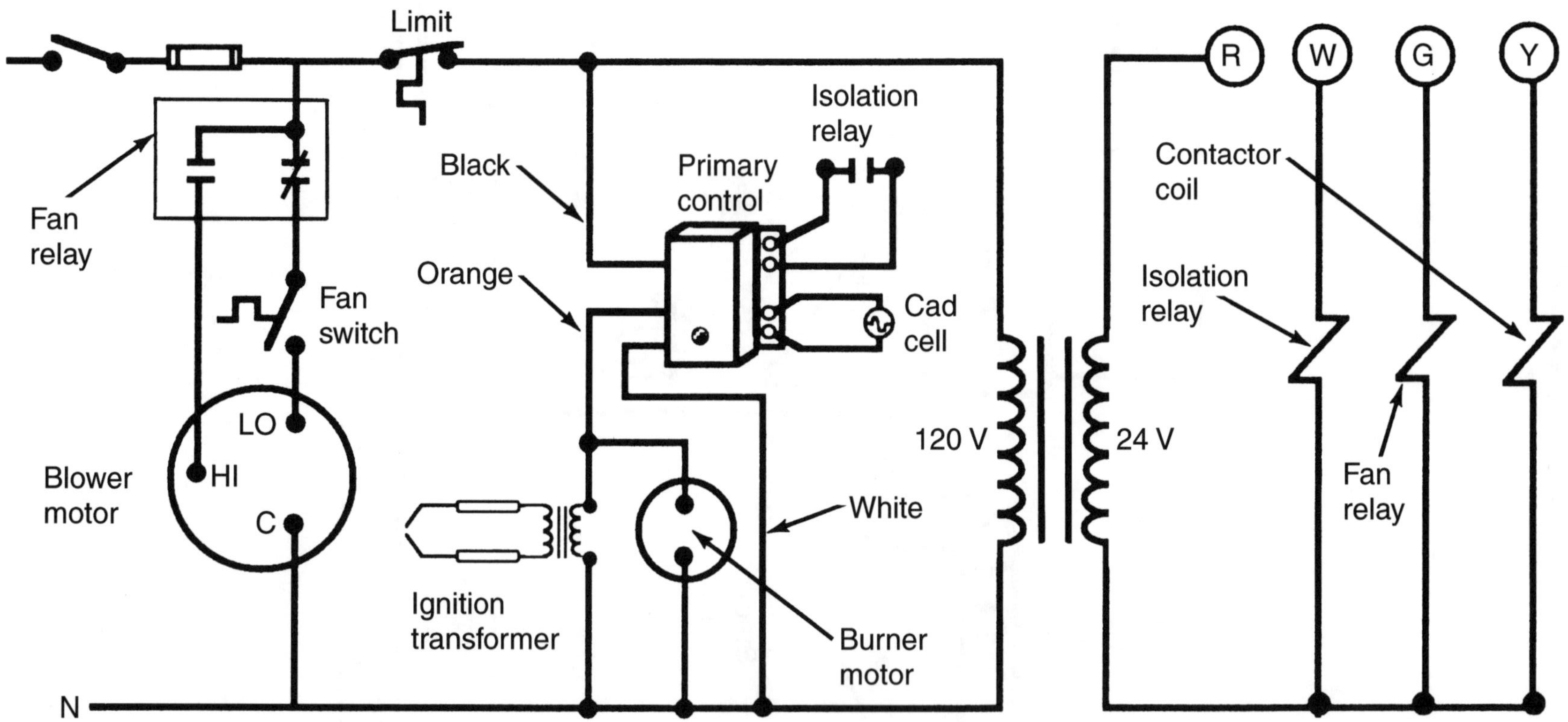

HC 30-4

Name ______________________________

Date ______________________________

Instructor ______________________________

Period______________________________

30

Oil Heat with Air Conditioning

Quiz

Instructions: Write your answers on the lines provided at the right-hand side of the page.

1. Fuel oil in the No. 2 grade is the most popular, and produces about _____ Btu/gal. 1. ______________

2. *True or false?* A draft control device is used to maintain a constant flue pressure. 2. ______________

3. The oil pump is used to increase oil pressure to ______ psig. 3. ______________
   a. 100
   b. 150
   c. 200

4. What is the purpose of the bleed port on an oil pump? 4. ______________

5. What are the two important ratings on a fuel nozzle? 5. ______________

   ____

6. How much gap is permitted between the electrodes? 6. ______________

7. The ignition transformer increases supply voltage to _____ V. 7. ______________

8. *True or false?* The primary control on an oil furnace has an automatic reset. 8. ______________

9. Flame safeguard is provided by a _____. 9. ______________
   a. thermocouple
   b. cadmium cell
   c. bimetal helix

10. The two wires from the cadmium cell are connected to the _____. 10. ______________

31

Electric Heat with Air Conditioning

Objectives

After studying this chapter in the text, the student will be able to:

- ❑ Identify components of an electric heating system and describe their purpose.
- ❑ Describe proper wiring connections for various heating system components.
- ❑ Recognize variations in wiring.
- ❑ Read schematics for electric heating systems.
- ❑ Describe the operation of two-stage heating systems.
- ❑ Discuss how to isolate control circuits from line voltage circuits.

Instructional Materials

Text: Pages 553–566
Lab Manual: Pages 239–246
Instructor's Manual:
Visual Masters, Pages 275–280
Chapter 31 Quiz, Page 281

Teaching Strategy

Text

1. Review objectives.
2. Discuss and assign chapter contents. Use these visual masters as appropriate:
   HC 31-1: Heat Relay and Sequencer
   HC 31-2: Two-stage Electric Heat
   HC 31-3: Outdoor Thermostat Configurations
   HC 31-4: Safety Controls
   HC 31-5: Two-stage Electrical Heat with Air Conditioning—Schematic
   HC 31-6: Two-stage Electric Heat with Air Conditioning–Pictorial
3. Direct students to complete Test Your Knowledge questions, page 566. After completion, discuss correct answers with students.

Lab Manual

1. Provide overview of Activities. Where necessary, stress safe laboratory practices and the importance of wearing protective clothing.
2. Assign Activities to individuals or pairs of students, as appropriate.
   Activity 31-1: Checking Components of an Electric Heat System
   Activity 31-2: Wiring an Electric Heat Circuit with Multiple Elements and Heat Relays
   Activity 31-3: Wiring an Electric Two-stage, Five-element Heating System
   Activity 31-4: Wiring a Two-stage Electric Furnace with Single-stage Air Conditioning
3. Observe student performance on Activities, providing guidance or correction as needed. After reviewing the student's work, initial and date the activity in the place provided.

Evaluation

1. Administer the quiz from this Instructor's Manual.
2. Grade the quiz, and return it with appropriate comments for the student.

Answers to Test Your Knowledge, Page 566

1. no
2. 240 V
3. It must be located within sight of the furnace. It must be capable of being locked in the open position.
4. nickel and chromium
5. Ceramic

6. for safety
7. temperature
8. fusible link or thermal fuse
9. heat relays and sequencers
10. true
11. capacity
12. switches (or bulbs)
13. second stage
14. blower relay
15. lack of condenser air; overcharge; air in the system

Answers to Lab Manual, Pages 239–246

Activity 31-1

1.–6. Student response (answers will vary).

7. The student's sketch should be similar to this sequencer drawing.

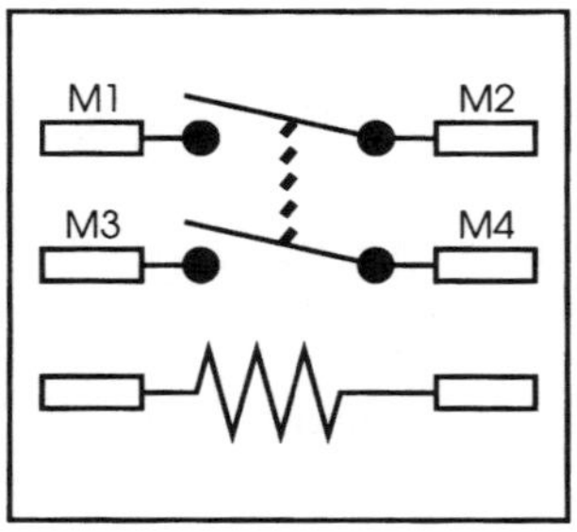

Activity 31-2

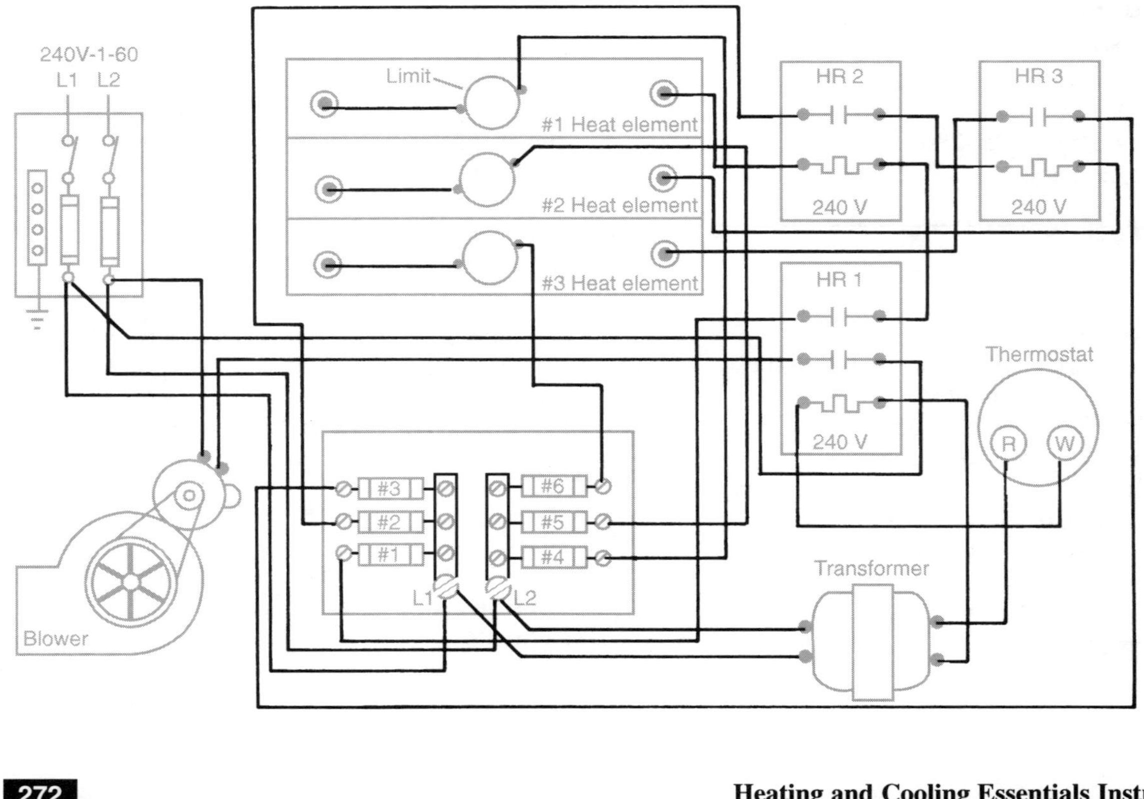

Activity 31-3 and Activity 31-4

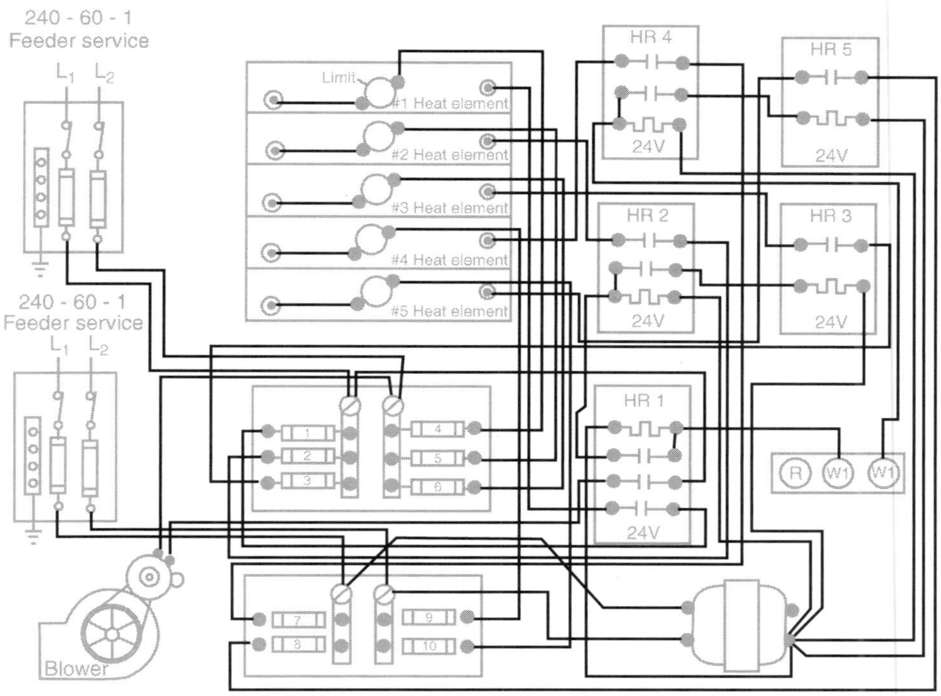

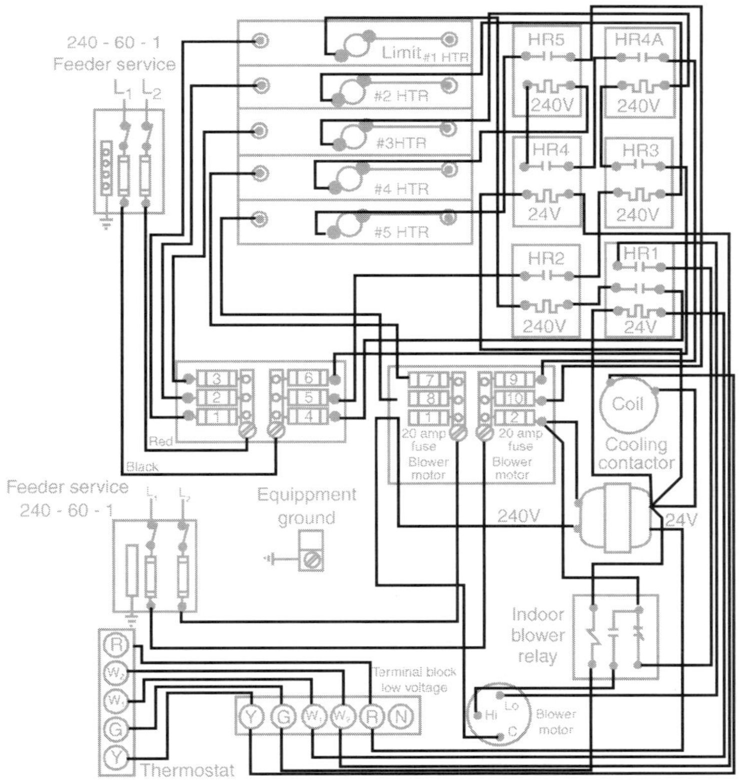

Answers to Quiz

1. 5 kW (5000 watts)
2. 240 V, three-wire, single-phase
3. dirty air filters; blower motor failure
4. nickel; chromium
5. heat relays; sequencers
6. capacity
7. b. W2
8. green
9. false
10. b. second-stage heat.

Heat Relay and Sequencer

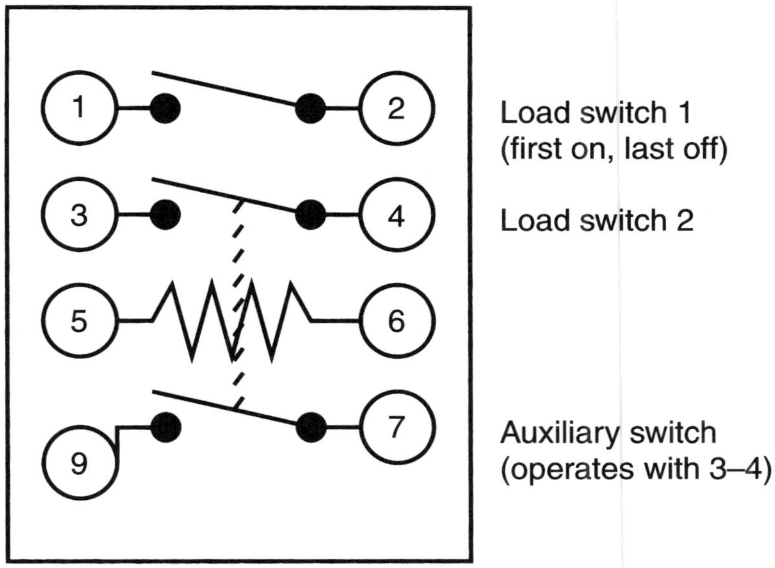

Heat Relay

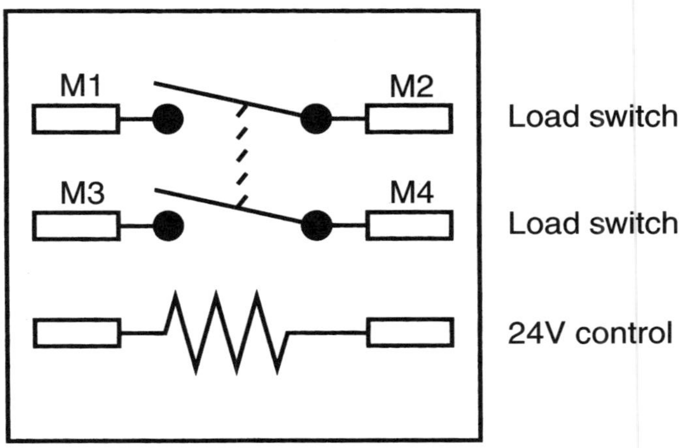

Sequencer

HC 31-1

Two-stage Electric Heat

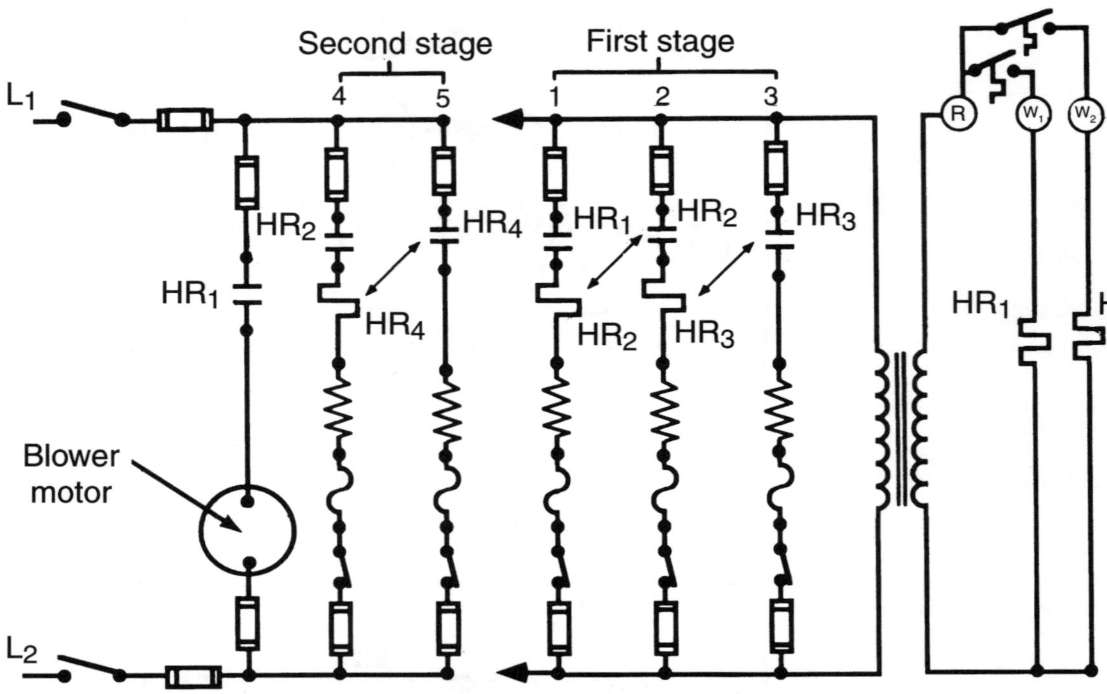

HC 31-2

Outdoor Thermostat Configurations

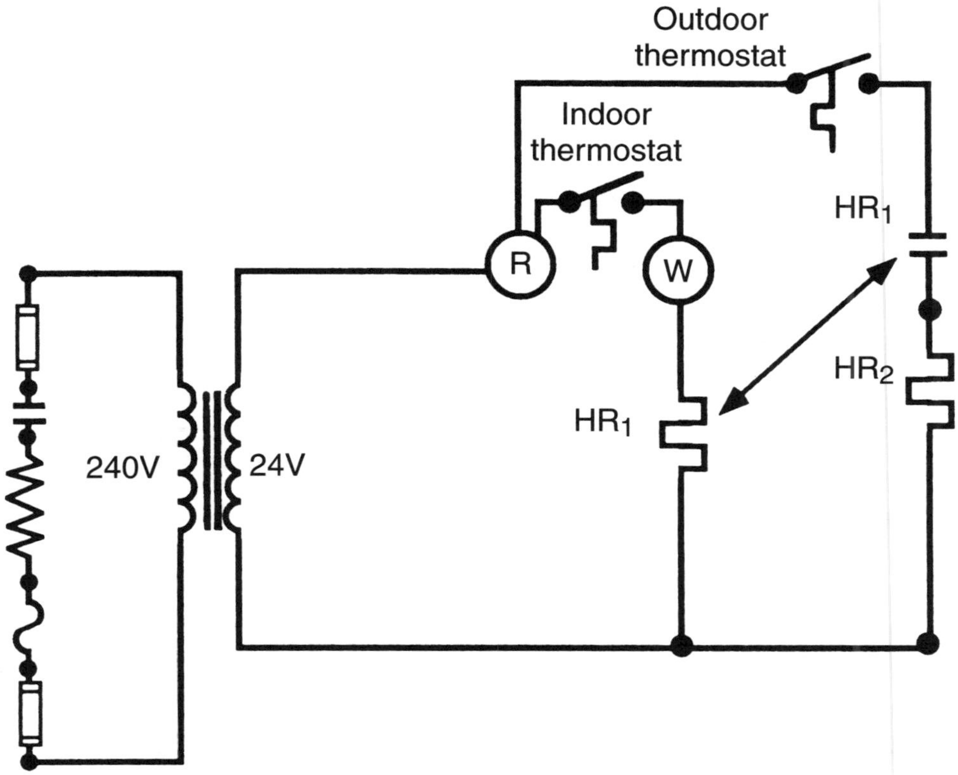

Circuit 1

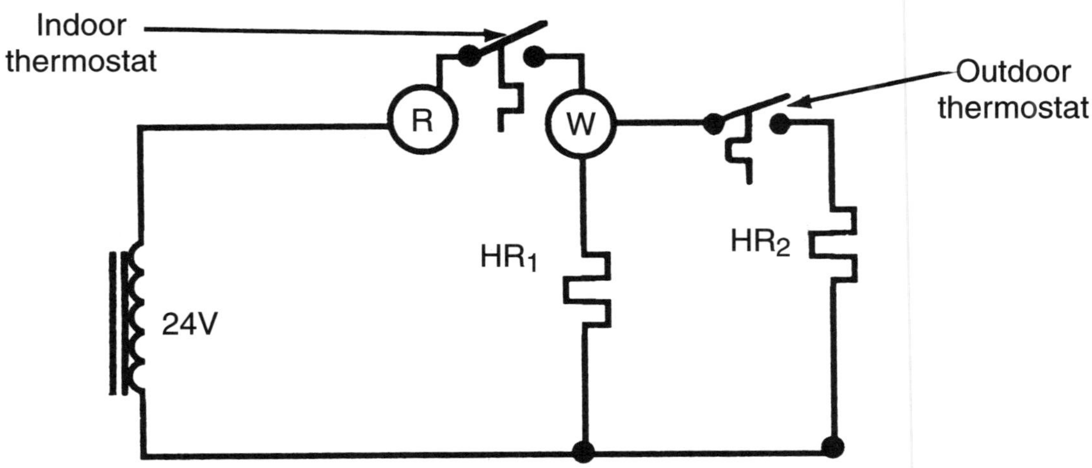

Circuit 2

HC 31-3

Safety Controls

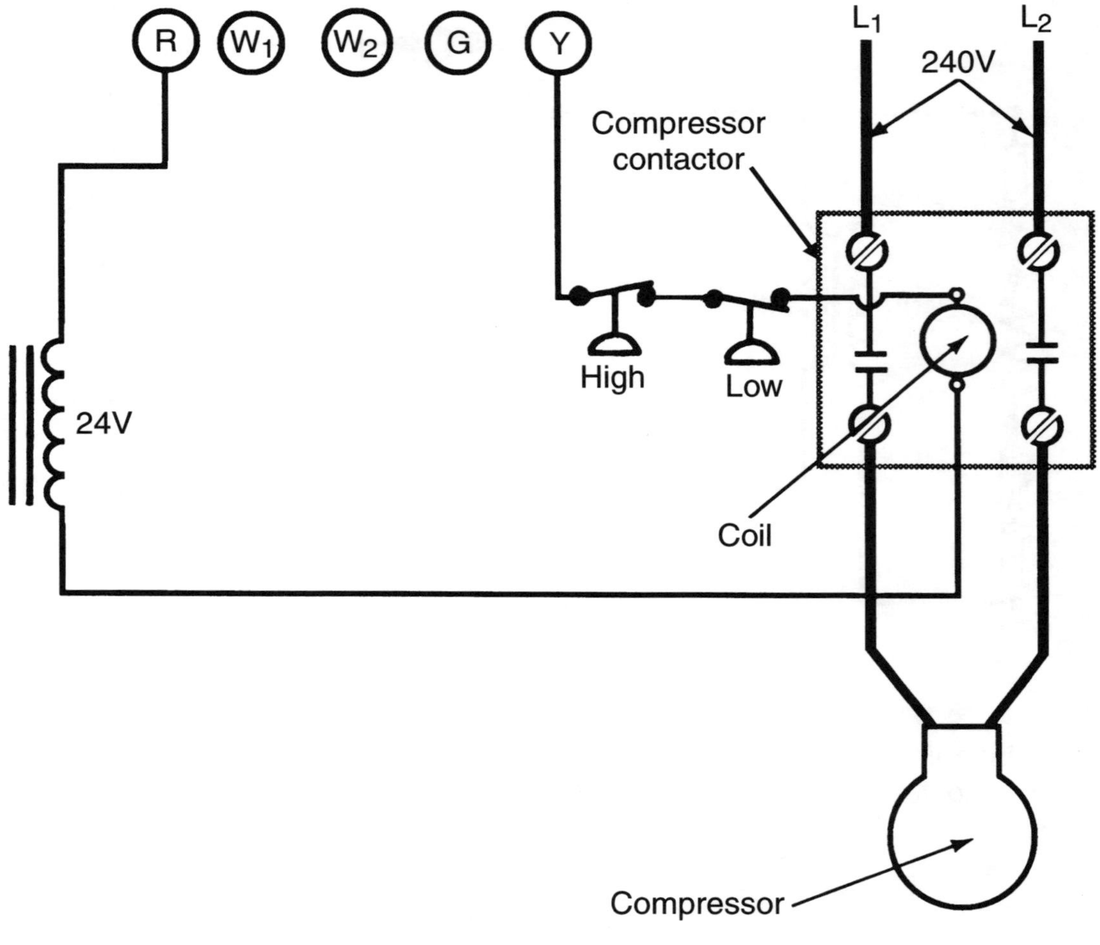

HC 31-4

Two-stage Electric Heat with Air Conditioning

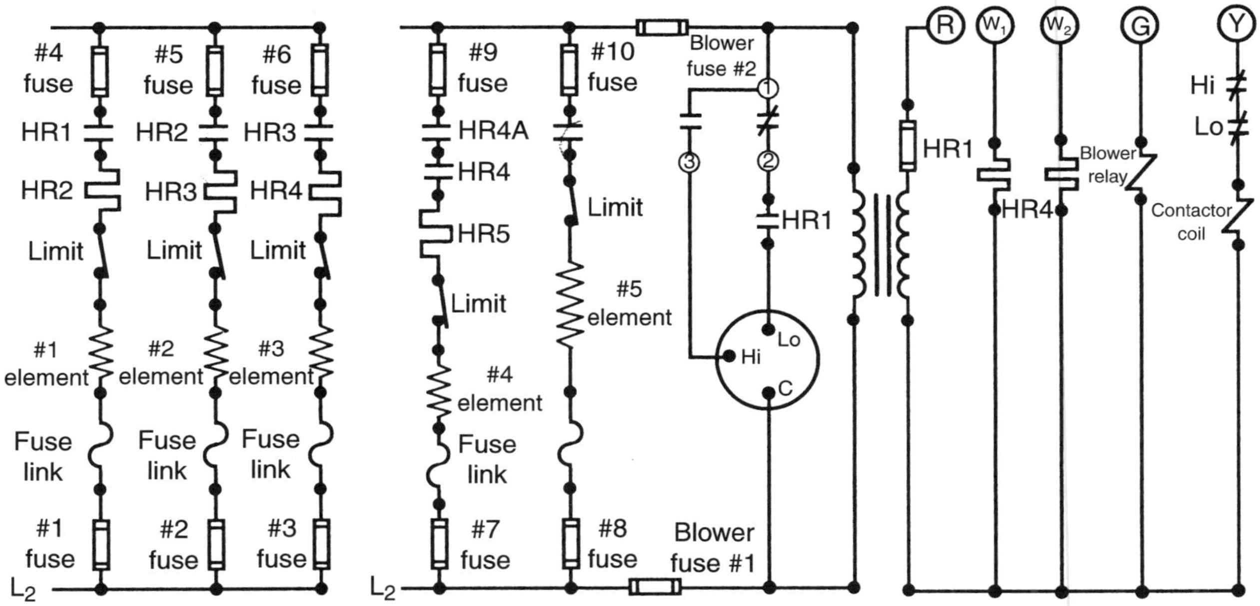

Schematic

HC 31-5

Two-stage Electric Heat with Air Conditioning

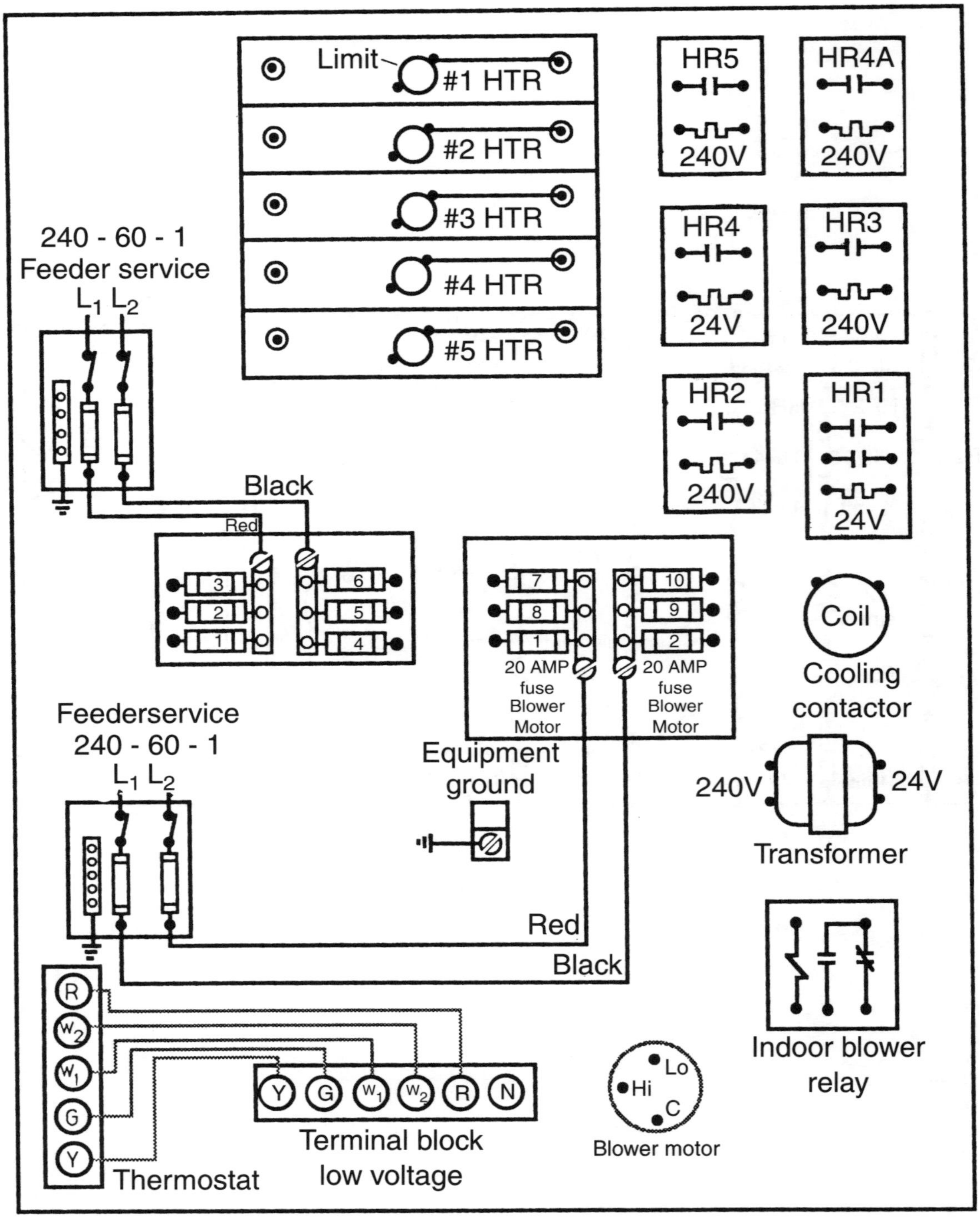

Pictorial

HC 31-6

Name ______________________________

Date ______________________________

Instructor ______________________________

Period ______________________________

31

Electric Heat with Air Conditioning

Quiz

Instructions: Write your answers on the lines provided at the right-hand side of the page.

1. Electric heating elements are normally sized at _____. 1. ______________________
2. What type of feeder service is required for a residential electric furnace? 2. ______________________
3. Name two primary causes of "cycling on the limit switch." 3. ______________________ ______________________
4. What two metals are used to manufacture heating elements? 4. ______________________ ______________________
5. Devices used to stage the heating elements on and off are called _____ and _____. 5. ______________________ ______________________
6. A two-stage heating thermostat is often used to control system _____. 6. ______________________
7. What terminal on the thermostat controls the second stage heat? 7. ______________________
   a. X2
   b. W2
   c. Y2
8. The wire from the thermostat that controls power to the blower relay coil is coded with what color? 8. ______________________
9. *True or false?* On a two-speed blower motor, high speed is used for heating. 9. ______________________
10. An outdoor thermostat is often used to control: 10. ______________________
    a. second-stage cooling.
    b. second-stage heat.
    c. first-stage heat.

32

Heat Pumps

Objectives

After studying this chapter in the text, the student will be able to:

- ❑ Describe the operating principles of a heat pump.
- ❑ Recognize and explain the purpose of major heat pump components.
- ❑ Define the terms *coefficient of performance, energy efficiency ratio,* and *balance point.*
- ❑ Describe the operating principle of the reversing valve.
- ❑ Identify and describe types of refrigerant flow controls.
- ❑ List types of defrost controls and describe their operation.
- ❑ Describe the operation of thermostats used on heat pumps.
- ❑ Read and interpret heat pump schematics.

Instructional Materials

Text: Pages 567–584
Lab Manual: Pages 247–252
Instructor's Manual:
Visual Masters, Pages 285–289
Chapter 32 Quiz, Page 290

Teaching Strategy

Text

1. Review objectives.
2. Discuss and assign chapter contents. Use these visual masters as appropriate:
   HC 32-1: Reversing Valve
   HC 32-2: Reversing Valves in Operation
   HC 32-3: Restrictor Flow Control
   HC 32-4: Four-bulb Thermostat
   HC 32-5: Heat Pump System
3. Direct students to complete Test Your Knowledge questions, page 583. After completion, discuss correct answers with students.

Lab Manual

1. Provide overview of Activities. Where necessary, stress safe laboratory practices and the importance of wearing protective clothing.
2. Assign Activities to individuals or pairs of students, as appropriate.
   Activity 32-1: Checking a Four-way Reversing Valve
   Activity 32-2: Checking a Defrost Timer
   Activity 32-3: Checking a Heat Pump System
3. Observe student performance on Activities, providing guidance or correction as needed. After reviewing the student's work, initial and date the activity in the place provided.

Evaluation

1. Administer the quiz from this Instructor's Manual.
2. Grade the quiz, and return it with appropriate comments for the student.

Answers to Test Your Knowledge, Page 583

1. reversing valve
2. b. Balance point.
3. solenoid coil
4. scroll (or compliant scroll)
5. TEV; capillary tube; restrictor-type flow control
6. outdoor
7. cooling
8. false
9. the outdoor defrost thermostat
10. 65°F (18°C)
11. by weight

12. four-bulb
13. auxiliary heat
14. low-ambient thermostat
15. R

Answers to Lab Manual, Pages 247–251

Activities 32-1 to 32-3

Student response (answers will vary).

Answers to Quiz

1. refrigerant
2. c. Reversing valve.
3. false
4. heat loss
5. true
6. cooling
7. a. outdoor thermostat
8. a. outdoor coil
   b. restrictor control
   c. filter-drier
   d. sight glass
   e. indoor coil
9. For fast defrost (keeps heat localized).
10. scroll

Reversing Valve

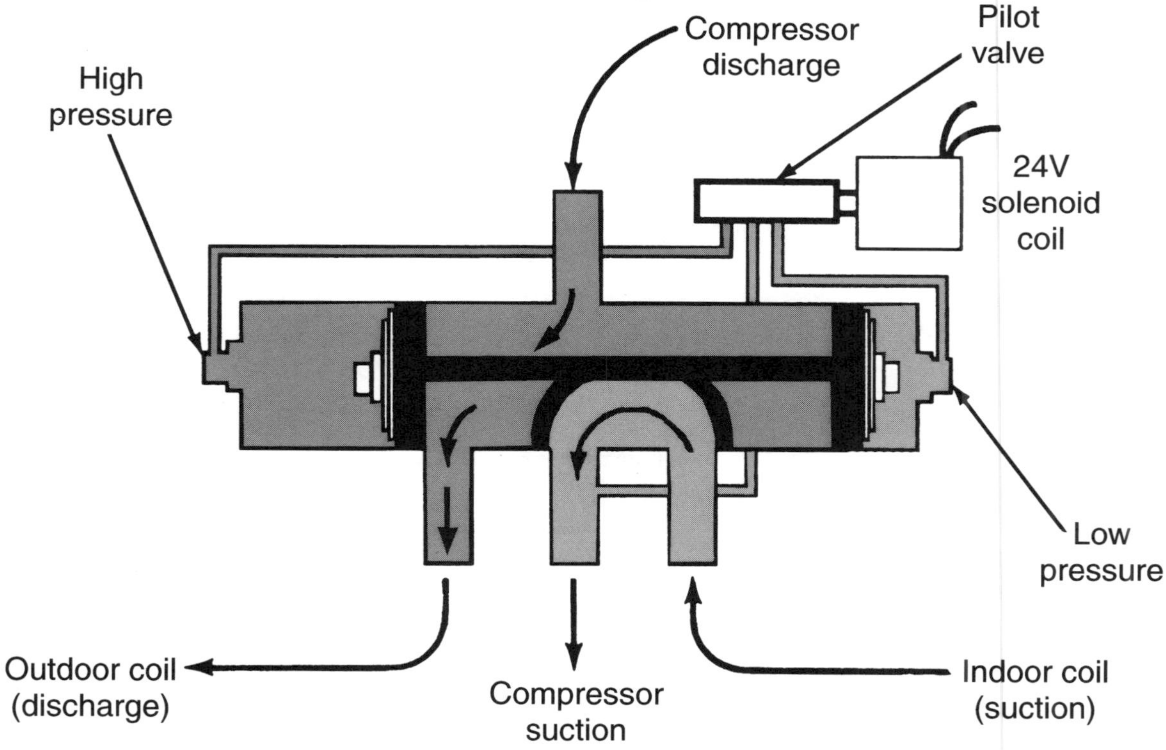

Cooling Cycle

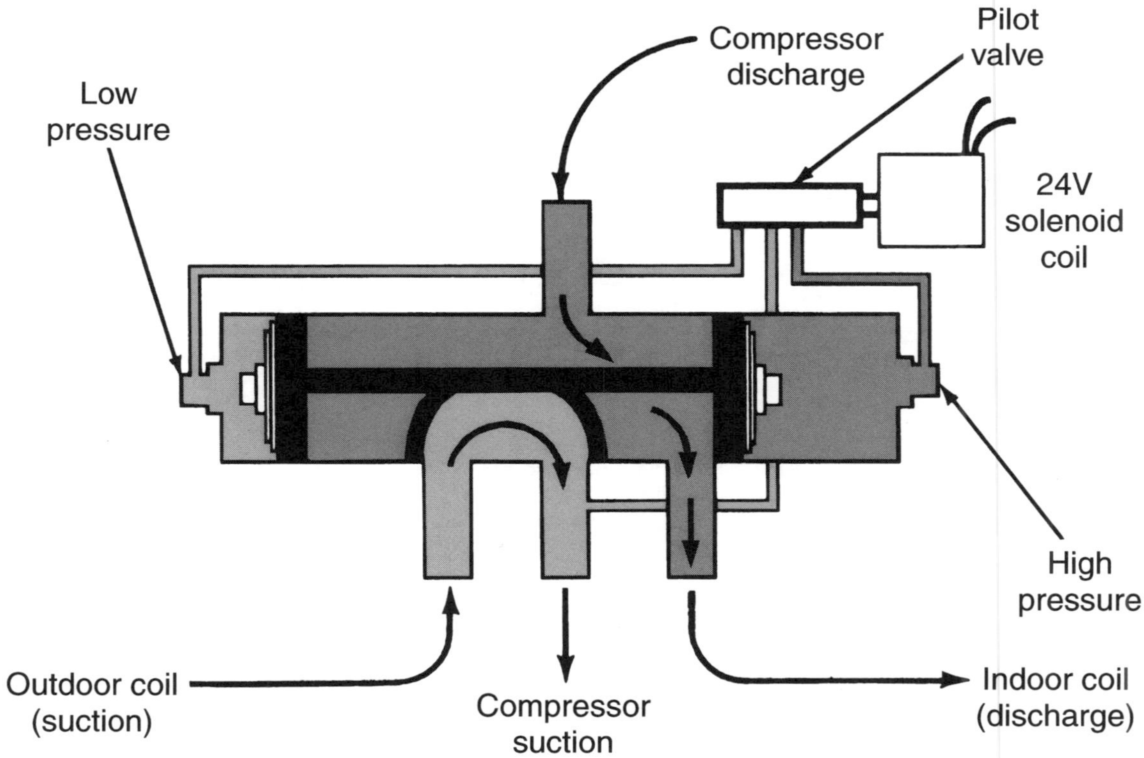

HC 32-1

Heating Cycle

Reversing Valves in Operation

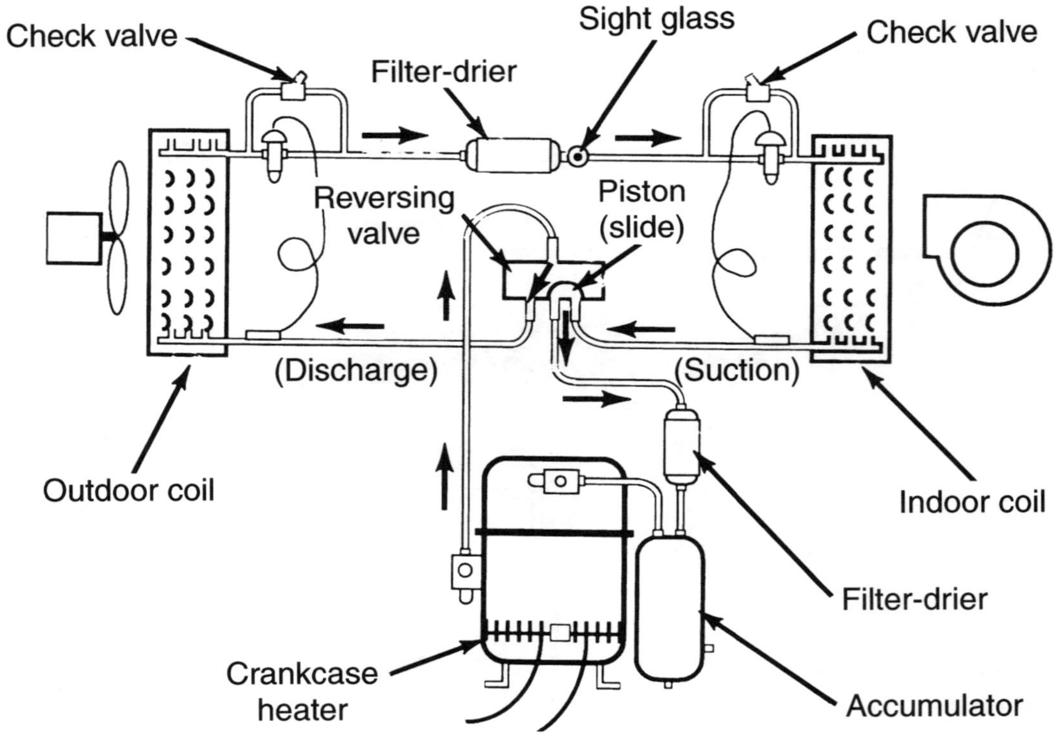

Cooling Cycle

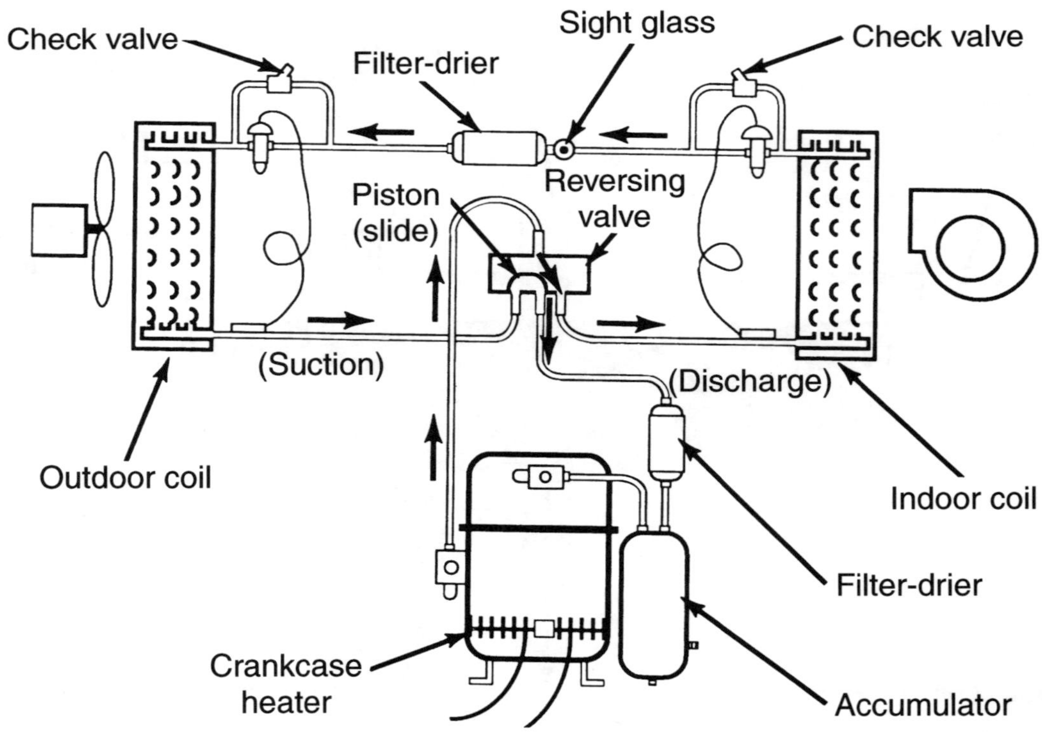

Heating Cycle

HC 32-2

Restrictor Flow Control

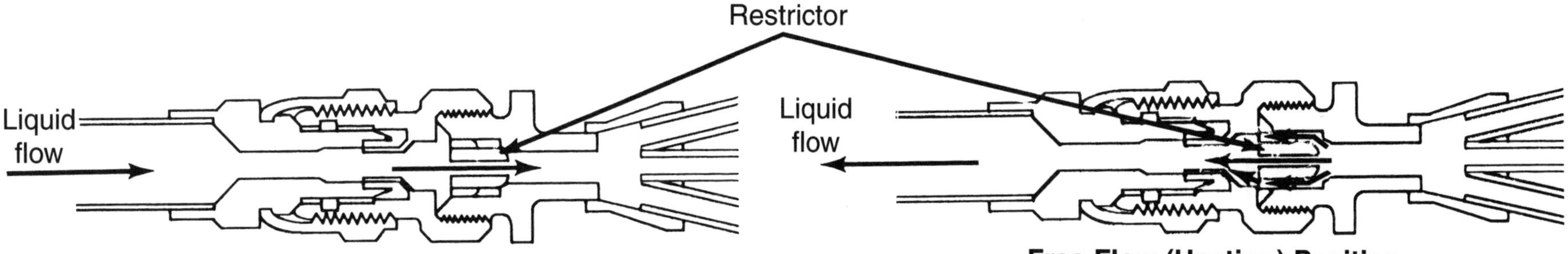

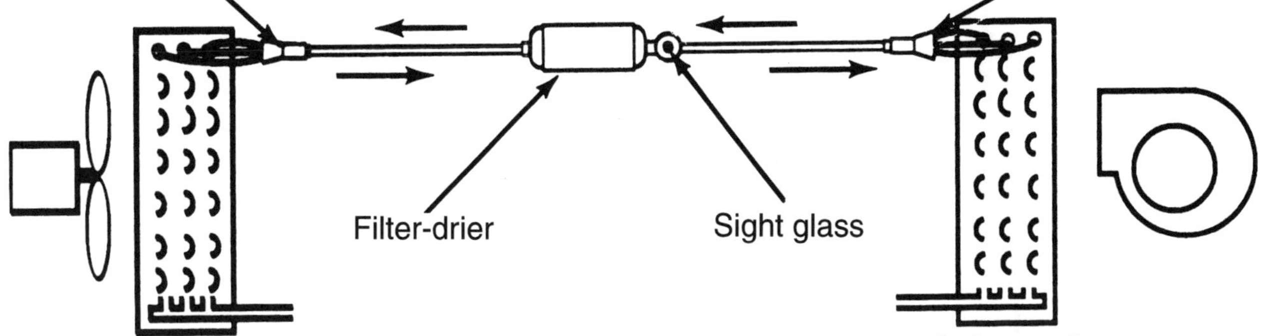

Four-bulb Thermostat

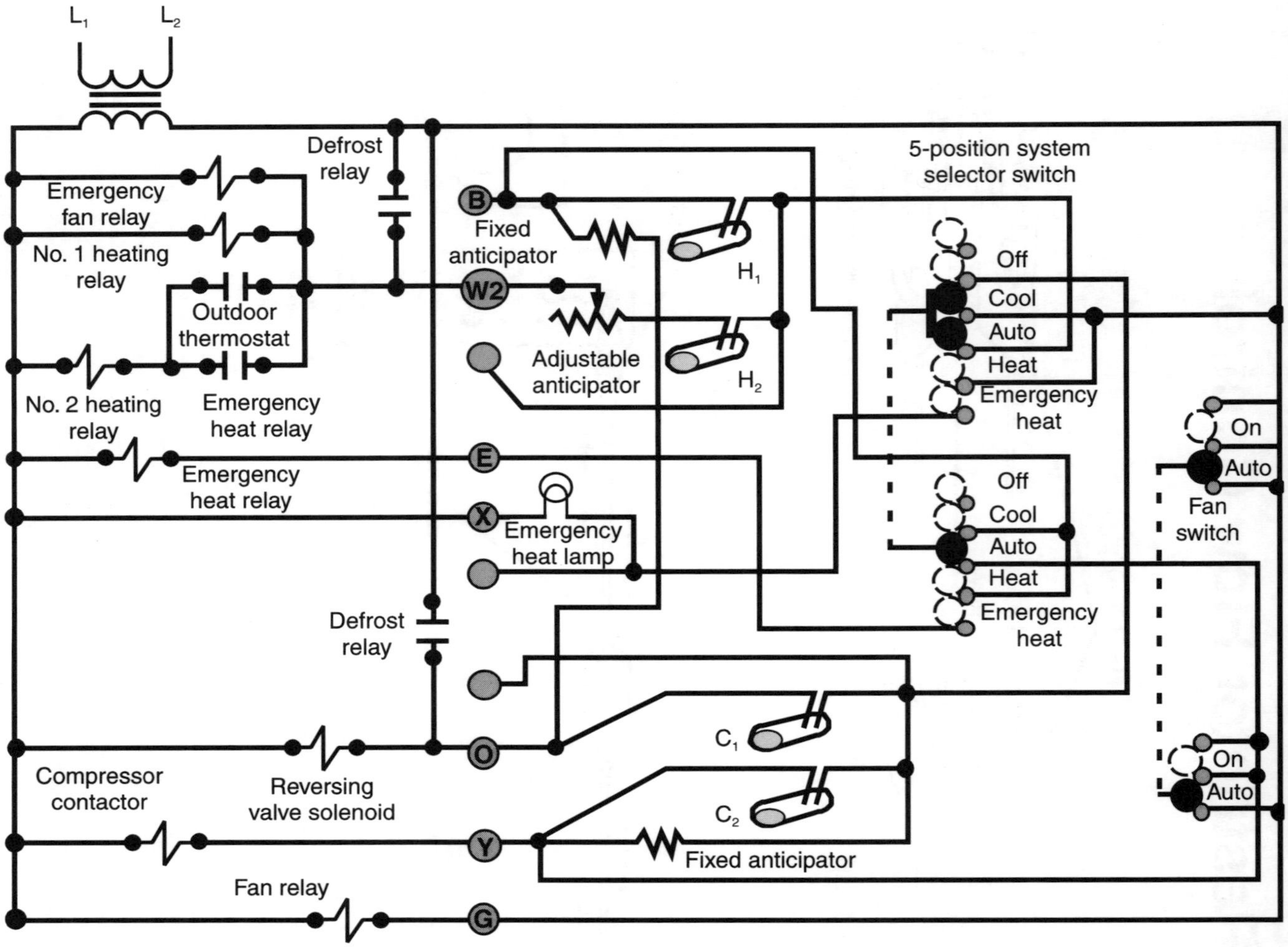

HC 32-4

Heat Pump System

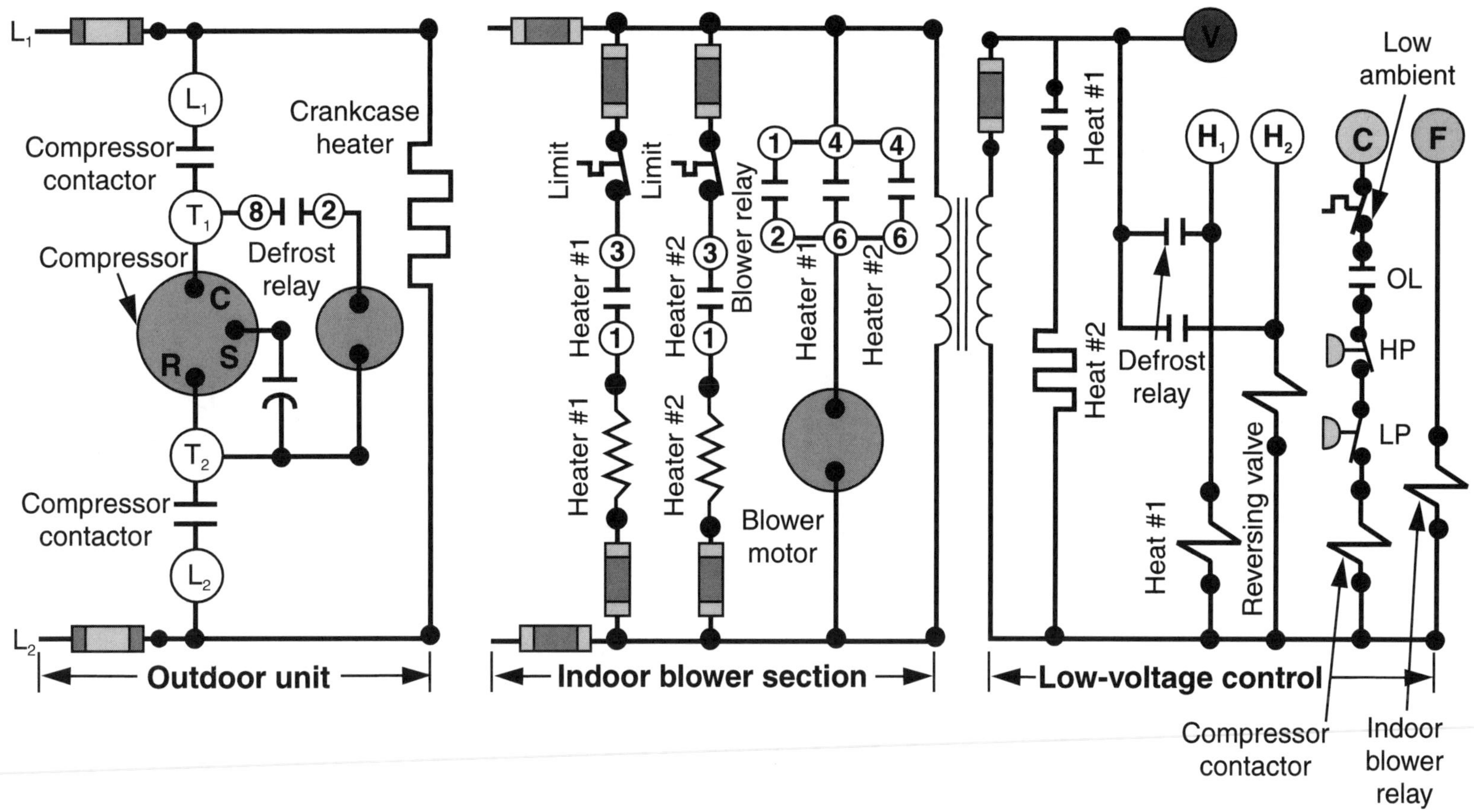

HC 32-5

Name ______________________________

Date ______________________________

Instructor __________________________

Period______________________________

32

Heat Pumps

Quiz

Instructions: Write your answers on the lines provided at the right-hand side of the page.

1. A reversing valve changes the direction of _____ flow. 1. ______________
2. What component of the heat pump allows it to provide either heat or air conditioning, depending on the season? 2. ______________
   a. Outdoor coil.
   b. Indoor coil.
   c. Reversing valve.
3. *True or false?* On a heat pump, the indoor coil always functions as an evaporator. 3. ______________
4. "Balance point" describes the point at which the heating ability of a heat pump equals the _____ of the building. 4. ______________
5. *True or false?* Auxiliary heat is required when the outdoor air temperature is below the balance point. 5. ______________
6. For a defrost cycle, the heat pump automatically switches to the mode. 6. ______________
7. The defrost cycle is normally terminated by the _____. 7. ______________
   a. outdoor thermostat
   b. defrost timer
   c. defrost relay
8. Identify the components of this heat pump system that uses restrictor-type flow controls. 8. a. ______________ b. ______________ c. ______________ d. ______________ e. ______________

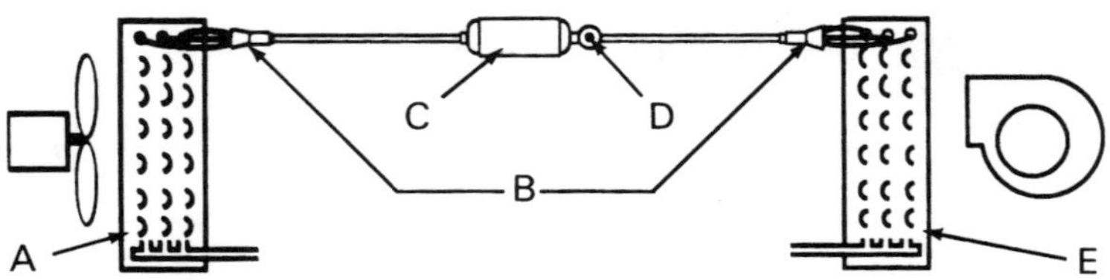

9. Why is the outdoor fan motor turned off during a defrost cycle? 9. ______________
10. What is the compressor of choice for heat pump applications? 10. ______________

33

Customer Relations

Objectives

After studying this chapter in the text, the student will be able to:

- ❑ Explain how to win arguments by avoiding them.
- ❑ Point out why it is important to learn and use people's names.
- ❑ Describe how to shake hands properly.
- ❑ Recognize the importance of good appearance.
- ❑ List ways of showing respect for others.
- ❑ Develop the skills needed to deal effectively with angry customers.
- ❑ Understand the value of admitting when you are wrong.
- ❑ Demonstrate how to begin and end service calls.

Instructional Materials

Text: Pages 585–590
Lab Manual: Pages 253–255
Instructor's Manual:
Chapter 33 Quiz, Page 293

Teaching Strategy

Text

1. Review objectives.
2. Discuss and assign chapter contents. (*Note:* There are no visual masters for this chapter.)
3. Direct students to complete Test Your Knowledge questions, page 590. After completion, discuss correct answers with students.

Lab Manual

1. Provide overview of Activities. Where necessary, stress safe laboratory practices and the importance of wearing protective clothing.
2. Assign Activities to individuals or pairs of students, as appropriate.
   Activity 33-1: Displaying Communication Skills
   Activity 33-2: Role-playing a Service Technician
3. Observe student performance on Activities, providing guidance or correction as needed. After reviewing the student's work, initial and date the activity in the place provided.

Evaluation

1. Administer the quiz from this Instructor's Manual.
2. Grade the quiz, and return it with appropriate comments for the student.

Answers to Test Your Knowledge, Page 590

1. mechanical
2. false
3. Agreeing with your opponent.
4. Changing the subject. Asking for an explanation.
5. Repeating
6. win-win
7. c. Admit your mistake.
8. false
9. sympathetic
10. safe; secure

Answers to Lab Manual, Pages 253–255

All activities are verbal; no written responses.

Answers to Quiz

1. people skills
2. d. all of the above
3. false
4. name
5. win-win
6. admit them
7. true
8. listen carefully
9. appearance
10. right

Name ______________________________

Date ______________________________

Instructor ______________________________

Period ______________________________

33 Customer Relations

Quiz

Instructions: Write your answers on the lines provided at the right-hand side of the page.

1. The road to success requires good mechanical skills combined with good _____. 1. ______________________

2. An important rule of using good people skills is to always avoid _____. 2. ______________________
   a. criticizing
   b. condemning
   c. complaining
   d. all of the above

3. *True or false?* The person who talks longest and loudest usually wins the argument. 3. ______________________

4. A person places an amazing amount of importance on his or her _____. 4. ______________________

5. In what type of situation do both parties benefit? 5. ______________________

6. Everyone makes mistakes, but few people have the courage to _____. 6. ______________________

7. *True or false?* When shaking hands, always make direct eye contact. 7. ______________________

8. The rule for dealing with angry people is to be sympathetic and _____. 8. ______________________

9. Your _____ conveys a strong message about your personality. 9. ______________________

10. At the end of the service call, it is important to assure the customer that you have fixed the _____ problem. 10. ______________________